BOLD VENTURES

Volume 3

Case Studies of U.S. Innovations in Mathematics Education

W0245874

Other Volumes in the Series: Bold Ventures

BOLD VENTURES

Volume 3

Case Studies of U.S.
Innovations in Mathematics
Education

edited by

Senta A. Raizen
Edward D. Britton

from

**The National Center for Improving
Science Education**

a division of
The NETWORK, Inc.

KLUWER ACADEMIC PUBLISHERS
DORDRECHT / BOSTON / LONDON

A C.I.P. Catalogue record for this book is available from the Library of Congress.

ISBN 978-0-7923-4237-3   ISBN 978-94-011-7111-3 (eBook)
DOI 10.1007/978-94-011-7111-3

Published by Kluwer Academic Publishers,
P.O. Box 17, 3300 AA Dordrecht, The Netherlands.

Kluwer Academic Publishers incorporates
the publishing programmes of
D. Reidel, Martinus Nijhoff, Dr W. Junk and MTP Press.

Sold and distributed in the U.S.A. and Canada
by Kluwer Academic Publishers,
101 Philip Drive, Norwell, MA 02061, U.S.A.

In all other countries, sold and distributed
by Kluwer Academic Publishers Group,
P.O. Box 322, 3300 AH Dordrecht, The Netherlands.

Printed on acid-free paper

The National Center for Improving Science Education

The National Center for Improving Science Education (NCISE) is a division of The NETWORK, Inc., a nonprofit organization dedicated to educational reform. The Center's mission is to promote change in state and local policies and practices in science curriculum, teaching, and assessment. To further this mission, we carry out research, evaluation, and technical assistance. Based on this work, we provide a range of products and services to educational policymakers and practitioners to help them strengthen science teaching and learning across the country.

We are dedicated to helping all stakeholders in science education reform, preschool to postsecondary, to promote better science education for all students.

Advisory Board, U.S. Case Studies

Contents

Contributors

The Case Study Teams

Brief biographical information about team members is found in their respective chapters. Team leaders are listed first. All three mathematics studies were coordinated by Norman Webb at the University of Wisconsin-Madison.

Setting the Standards: NCTM's Role in the Reform of Mathematics Education

Douglas B. McLeod
Bonnie P. Schappelle
Melissa Mellissinos

San Diego State University

Robert E. Stake
Mark J. Gierl

*University of Illinois at
Champaign-Urbana*

Teaching and Learning Cross-Country Mathematics: A Story of Innovation in Precalculus

Jeremy Kilpatrick
Lynn Hancock
Denise S. Mewborn
Lynn Stallings

University of Georgia

The Urban Mathematics Collaborative Project: A Study of Teacher, Community, and Reform

Norman L. Webb
Daniel J. Heck
William F. Tate

University of Wisconsin-Madison

The Editors

Senta A. Raizen, Director of The National Center for Improving Science Education, is principal investigator and editor of the U.S. case studies discussed in this and two companion volumes. Raizen is the primary author of a number of books, reports, and articles on science education in elementary, middle, and high school; indicators in science education; preservice education of elementary school teachers; and technology education. Her work also includes educational assessment and program evaluation, education policy, reforming education for work, and linking education research and policy with practice. She is principal investigator for NCISE research for the

ix

Third International Mathematics and Science Study (TIMSS) and serves on the TIMSS International Steering Committee. Raizen directs NCISE evaluations of several federal programs that support science education. She serves in an advisory capacity to—among others—the National Assessment of Educational Progress, the National Goals Panel, the National Institute for Science Education, and the National Research Council.

Edward D. Britton, Associate Director of NCISE, serves as project director for several international studies, including the work presented in this volume. He was lead editor of *Examining the Examinations: An International Comparison of Science and Mathematics Examinations for College-Bound Students*. Britton also works on several aspects of TIMSS, including the U.S. and international curriculum analyses and the international teacher and student questionnaires. In addition, he has managed development of CD-ROM disks and videotapes designed to help elementary teachers enhance their science knowledge and pedagogy. Britton has written on indicators for science education, dissemination of innovations, and evaluation.

Preface

This book presents comprehensive results from case studies of three innovations in mathematics education that have much to offer toward understanding current reforms in this field. Each chapter tells the story of a case in rich detail, with extensive documentation, and in the voices of many of the participants—the innovators, the teachers, the students. Similarly, Volume 2 of *Bold Ventures* presents the results from case studies of five innovations in science education. Volume 1 provides a cross-case analysis of all eight innovations.

Many U.S. readers certainly will be very familiar with the name of at least one if not all of the mathematics innovations discussed in this volume—for example, the NCTM *Standards*—and probably with their general substance. Much of the education community's familiarity with these arises from the projects' own dissemination efforts. The research reported in this volume, however, is one of the few detailed studies of these innovations undertaken by researchers outside the projects themselves.

Each of the three studies was a large-scale effort involving teams of researchers over three years. These teams analyzed many documents, attended numerous critical project meetings, visited multiple sites, conducted dozens of individual interviews. The team leaders (Kilpatrick, McLeod, Webb), having spent much time with mathematics education over long careers, looked at these innovations through many lenses. It was a daunting task for each team to sift through the mountains of detail in order to bring the most compelling themes to the surface. But through some exciting—if exhausting—meetings two or three times each year as well as ongoing exchanges, the key storylines did become clear. Deliberately, so as to let the stories emerge for the reader as they did for the researchers, we have not enforced a uniform format on the case study reports. In fact, at the request of some of its authors, chapter 1 on "Setting the Standards" did not undergo the same editorial treatment as the other chapters and has been included as submitted.

The Introduction gives a more detailed overview of the eight U.S. innovations we studied and the resulting case studies. It is important to note here, however, that this substantial case study effort was part of a larger international endeavor. Both the mathematics and science cases constituted the participation of the United States in a study of educational innovations in science, mathematics and technology education that was undertaken by 13 member countries of the Organisation for Economic Co-operation and Development (OECD). The genesis of the case study project was member countries' shared dissatisfaction with the state of education in these school subjects; this led them to seek a contextualized understanding of what innovations were under way to address this concern. The result was the largest qualitative research project ever undertaken across countries.

Just as we in the United States are learning from reform efforts in other countries, so too will the stories and findings of the U.S. projects be beneficial for the educational systems of the other OECD member nations. Innovation in a large, decentralized, and diverse country such as the United States is of particular interest for countries in which centralized government bureaucracies experiment with decentralized curriculum reform on the local or regional level, as several of the industrialized countries are now doing. Educators in other countries will be able to compare policy priorities in the United States to issues considered important in their own countries. Also, the documentation of these reform efforts available through the three volumes in this series will complement and help interpret both for this country and for an international audience the quantitative results generated on student achievement, current curriculum, and teacher and student background by the Third International Mathematics and Science Study (TIMSS).

Audiences

Bold Ventures is intended primarily for people working to improve science and mathematics education. We include in this audience:

- policymakers in a position to influence schools and educational systems;

- teacher educators and staff developers working with prospective or already practicing teachers;

- science, mathematics, and engineering professionals;

- school administrators; and

- teachers of science and mathematics.

The three volumes in the series unravel the origins, development, and implementation stories of eight of this country's major reform initiatives. They address questions surrounding the "what" as well as the "how" of these innovations. Curriculum specialists and teachers of mathematics in particular will find the detailed case studies in this volume of practical utility in the teaching and learning of mathematics. Having the case stories before them should help educators reflect on their own practices and programs.

Acknowledgments

We gratefully acknowledge the many individuals who contributed to the research effort reported in this and the two companion volumes. Just counting the researchers, advisors, and support staff, over 40 U.S. professionals invested substantial amounts of their time in this effort over the last four or more years. Most of these individuals are cited on other pages of this volume: advisors are listed just before the contents page, and short biographies of research team members are included in each of the individual studies. We thank the indispensable administrative assistants to this project, especially those who spent large amounts of time organizing meetings, preparing briefing books, and working with manuscripts: Susan Callan and LaDonna Dickerson at NCISE, Sally Lesher at the University of Wisconsin, and Sunny Toy at Stanford University. Nita Congress and her colleagues toiled to help us line edit this rather large body of work; Shelley Wetzel of Marketing Options handled the manuscript layout.

The Department of Education and the National Science Foundation have supported this work on an equal basis, as administered by NSF under grant number RED-9255247. We are grateful for their funding and supportive monitoring. The content of this report does not necessarily reflect the views of the Department or NSF. Their funding extended beyond the research to support dissemination efforts including a 1993 workshop to acquaint potential audiences with the goals of the U.S. case studies; and a 1996 conference in Washington, "Getting the Word Out," to release the U.S. and international results. Eve Bither, Acting Director of the Office of Reform Assistance and Dissemination, spearheaded the Department of Education's support of this project and has steadfastly advanced our work. Several current and former officers at NSF have generously assisted us over the years: Daryl Chubin, Director of Research, Evaluation, and Communication; David Jenness, now an independent consultant; Conrad Katzenmeyer, senior program officer; Iris Rotberg, program officer; and Larry Suter, deputy division director.

In closing, we wish to applaud and thank those who have made our work possible—the reform-minded professionals who worked in and out of the classroom to make their visions of better science and mathematics education come to life. Many people were gracious enough to let us witness their endeavors again and again and to speak with them at length. When busy educators and innovators welcome researchers into their world, it behooves these researchers to take the greatest care in depicting their efforts. We and our colleagues hope the many individuals who generously gave of themselves and their time will find *Bold Ventures* to be respectful, accurate, and helpful in advancing their important work.

Senta A. Raizen   *Washington, D.C.*
Edward D. Britton   *May 1996*

Introduction
Study Background

Senta A. Raizen

The last dozen years have been ones of both challenge and excitement for mathematics education—challenge because, on the whole, students' mathematics achievement continues to be disappointing, excitement because of the significant reform efforts that have been undertaken. This volume of *Bold Ventures* presents case studies of three such reform efforts: the Contemporary Precalculus course, the National Council of Teachers of Mathematics (NCTM) *Standards*, and the Urban Mathematics Collaboratives (UMC). Because of their promise for improving student learning in mathematics, while representing different approaches to reform, these three innovations were selected for intensive study. Along with the five science innovations discussed in volume 2, they comprise the participation of the United States in the largest cross-national case study project ever undertaken, the international context of which is described later in this chapter. This volume provides the full, rich detail of the stories these mathematics innovations have to tell and what lessons can be learned from them. Volume 1 of *Bold Ventures* presents our cross-case analysis of all eight studies.

The U.S. Studies

In selecting the project's mathematics and science innovations for intensive study, we aimed for a varied and balanced set with respect to subject areas, grade levels, nature of the interventions demanded by the innovation, and the origins and main driving force behind the innovation. Table 1 provides an overview of the range of the selected innovations.

As noted, five of the innovations are in science education: the California Science Education Reform initiative, the Chemistry in the Community (ChemCom) course, the Kids Network units, Project 2061, and the Voyage of the Mimi materials.[1] Two of the innovations (Kids Network and Voyage of the Mimi) use communication technology and multimedia approaches for instruction. Four of the choices are focused on innovative curriculum materials; four are more comprehensive in nature. Two of the materials development projects are aimed at high school (ChemCom, Contemporary Precalculus); one is intend-

---

[1]The Voyage of the Mimi was designed to be used as a tool in both mathematics and science instruction, but in practice teachers typically have emphasized science much more than mathematics.

1

Table 1.   Characteristics of U.S. Innovations Studied

	Curriculum Development		Comprehensive		
	K-8	9-12	District	State	National
Mathematics		Contemporary* Precalculus	Urban Math Collaborative		NCTM Standards
Science	Kids Network* Mimi*	ChemCom		California Science Ed. Reform	Project 2061

*While the primary focus of these three innovations was new approaches to science or mathematics, they also developed new uses of technology in instruction.

ed for middle school and upper elementary (Voyage of the Mimi) and one for upper elementary (Kids Network).

The origins of the innovations also are varied: two are large national initiatives (NCTM *Standards*, Project 2061); one was developed as a result of a National Science Foundation (NSF) grants competition (Kids Network); two came about through dynamic leadership within a professional school or organization and subsequently received federal and foundation funding (Voyage of the Mimi and ChemCom); one represents a state initiative (California Reform of Science Education); one was conceived by a private foundation and focuses on large urban school systems (UMC); and one came into existence through the initiative of teachers at a magnet science and mathematics school (Contemporary Precalculus). The selection of the U.S. cases was made by the case study researchers based on advice from a widely representative group of experts, i.e., the U.S. project's advisory board convened by The National Center for Improving Science Education (NCISE). (Members are listed in the front of this volume.)

In what ways does each of the selected innovations represent a change? How notable are these changes for U.S. education? Why is the innovation worth studying? What can be learned from each case? These and many other questions find responses in the richly told and documented stories of the innovations in this volume. To provide the necessary background for the rest of this book, we briefly summarize below some of the most innovative features of the mathematics cases.

Setting the Standards: The National Council of Teachers of Mathematics and the Reform of Mathematics Education. In 1989, NCTM published *Curriculum and Evaluation Standards for School Mathematics*; in 1991, *Professional Standards for Teaching Mathematics;* and in 1995, *Assessment Standards for School Mathematics*. With these three documents as a guide, NCTM is promoting a systematic program to reform mathematics education. The standards documents emphasize a new conception of mathematics as problem solving, reasoning, communication, and connection. An extensive dissemination effort has brought the mathematics standards to national attention, one result being that they have served as a model for the development of standards documents in science and other school subjects. NCTM is the first subject-area association to create a comprehensive series of standards documents. Further, NCTM did this on its own initiative, over a long period of time, without government resources. The case study tells the story of how a professional organization of mathematics teachers assumed national leadership in the field of mathematics education and influenced national and state policy in the movement to develop high educational standards.

Teaching and Learning Cross-Country Mathematics—A Story of Innovation in Precalculus. This "grassroots" effort differs markedly from most curriculum projects. A group of teachers at a science and mathematics magnet high school in North Carolina have developed a precalculus course based on the modeling of real phenomena. It emphasizes applications, data analysis, and matrices; the use of graphing calculators is integral to the course. Through word of mouth, teachers in other locales have learned of the course and begun to use it. This in turn has enabled the North Carolina teachers to obtain NSF funding for polishing the materials and also to secure a commercial publisher. The case study describes how the original group of teachers became involved in developing the course, holding workshops for other teachers interested in changing their mathematics instruction, and thus spreading their innovative approach to high schools across the country.

The Urban Mathematics Collaborative Project: A Study of Teacher, Community, and Reform. The UMC project is aimed at improving mathematics education in inner-city schools, while at the same time identifying new models for meeting the ongoing professional needs of teachers. Funded by the Ford Foundation, mathematics teachers in 16 school districts and others in their communities have worked together to improve the quality of mathematics teaching in the districts' middle schools. The case study reports on the interaction among business, industry, higher education, school systems, and the teachers to create change in the teaching of mathematics. It chronicles the personal growth of teachers as they assumed leadership roles as well as the new conceptions and practices of classroom mathematics that emerged. It analyzes the factors that facilitate—and those that impede—institutionalization of this approach to continuous teacher growth; for example, the problems associated with using a volunteer model.

To give a little more comprehensive view of the U.S. research, following are short descriptions of the five case studies of innovations in science education.

Building on Strength—Changing Science Teaching in California Schools. California has led the way in systemically influencing science instruction by concerted, congruent actions in four policy arenas: reconceptualizing the science curriculum in an innovative state framework, developing new statewide assessments that support appropriate science instruction, creating means for helping teachers change their science instruction, and requiring state-approved instructional materials to be consistent with the framework. The case study analyzes the changes in approach to science content and teaching required by the reform and describes in detail California's reliance on teacher networks to effect changes.

Chemistry in the Community: A Science Education Curriculum Reform. This course, popularly known as ChemCom, is an alternative to traditional high school chemistry. The curriculum focuses on chemistry as needed to understand sociotechnological problems, and attempts to attract to the study of chemistry nontechnical college-bound students as well as students not planning to attend college. The case study details the development of the course; its implementation; and the continuing commitment to ChemCom of the American Chemical Society, the sponsoring society.

The Case Study of the Kids Network. Through telecommunications, Kids Network enables upper elementary school students across the country—and in other countries—to exchange scientific data they collect during investigations of real-world environmental problems. The Kids Network was developed by Technical Education Research Centers (TERC) and is distributed by the National Geographic Society. The case study describes how the project provides a medium of change for teachers and classrooms, how teachers implement the curriculum, the reaction of students, and incentives and barriers related to the use of the technology.

The Different Worlds of Project 2061. This initiative, led by the American Association for the Advancement of Science, endeavors to fundamentally restructure science, mathematics, and technology education across all of precollegiate schooling. *Science for All Americans*, the project's first publication, set out a vision for what science was important for all students to know. The second publication, *Benchmarks*, suggests in detail what students should know and be able to do at various grade levels. The case study examines the development of the publications and their national impact, and emphasizes their relationship to curriculum designs formulated in six partnering school districts.

The Case Study of the Voyage of the Mimi. The Voyage of the Mimi is a curriculum program for 9- to 14-year olds that combines videos or videodiscs, computer software, and print materials to present an integrated set of concepts in mathematics, science, social studies, and language arts. The project uses a storyline of expeditions involving a research ship called Mimi. The case study relates how the course units were introduced into schools, how teachers collaborated to teach them, and how they adapted these materials. It describes teachers' reactions to the materials and analyzes factors that affected implementation.

The International Project

The origin of the international Science, Mathematics and Technology Education case study project dates back to 1989, when a meeting of interested researchers resulted in an issues paper, further refined during a 1990 meeting of experts from 18 member countries (including the United States) of the Organisation for Economic Co-operation and Development (OECD). The idea for conducting international case studies developed out of a growing concern shared by OECD member nations for more effective programs in science, mathematics and technology education to serve their populations. Moreover, the participating countries recognized the need to have a better understanding of the policies, programs, and practices likely to lead to more successful outcomes in these educational fields. What is the nature of the changes being formulated? How are such programs, policies, and practices developed? How are they implemented in settings where they are successful? What processes support implementation? What roles are played by whom? What outcomes are attained?

The desire to have answers to these questions led 13 countries to carry out one or more case studies of innovations in science, mathematics, and technology education.[2] The participating countries included, in addition to the United States, Australia, Austria, Canada (two studies), France, Germany, Ireland, Japan (two studies), The Netherlands, Norway (two studies), Scotland, Spain, and Switzerland.

Countries agreed on main research themes that each case study should seek to address. These themes are:

- context (historic, social, political, educational) within which the innovation was formulated;

- processes by which change was implemented, both as envisioned in planning and as experienced in reality;

- goals and content of the innovation;

- perspectives of the students participating in the innovation;

- methods, materials, equipment, and settings for learning;

- teachers and teacher education; and

- assessment, evaluation, and accountability.

---

[2]Three of these countries studied the introduction into precollegiate education of technology as a school subject. As the U.S. work included no such case—though three of the cases involve the use of educational technology in teaching science or mathematics—we generally refer to the U.S. cases as innovations in science and mathematics education.

In terms of the three subject areas, researchers were to address:

- how the scope and structure of science content was being redefined;

- the role of problem-solving approaches and applications in building mathematical knowledge and skills;

- the extension of science and mathematics education to more diverse students;

- ways in which technology education is being implemented in the general curriculum of elementary and secondary schools; and

- the interrelationships among science, mathematics, and technology in the school curriculum.

The outcome of the international project is *Changing the Subject: Innovations in Science, Mathematics and Technology Education* (Black and Atkin 1996). This book distills the major findings from all 23 cases, concentrating on the themes of greatest concern to the participating countries. The insights provided complement those of the Third International Mathematics and Science Study (TIMSS) which has collected data on a number of contextual factors as well as data on student achievement in science and mathematics in over 40 countries. TIMSS was not able, however, to examine in detail how these factors operate at specific sites.

Below are very brief descriptions of those countries' studies that focused on mathematics education only, or in combination with science or technology education. More detailed descriptions are available in the appendix of Volume 1 of *Bold Ventures*.

The state of Tasmania in **Australia** initiated the introduction of technology education, and concomitant changes in science and mathematics instruction for pupils from ages 4 to 16. The case study report, Science, Mathematics and Technology in Education (SMTE) Project, notes that the enthusiasm and leadership of a key teacher were imperative for successful introduction; other factors that were important included using approaches that emphasized student-centered learning and encouraged collegiality among teachers. The study was led by John Williamson of the School of Education at the University of Tasmania.

In **Austria**, a small group of experienced teachers was established by the education ministry to design computer-based materials for mathematics teaching in secondary technical colleges, for students 15 to 19 years old. Modern Mathematical Engineering Using Software-Assisted Approaches tells of the resulting countrywide interest and of the greater demands the course made on students. Peter Schüller of the Federal Ministry of Education led this case study.

From **Canada**, A Case Study of the Implementation of the Ontario Common Curriculum in Grade Nine Science and Mathematics, looks at a system undergoing extensive change. Two statewide requirements were introduced simultaneously: the integration of courses in science and mathematics, and the de-tracking of ninth grade classes. Detailed learning outcomes were specified. The case study team was led by John Olson of Queen's University.

National mathematics assessments have been developed in **France** for ages 8, 11, and 15 years to help teachers identify the needs of their students and adapt their instruction accordingly. The case study, The Impact of National Pupil Assessment on the Teaching Methods of Mathematics Teachers, identifies the benefits of the assessments for the diagnosis of students' needs and for dialogue between teachers and parents; it also documents teachers' wishes to receive more in-service training. This study was headed by Claudine Peritti, Ministry of Education.

In **Japan**, schools have introduced a revised course of study for mathematics in grades 1-9, though with less teaching time available, as reported in A Case Study of Teacher/Student Views About Mathematics Education in Japan. The desire is to promote problem solving and individuality, and to deal with advances in technology. Keiichi Shigematsu of the Nara University of Education led this research.

Five schools in **Norway** were asked to develop approaches to mathematics teaching for grades 7 and 8 that would involve students more actively and responsibly. This is reported in Assessment as a Link Between Instruction and Learning in Mathematics, Especially Focusing on Pupil Self-Assessment. Good communication among teachers, parent,s and students was found to be important to success. The case study was carried out by Sigrun Jernquist of the National Examination Board in Oslo.

The study from **Switzerland** tells of the use of computers as the medium whereby 13- to 16-year-old students gain understanding of real-life phenomena. The innovations of a small group of teachers and researchers are reported in The Representation, the Understanding and the Mastering of Experience—Modelling and Programming in a Transdisciplinary Context. Bruno Vitale of the Centre de Recherches Psycho-Pédagogiques in Geneva conducted this research.

Methodology

We chose the case study approach as our main methodology for studying the innovations because it is the only method that allows a close-up, in-depth understanding of their origins, progress, and implementation. Case studies also allow researchers to deal with latent or underlying issues that would escape the survey analyst or the "hit-and-run" field observer. For example, an understanding of site dynamics can evolve only gradually, from the time when a project is just under way to the time when it has spread widely enough to allow the formulation and testing of hypotheses about what facilitates or inhibits success. All the same, case study research admits that there can be several, varying interpretations, both among actors at the research sites and between these actors and the field researchers. We have tried to reflect these differences in interpretation in our individual case reports.

Each case report represents a *snapshot in time*. The case study researchers summarized, to the extent possible based on documents and interviews, critical events that occurred before they picked up the case; they were in the field for a year (and in some cases longer) following project activities and developments as they took place that year at selected sites and events associated with the project. Nevertheless, each of the case reports documents a "work in progress"—the situation and context as they were observed and recorded during a specific year in the life of each project.

The development of the U.S. case studies occurred in two phases, both funded by the U.S. Department of Education and the National Science Foundation. In the first phase, NCISE and the National Center for Research in Mathematical Sciences Education selected eight innovations for case study development and prepared 20-page descriptions of each. These descriptions were published by OECD (1993) under the title *Science and Mathematics Education in the United States: Eight Innovations.* The second phase consisted of the in-depth study of each of the innovations as reported in Volumes 2 and 3 of *Bold Ventures* and the cross-case analyses summarized in Volume 1.[3]

The Research Questions

Each of the three case studies in this volume addressed some general research questions; a second group of questions was particularly pertinent to the more comprehensive efforts represented by the NCTM standards project and the Urban Mathematics Collaboratives; a third group consisted of some more detailed questions applicable to implementation issues—for example, regarding the Contemporary Precalculus course. The three groups of questions are given below. Specific questions addressed by each case study are elaborated in the individual chapters.

General Questions .

- What has motivated the genesis of the innovation? Who were and are the key movers? What were the contexts of development—social, historic, institutional, political?

- What is the problem or dilemma addressed by the innovation?

---

[3]Seven of the innovations studied in depth are the same as those summarized in the OECD 1993 report; for the eighth innovation, the precalculus project was substituted for the California Restructuring of Mathematics Education (Webb 1993). The reason for the substitution was that the precalculus project represents a unique grassroots effort by teachers, whereas a state initiative already was represented in the California Science Education Reform case study.

- What are the content, design and underlying assumptions of the innovation? How much variation in point of view is there—or has there been—in different settings?

- For each of the innovations, how is progress or success defined, assessed, and known?

- Which criteria for success are used by different key actors or institutions?

- What were—or are—the principal facilitators and constraints in the design and execution of the innovation?

- What is the involvement of teachers in each innovation? How does teacher involvement affect the introduction and acceptance of innovations?

Additional Questions for Comprehensive Efforts. This next set of questions is particularly appropriate for such comprehensive efforts as represented by the UMC and the NCTM standards projects.

- What policies are being advocated? What is the rationale for these policies? Which of the policies have been instituted?

- For UMC's district-focused efforts, what direct funds or indirect resources have been allocated to the reform?

- Has the effort generated a favorable climate for mathematics or science education reform—nationally, in states, in districts?

- What roles are envisioned in the effort for business and industry, institutions of higher education, and other nonschool organizations?

- What collaboration occurred among professionals working in different contexts (scientists or mathematicians in higher education or industry, teachers, school administrators)?

- How have local users modified the intended innovations and for what reasons?

Additional Implementation Questions. These questions are relevant to the Contemporary Precalculus project, but also to the implementation of the NCTM *Standards*.

- What are the key marker events of each phase of implementation, and how have these events affected the outcomes of local implementation?

- What resources and working conditions have schools or districts provided to facilitate use of project materials?

- What do the innovations "look like" on a daily basis? Do teachers' and students' perceptions of what goes on in the classroom match or vary?

- What changes do teachers make in their instructional practices during successive uses of materials? What training and other support are available?

- Are schools and classrooms equipped technologically and with needed supplies to carry out the new curriculum or the innovation's teaching strategies?

The Case Study Research Cycle

The overall work of studying the eight U.S. innovations took three years, as summarized in table 2.

The first year was devoted to three tasks: The first task was to identify, within the OECD framework, the common research questions covering all the cases as well as project specific ones. The second task was to identify potential sites for in-depth study; quite different approaches were needed for the individual cases, ranging from developing a sampling plan for innovations already in many schools (as was the case for several of the curriculum materials projects) to selecting from a few extant sites (for example, for the Urban Mathematics Collaboratives). The third task consisted of developing specific methodology:

Table 2.   Research Cycle for Eight Individual Cases

1989	OECD meeting: design international study, including major research themes.
1991-1992	U.S. pilot study: Researchers identify and describe potential U.S. cases; Advisory board selects eight innovations for extensive case studies; brief descriptions of U.S. innovations published by OECD (1993).
1992-1993	Refine research questions, analyze documents, develop instruments, explore and sample sites, begin field work.
1993-1994	Main data collection and analysis: observe classrooms and events, interview key individuals including initiators, developers, and teachers.
	ED and NSF sponsor an NCISE "dissemination" workshop to alert potential audiences to the research and elicit suggestions for directions of special interest and importance.
1994-1995	Complete data analysis, draft reports.
1995-1996	Edit and publish *Bold Ventures* in three volumes.
	Release synopsis of results in *Education Week's* "Forum" series, April 10, 1996; ED, NSF, OECD and NCISE conduct "Getting the Word Out," a conference in April 1996 to highlight international and U.S. results.

identifying key project informants to be interviewed, formulating appropriate surveys and questionnaires, and creating interview and observation protocols. Each of the case study reports provides further methodological details germane to the individual case.

The second year was devoted to intensive field work, development of analytical methods including coding of observations and interviews, analysis of critical documents, preliminary analyses of field data to generate further field work, and interaction with potential audiences to generate dissemination plans for the study results. The Introduction to Volume 1 illustrates the scale and nature of activity that went into data collection and analysis. Intermediate analyses, reviews, and feedback from the projects and sites as well as from outside consultants, final analyses, and writing of each case consumed the third year.

The entire effort was characterized by close collaboration among the research teams. All the teams met together several times each year, first to decide on the common research questions and general approaches to the field work and site selection; later to exchange early findings, develop common themes for the cross-case analysis, and decide on the length and style of the case reports; still later to discuss key chapters of the cross-case volume and address dissemination issues. Several of the meetings also involved reviews and critiques by the project advisors (listed in the front of this volume). We know that this unusual collaboration among eight research teams proved highly motivating and intellectually stimulating to all involved, especially to the graduate students serving as research associates. We hope also that this mode of working served to enrich both the cross-case volume and the individual case studies reported in the following chapters.

References

Black, P. and J.M. Atkin, eds. 1996. *Changing the Subject: Innovations in science, mathematics and technology education.* London and New York: Routledge.

Organisation for Economic Co-operation and Development (OECD). 1993. *Science and mathematics education in the United States: Eight innovations.* Paris.

Webb, N. 1993. *State of California: Restructuring of Mathematics Education.* In Science and Mathematics Education in the United States: Eight Innovations. Paris: Organisation for Economic Co-operation and Development.

Chapter 1

Setting the Standards
NCTM's Role in the Reform of Mathematics Education

Douglas B. McLeod   San Diego State University
Robert E. Stake   University of Illinois at
                             Champaign-Urbana

Bonnie P. Schappelle   San Diego State University
Melissa Mellissinos   San Diego State University
Mark J. Gierl   University of Illinois at
                             Champaign-Urbana

Contents

Setting the Standards
NCTM's Role in the Reform of Mathematics Education

Introduction

On March 21, 1989, the National Council of Teachers of Mathematics (NCTM) held a press conference in Washington, DC, to release the *Curriculum and Evaluation Standards for School Mathematics* (NCTM, 1989). The *Curriculum and Evaluation Standards* had been under development for almost five years (Crosswhite, Dossey, & Frye, 1989), and public interest was high; the press conference, which some had expected to be quite small, was moved to the larger venue of the Willard Hotel in downtown Washington. Since NCTM had not organized such a large media event before, it was an exciting time for the organization. As Marilyn Hala, an NCTM staff member, recalled:

> This is the first time I ever remember that we had a bank of six or seven video cameras in the back of the room. There were probably 200 people at the press conference in the Willard Ballroom. It was quite impressive! You had to walk by these members of the press—these cameras—to sit down.[1]

NCTM leaders John Dossey,[2] Shirley Frye, and Tom Romberg were there to answer the calls that came in from people in the media. The public relations firm hired by NCTM, Burson-Marsteller, provided the expertise that was needed for such an event. As John Dossey put it:

> That's why I was on the *Today Show* [on NBC-TV], why Shirley Hill[3] was on CBS or ABC during that period of time. That's why Sally Ride4 and Bud Wonsiewicz were at the press conference. We gave the PR firm some ideas,

---

[1]Quotations were usually taken from transcripts of interviews with sources; some sources provided written responses. We identify sources by name with their permission.
[2]John Dossey was immediate past president of NCTM, Shirley Frye was president, and Tom Romberg had served as chair of the Commission on Standards for School Mathematics.
[3]Shirley Hill was chairman of the Mathematical Sciences Education Board (MSEB) and a past president of NCTM.

they came back with ideas, and it all came together with a lot of their leadership. They arranged and took me on a tour of New York City one day; I spent the morning talking with the editorial board of the *Wall Street Journal*, the afternoon with the editorial board of the *New York Times*, and the next day on the telephone with the *Los Angeles Times*, the *Chicago Tribune*, and a few others. Those were not things that we were used to doing.

The *New York Times* and the *Washington Post*, along with many other newspapers, reported on the release of the *Curriculum and Evaluation Standards* (NCTM, 1989) in their issues of March 22, 1989. NCTM President Shirley Frye and Past-President John Dossey were quoted on the need for reform in mathematics education; the newspaper reports focused on the need for reasoning over rote learning, the usefulness of calculators and computers, and the emphasis on applications. Critics of NCTM, like textbook author John Saxon, were also quoted on the dangers of technology and the need for drill and practice. A "strong national debate" was expected, according to the *New York Times*.

NCTM, a professional organization of mathematics teachers and others concerned with mathematics education, had taken an enormous step with its release of the *Curriculum and Evaluation Standards* (NCTM, 1989), a step that led to a national effort at systemic change and to major revisions in government educational policy.[5] NCTM had made recommendations about curriculum in the past (e.g., NCTM, 1980), but never before did an NCTM initiative have such a broad impact on education at the national level. In the years following the publication of the *Curriculum and Evaluation Standards*, governmental leaders from both major political parties agreed that the US needed to develop *standards* in all the core areas of the school curriculum, a significant shift in federal policy (Massell, 1994). Educational leaders from outside of mathematics described the NCTM effort as "a prime factor in inspiring and shaping today's standards movement" (Atkin, 1994, p. 64) and an "example for emulation" (Ravitch, 1995, p. 49) in other subject-matter areas.

What are the NCTM *Standards*?[6] What are their origins? What propelled and what shaped their development? How did they become known? How are they being interpreted? What is their standing? How did the NCTM *Standards* themselves become a standard for educational change in other disciplines? These questions were the focus of our case study.

---

[4]Sally K. Ride was an astronaut with NASA, and Bud C. Wonsiewicz was vice president of US WEST Advanced Technologies.

[5]In spite of its name, NCTM is an international organization that is independent of the US government.

[6]Most sources used the phrase "NCTM *Standards*" to refer to the *Curriculum and Evaluation Standards for School Mathematics* (NCTM, 1989), and we follow that usage here.

The NCTM *Standards*

Romberg and Webb (1993) have described the NCTM *Curriculum and Evaluation Standards* as a "258-page document [that] presents a vision of a mathematics program for Grades K-12" (p. 145). The document begins by describing itself as "one facet of the mathematics education community's response to the call for reform" (NCTM, 1989, p. 1), citing *A Nation at Risk* (National Commission on Excellence in Education, 1983) and *Educating Americans for the 21st Century* (National Science Board, 1983) as the source of that call. The NCTM *Standards* continue with a statement of new goals for education in the information age, including societal goals[7] and goals for students.[8] After describing these goals, the document outlines the content that should be included in the mathematics curriculum through its specification of 13 standards for Grades K-4, 13 for Grades 5-8, and 14 for Grades 9-12. In each of these three levels, the first four standards deal with mathematical problem solving, communication, reasoning, and connections.[9] The other standards in each level specify various mathematical topics like number sense, geometry, and algebra. A sample of a standard (Statistics, Grades 5-8) is included below.

Standard 10: Statistics
In grades 5-8, the mathematics curriculum should include exploration of statistics in real-world situations so that students can—
- **systematically collect, organize, and describe data;**
- **construct, read, and interpret tables, charts, and graphs;**
- **make inferences and convincing arguments that are based on data analysis;**
- **evaluate arguments that are based on data analysis;**
- **develop an appreciation for statistical methods as powerful means for decision making.** (NCTM, 1989, p. 105)

All standards are presented in the same way, with a concise specification of what students should be able to do (e.g., the "bullets" listed above), followed by two or three pages describing student activities that are meant to convey the "vision [of] both mathematical content and instruction" (NCTM, 1989, p. 10). The document also includes 14 evaluation standards related to general assessment principles (e.g., alignment of tasks and curriculum, multiple sources of information), assessment for instructional purposes, and program evaluation.

Two years after the appearance of the *Curriculum and Evaluation Standards* (NCTM, 1989), a companion volume (the *Professional Standards for Teaching*

---

[7]Societal goals include mathematically literate workers, a workforce capable of lifelong learning, equal opportunity for all, and an informed electorate (NCTM, 1989, pp. 3-5).

[8]New goals for students deal with valuing mathematics, confidence, problem solving, communicating mathematics, and reasoning.

[9] Some sources referred to these four as the "process" standards.

Mathematics [NCTM, 1991]) was produced to "provide guidance to those involved in improving mathematics teaching" (p. 2). Building directly on the earlier conceptualizations of curriculum and evaluation, this second document is divided into sections that present standards for teaching mathematics, for evaluating the teaching of mathematics, and for the professional development and support of mathematics teachers. The pedagogical goals of the *Teaching Standards* (NCTM, 1991) are illustrated through its extensive use of vignettes about classrooms.

More recently NCTM published *Assessment Standards for School Mathematics* (NCTM, 1995) to complement the two earlier documents (NCTM, 1989, 1991) and to help develop assessment strategies that are "consistent with NCTM's vision of school mathematics" (NCTM, 1995, p. 1). For the purposes of our case study, however, we focused on the first two documents, the *Curriculum and Evaluation Standards* (NCTM, 1989) and the *Teaching Standards* (NCTM, 1991).

NCTM—The Organization

The National Council of Teachers of Mathematics identifies itself as the world's largest organization in the field of mathematics education. Its membership includes 118,000 individuals and institutions, most of them in the US and Canada. NCTM serves the mathematics education community through its publications,[10] conferences,[11] and 260 affiliated groups.[12] Although the development of the three *Standards* documents is the largest and most significant project ever undertaken by NCTM, the organization has a long history of involvement in advocacy and reform.

NCTM was founded in 1920 (Osborne & Crosswhite, 1970), partly as a response to "educational reformers" in the early part of this century who wanted to reduce the role of mathematics in the secondary school curriculum (Kilpatrick & Stanic, 1995). As a new organization, NCTM was an active participant in discussions surrounding the "1923 Report" on secondary mathematics curriculum, but the leadership for that report came mainly from college mathematicians (Osborne & Crosswhite, 1970). Similarly, from the 1920s through the 1960s, NCTM played an important but usually secondary role in curriculum recommendations. For example, the College Board's influential Commission on Mathematics (College Entrance Examination Board, 1959) was dominated by mathematicians from universities, not by leaders of NCTM.

Within NCTM a tradition had developed that worked against having the organization take a leading role in policy recommendations. Up until the 1970s,

---

[10]NCTM publishes four journals, yearbooks, and other professional materials.

[11]Attendance at the annual meeting has been about 20,000 in recent years; there are also eight or nine regional conferences in North American cities each year.

[12]Most states and large cities have such groups, as do national groups with a common interest, such as the Association of State Supervisors of Mathematics.

many leaders thought that NCTM should not take positions that might be opposed by some of its members. The publication of *An Agenda for Action* (NCTM, 1980), with its brief recommendations on curriculum and teacher professionalism, was a significant change. James Gates, NCTM Executive Director, noted:

> Many of us felt that *An Agenda for Action* was a set of bones without much meat. But it was a rather bold step compared to what we had done in the past. [Past President] Joe Crosswhite may discuss recommendations that were in NCTM publications in the early years, but he will also point out that those were not official positions.

NCTM has for some time routinely reminded the reader that its yearbooks and other publications present "a variety of viewpoints ..., [which] should not be interpreted as official positions of the Council" (e.g., Webb, 1993a, p. ii). Gates continued:

> We didn't take positions. We were a forum, a place where you got a lot of information. I believe that changed with *An Agenda for Action*. We thought at the time that it was a big change for NCTM to be taking such a position, and publication of the *Standards* was a major jump from that. We started making position statements when basic skills came along as a hot topic in the 1970s. The first position statement was on basic skills. Some people would say that taking such positions isn't the role of NCTM.

During the late 1970s, NCTM became much more active in trying to influence public policy. Some people associate the change with the activist role taken by Shirley Hill as president of NCTM during 1978-80. Joe Crosswhite noted:

> Prior to Shirley's time, you couldn't interest an NCTM president in having a national presence in Washington—an NCTM presence. And Shirley was the first, to my knowledge, that was sympathetic to the notion that NCTM should become politicized.

Shirley Hill described some factors that influenced her thinking about NCTM's role:

> Although I could certainly agree that sufficient motivation for belonging to NCTM might be found in the interaction with colleagues and the benefits of publications and conferences, I had also felt a certain frustration that we weren't being listened to seriously enough outside our own circles. In one of my newsletter columns as president, I noted that an important reason for my own membership had always been the opportunity to have a collective voice representing our professional community. I remember attending some meetings of the presidents of like organizations in Washington, DC, in the 1970s and noticing the frequent absence of the president of one of our sister organizations. It turned out that he was being escorted by his staff government relations expert in visits to members of Congress. At that time his organization seemed to be

very influential in the establishment of federal programs. I thought that we in NCTM should be doing more of these things. I thought that we and most of our sister organizations were being a little naive about government relations and public relations at that time.

Although NCTM became a more "activist" organization in the 1970s, that decade was also a time of decline for the organization. NCTM membership went from a high of over 82,000 in 1968 to a low of 56,000 in 1983, a drop of over 30% in 15 years (Willoughby, 1984). As Joe Crosswhite noted, his term as president (1984-86) began at a difficult time:

> We were at the bottom of a fairly long slide in both membership and finances. We had reached the point where we had considered—and actually had a motion made to the Board—to discontinue NCTM's publication program.

Although some members of the NCTM Board of Directors were ready to cut back on programs, others were determined to push ahead. As President Shirley Hill (1980) put it in her presidential address, NCTM could improve mathematics education if it would speak out with a united voice to those who influence educational policy. Speak out it did, with results that have surprised NCTM leaders and members alike. By 1995, membership in the organization had increased to 118,000, and NCTM presidents were regularly asked to advise governmental leaders on educational policy. Jim Gates reported that contacts with federal officials were not that common a decade or two ago, but now such contacts are frequent and often helpful to the organization:

> Next week the NCTM president will meet with a Senate committee on the Eisenhower legislation, and recently we visited some of the key members of the House of Representatives about that legislation. Ten or twenty years ago, would they have let us help them write their legislation?

There are a number of reasons one could hypothesize for the success of NCTM as an organization during the 1980s. First, there was a renewed interest in the public in educational change, including change in mathematics education; this interest was reflected in federal funding for programs that encourage teachers to be professionally active (e.g., the Eisenhower program, the Presidential Awards for mathematics teachers). Second, the continuity of NCTM's administrative leadership is unusual; Executive Director Jim Gates served NCTM for 34 years before his retirement in 1995. Third, NCTM presidents (see Table 1) who had been serving part-time (with partial salary supplied from NCTM) became full-time employees during the 1980s. Joe Crosswhite took early retirement from Ohio State University and devoted himself full-time to his duties as president of NCTM during 1984-86. John Dossey was the last part-time president (1986-88), but he invested tremendous energy in the position, especially the *Standards* project. After Dossey's term, NCTM began to fund the presidency as a full-time appointment.

Table 1. **Recent Presidents of the National Council of Teachers of Mathematics**

Term	President*
1978-80	Shirley A. Hill, University of Missouri-Kansas City
1980-82	Max A. Sobel, Montclair (New Jersey) State College
1982-84	Stephen S. Willoughby, New York University
1984-86	F. Joe Crosswhite, Ohio State University
1986-88	John A. Dossey, Illinois State University
1988-90	Shirley M. Frye, Scottsdale (Arizona) School District
1990-92	Iris M. Carl, Houston (Texas) Independent School District
1992-94	Mary M. Lindquist, Columbus (Georgia) College
1994-96	Jack Price, California State Polytechnic University at Pomona
1996-98	Gail Burrill, Whitnall (Wisconsin) High School

*The NCTM president and Board of Directors are elected by the members.

In part the story of the NCTM *Standards* is the story of NCTM as an organization. NCTM went from being a professional organization where issues of teaching and learning were discussed and debated to an organization with an activist and political orientation. NCTM's new mission focused on providing national leadership in mathematics teaching and learning.

Origins of the NCTM *Standards*

The story of the creation of the NCTM *Standards* (NCTM, 1989) was told somewhat differently by different groups within NCTM. NCTM presidents and other national leaders tended to focus on major conferences and the broader social context of reform. They talked about the national conferences that they had attended and their attempts to influence government policy. NCTM committee members, by contrast, emphasized the internal workings of NCTM as an organization, emphasizing the tasks that their committees had been engaged in. A third group, the writers who actually put words to paper, talked more about the specifics of the discipline (e.g., the impact of graphing calculators on secondary mathematics). As they wrote the *Standards*, they thought about their own state's curriculum guidelines and their own work in curriculum development, rather than national conferences or government policies. The comments from each of these groups reflect how their individual experiences shaped their view of the NCTM *Standards*.

Our interviews with these three groups were the main sources of data[13] for this part of our report. We began by interviewing writers, NCTM staff, and NCTM presidents, and followed the recommendations of those sources in choosing additional leaders to interview. We sent questionnaires to the writers of the *Curriculum and Evaluation Standards* that we were unable to interview. We also gathered questionnaire data from the State Supervisor of Mathematics in each state and interviewed a sample, including several who had been directly involved in the development of the NCTM *Standards*.

Views from the Leaders of NCTM

The origins of the NCTM *Standards* have been outlined by NCTM leaders themselves (e.g., Crosswhite, 1990; Crosswhite et al., 1989; Dossey, 1990; Romberg & Webb, 1993). These authors have documented the major conferences and reports that led to the publication of the *Curriculum and Evaluation Standards* in 1989. We summarize that part of the *Standards'* story here (see Table 2) and supplement it with information from NCTM committee members and others who were involved in the organizational work that preceded the decision to produce the NCTM *Standards*.

In the mid-1970s, the public perception of mathematics education in the US was reaching a low point. The reforms of the 1960s were perceived as failures, and a renewed emphasis on basic skills was the focus of media attention. Crosswhite (1990) recalled the context:

> There was a growing concern among professionals in mathematics education that the school curriculum was being narrowed by what has been called the Back to Basics movement. That movement seemed to continue a cyclic pattern of overreaction that has characterized the history of school mathematics in this country. (Crosswhite, 1990, p. 454)

Crosswhite argued that such a cyclic pattern was damaging to mathematics education. He noted that the National Advisory Committee on Mathematical Education (NACOME, 1975) had warned against "false dichotomies"—those

---

[13]We interviewed more than 40 people, including 6 NCTM presidents, 3 NCTM staff members, 15 (of 26) writers of the *Curriculum and Evaluation Standards,* and 11 past or present State Supervisors of Mathematics, as well as additional leaders of NCTM committees. Interviews (most longer than an hour) were usually audiotaped and transcribed; follow-up interviews were conducted for clarification when necessary. Questionnaires were not returned by 3 (of 26) writers and 9 (of 50) State Supervisors of Mathematics.

**Table 2.   Notable Documents Related to the Origins of the
                 NCTM *Standards***

Documents	Description
The National Advisory Committee on Mathematical Education Report (NACOME, 1975)	Analysis of trends in mathematics education at the end of the New Math era
Position Statement on Basic Skills (National Council of Supervisors of Mathematics, 1978)	Mathematics supervisors re-define basic skills to include problem solving
Case Studies in Science Education (Stake & Easley, 1978)	Field observations of mathematics and science classrooms; see also related surveys discussed by Fey (1979)
An Agenda for Action (NCTM, 1980)	A major policy document on curriculum, evaluation, and teacher professionalism
What Is Still Fundamental and What Is Not (Conference Board of the Mathematical Sciences, 1983)	Report of a conference making recommendations to the National Science Board
A Nation at Risk (NCEE, 1983)	Report of a "rising tide of mediocrity" and the need for high standards in all subjects
Educating Americans for the 21st Century (NSB, 1983)	A plan for making US mathematics and science education best in the world by 1995
New Goals for Mathematical Sciences Education (CBMS, 1984)	Report of a conference responding to NSB (1983) and related publications
School Mathematics: Options for the 1990s (Romberg 1984b)	Report of a conference responding to NCEE (1983) and related publications
The Underachieving Curriculum (McKnight et al. 1987)	Report of weak US performance in the Second International Mathematics Study (SIMS)
Curriculum and Evaluation Standards for School Mathematics (NCTM 1989)	A 1987 draft was widely reviewed and revised before final publication in 1989

"false choices between the old and new in mathematics, skills and concepts, the concrete and the abstract" (NACOME[14], 1975, p. 136). The call from politicians and the press, however, was for skills, the so-called "basics." In response, the National Council of Supervisors of Mathematics (NCSM, 1978) issued a position paper that re-defined basic skills in mathematics to include problem solving and applications, and NCTM appointed a new committee on mathematics curriculum for the 1980s. Aware of the importance of public perceptions of mathematics curriculum, NCTM also obtained funding from the National Science Foundation (NSF) to conduct the Priorities in School Mathematics Project (NCTM, 1981a), an extensive survey of professional and public opinion on content and pedagogical issues.

The discussion of mathematics education in the public arena was dominated by the decline in test scores (NCTM, 1980), although the extent and importance of the decline was much debated in the mathematics education literature (Price & Carpenter, 1978). Nevertheless, NCTM leaders were very much aware that past efforts to improve the mathematics curriculum had not had much effect. Stake and Easley's (1978) case studies of mathematics classrooms (along with other studies funded by NSF) provided a vivid description of the unlively state of mathematics education during the 1970s. Fey (1979), in his response to the Stake and Easley (1978) case studies and related research, spoke for many in mathematics education when he asked: "What role should the NCTM play in future developments to improve school mathematics?" (p. 504).

Setting an Agenda. One answer to Fey's question came the following year. The report, *An Agenda for Action* (NCTM, 1980), made a number of recommendations[15] for school mathematics and indicated that NCTM wanted to provide direction to the field, to assert its authority and share its expertise with a higher level of intensity than had been its custom. This *Agenda* provided a focus for NCTM activities and an organizational direction that had not been present in previous years. Shirley Hill, in her presidential address of 1980, gave some of the background for the *Agenda*:

> In the 1960s we learned that curriculum change is not a simple matter of devising, trying out, and proposing new programs. In the 1970s we learned that many pressures, from both inside and particularly outside the institution of the school, determine goals and directions and programs. ... A major obligation of a professional organization such as ours is to present our best knowledgeable advice on what the goals and objectives of mathematics education ought to be. ... In my opinion, we are approaching a crisis stage in school mathematics.

---

[14] A list of acronyms is presented in the appendix of this volume.

[15] The first five *Agenda* items recommended that problem solving be the focus of school mathematics, basic skills be defined as more than computation, calculators and computers be used at all grade levels, stringent standards of effectiveness be applied to teaching, and student learning be evaluated by a wider range of measures than conventional testing (NCTM, 1980).

> Policy makers in education are not confronting the deepest problems because the public and its representatives have been diverted by a fixation on test scores. ... We are still battling against an excessive narrowing of the curriculum in the name of "back to basics." (Hill, 1980, pp. 473-476)

During the early 1980s the various NCTM conferences (both national and regional) and committees gave much attention to the *Agenda* and its emphasis on problem solving, its broad definition of basic skills, and its recommendations on using calculators and computers at all grade levels. The *Agenda* was certainly a useful and even powerful policy document, but it was only about 30 pages long. When an NCTM committee tried to evaluate the success of the *Agenda*, they had considerable difficulty. As Joe Crosswhite put it, "The *Agenda* lacked the kind of specificity that would enable one to evaluate its impact." The *Agenda* was generally seen as a good policy document, but more was needed. As Crosswhite noted, "It was a framework, a pretty loose framework. We weren't able to invest the time or resources to flesh it out. It was always intended to be just a framework, something that people could work from." A state supervisor of mathematics confirmed the need for more direction[16] than was given by the *Agenda*, noting that "The *Agenda for Action* in 1980 was the best known document, literally a little pamphlet, 5×7. It was nice, but ... it didn't have a lot of heft, and people weren't paying much attention."

As NCTM's first major effort to influence public policy, however, the *Agenda* was an important activity. The planning that went into the *Agenda* was relatively sophisticated, foreshadowing the later work on the 1989 *Curriculum and Evaluation Standards*. Hill (1981), in an NCTM publication on the nature of educational change, indicated the thinking of the NCTM leadership on the change issue. She noted the complexity of the change process, where on the one hand, the school "appears to favor the status quo"; on the other hand, "it is ... particularly vulnerable to curricular fads" (p. 4). She also observed that providing guidance on curriculum is "an obligation" of professionals, and recommended increased efforts in public relations and political action. She concluded with an optimistic note: "The decade of the 1980s is a decade for mathematics" (p. 10).

Politics and the National Science Board.[17] Politically, the *Agenda* was an important force for unifying the NCTM membership, but other political forces were also at work. Shortly after the publication of the *Agenda*, the Reagan Administration eliminated all funding for mathematics and science education in Grades K-12 from the 1982 budget of NSF.[18] To the dismay of those

---

[16]Others were wary of over-specifying what teachers should do (Hill, 1980). This issue comes up again in discussions of the "grain size" of standards.

[17]The National Science Board (NSB) is the governing body of the National Science Foundation (NSF), the federal agency charged with supporting basic research and education in the sciences.

[18]See NCTM (1981b) for one NCTM response.

who worked in the Science and Engineering Education Directorate at NSF, some of their leaders capitulated easily to the demands of the Reagan Administration and its Office of Management and Budget. The NSF leadership took as its first priority the preservation of the research programs in the sciences and engineering. Although social science research at NSF was also targeted for elimination, professional societies in the social sciences were very effective in lobbying the Reagan Administration and were able to save substantial amounts of funding for the programs that were of most interest to the social sciences.

Mathematics and science education were not successful in opposing the Reagan Administration's cuts. It is ironic that the Reagan presidential campaign had emphasized its plans to cut the Education Department, which had only recently been raised to cabinet status during the term of President Jimmy Carter (Kaestle, 1992). However, the Education Department was saved by its friends in Congress, while the Science and Engineering Education directorate at NSF was lost. It seems clear that mathematics and science education were not effective enough in their lobbying efforts. The weakness of the lobbying effort, along with the lack of support for K-12 education from NSF leaders, was a significant factor in the loss of federal support.

Although the National Science Board was not effective in preserving mathematics and science education at NSF, they did negotiate with the Reagan Administration to preserve the graduate student fellowship program (a part of Science Education at NSF) and to establish a Commission on Precollege Education in Mathematics, Science, and Technology. The co-chair of the Commission was William T. Coleman, a prominent Republican who had served as Transportation Secretary in the Ford Administration.

The Commission began hearings in 1982 and invited Henry Pollak, then chairman of the Conference Board of the Mathematical Sciences, to conduct a small conference in September 1982 to advise the Commission on "The Mathematical Sciences Curriculum K-12: What Is Still Fundamental and What Is Not" (CBMS, 1983). Coleman, among others, was not at all convinced that curriculum change in mathematics was necessary or advisable. According to a source within the Commission, he held very traditional views of the mathematics curriculum, wanted a strong emphasis on computation, and was particularly resistant to the use of calculators in the schools. Apparently the efforts of Pollak and others were convincing; when the Commission presented its conclusions (NSB, 1983) the next year, the report of the Pollak conference was the first document in its set of source materials, and Coleman had become a stalwart supporter of educational change in mathematics and science. Under his leadership, the Commission even recommended a plan for providing additional federal financing of educational reform (NSB, 1983). Unfortunately, such funding never materialized.

Among the Commission's many recommendations was the call for developing a consensus on new objectives for the precollege curriculum and "guidelines for new curricula" (NSB, 1983, p. 46). They also recommended that professional organizations should take responsibility for directing educational change in their fields (p. 47). However, government funding did not become available in these areas either, and no federal support was forthcoming for NCTM's development of the *Curriculum and Evaluation Standards*.

The NSB (1983) report also referred to the National Commission on Excellence in Education (NCEE, 1983), which had published *A Nation at Risk* earlier that same year. The dramatic language of *A Nation at Risk* had received a great deal of media attention; all America had heard that "Our Nation is at risk ... the educational foundations of our society are presently being eroded by a rising tide of mediocrity that threatens our very future as a Nation and a people ... others are ... surpassing our educational attainments" (NCEE, 1983, p. 5). Many national leaders give credit to *A Nation at Risk* for helping to establish a climate that would support change in education.[19] As Tom Romberg noted:

> I think *A Nation At Risk* (NCEE, 1983) served primarily as a spark plug, a starting point for people. One of the concerns was that people in the states were requiring a third year of high school mathematics and making other political decisions without ever talking to the mathematics education community. That started a lot of people talking about the need for reform.

The publication of these two 1983 reports (NCEE, 1983; NSB, 1983) did much to spur the nation's concern about education. In particular, the response to *A Nation at Risk* was unprecedented (Ravitch, 1995). For example, business and academic leaders began to show increased interest in the problems of education (Lund & Wild, 1993). Following in the wake of these two 1983 reports, two meetings held in the fall of 1983 were particularly central to the development of the *Curriculum and Evaluation Standards for School Mathematics* (NCTM, 1989).

Responding to *A Nation at Risk*. In November 1983, the Conference Board of the Mathematical Sciences (CBMS), under the leadership of chairman Henry Pollak and administrative officer Marcia Sward, organized a two-day meeting at Airlie House in Warrenton, Virginia. NSF funded the meeting, which began with a discussion of the NSB (1983) recommendations. The report of the meeting, *New Goals For Mathematical Sciences Education* (CBMS, 1984), presented a broad range of recommendations on curriculum, instruction, and teacher education in mathematics.

One of the recommendations included in the CBMS (1984) report concerned the "communication of standards and expectations" (p. 5), especially "the expectations of colleges and universities relating to mathematics achievement" (p. 6). Lynn Steen, who attended the CBMS meeting in his role as President-Elect of the Mathematical Association of America, recalled:

---

[19]See Porter, Smithson, and Osthoff (1994) for a discussion of this point and the results of a related study.

In 1983, after the *Nation at Risk* (NCEE, 1983) report came out, the Conference Board of the Mathematical Sciences (CBMS) organized a retreat at Airlie House in Virginia and published a brief report. It was at this meeting that Joe Crosswhite introduced a motion that the group endorsed: that there should be a set of standards for school mathematics prepared by NCTM and that he wanted to get the whole group behind him to support this.

Several other recommendations were emphasized in the report (CBMS, 1984), including improvements in curriculum, teacher support networks, remediation, and faculty renewal. These topics had been addressed before in other conferences. The last of the recommendations in this report, however, was something new: the establishment of the Mathematical Sciences Education Board. MSEB was strongly recommended as an umbrella organization (with representatives from the mathematical sciences, education, and industry and business) that could provide overall direction for a broad reform movement in mathematics education.

In December 1983, a month after the CBMS meeting in Virginia, the US Department of Education, through its Office of Educational Research and Improvement (OERI), sponsored a meeting at the University of Wisconsin: "School Mathematics: Options for the 1990s." The reports of this second meeting (Romberg, 1984b; Romberg & Stewart, 1984) showed a direct link to the origins of the NCTM *Standards* (Romberg & Webb, 1993); the recommendations for a task force to develop guidelines for mathematics curriculum were explicit (Romberg, 1984b). Other recommendations dealt with teacher education, research, and, as at the Airlie House conference (CBMS, 1984), the need for a national steering committee like MSEB. Romberg and Webb (1993) have provided more detail on how the subsequent reform efforts have addressed the ten recommendations of the "Options for the 1990s" conference.

Lynn Steen described the role of the two conferences in the reform effort this way:

> What I specifically remember—which from my point of view was a fairly dramatic event—was the wrap-up session of the Airlie House Conference (CBMS, 1984). At that session, Joe Crosswhite introduced two motions that led to the *Standards* and to MSEB. A month later, Tom Romberg had a meeting at Wisconsin where the same issues were discussed. I know from talking to people who attended the Wisconsin meeting that they think of that meeting as the origin of the *Standards*, whereas those who attended the Airlie House meeting think of it as the origin. They just happened to be at different meetings, but both were within a month of each other.

The two conferences (CBMS, 1984; Romberg, 1984b) represented in large part the response of the mathematics education community to the passionate rhetoric of *A Nation at Risk* (NCEE, 1983) and the subsequent national discussion about education. The 68 participants who are listed in the two reports include six people who attended both meetings: Joe Crosswhite of Ohio State University (and NCTM President-Elect), Barbara Nelson of the Ford Foundation,[20] Henry Pollak of Bell Communications Research (and chair of CBMS), Tom Romberg of the University of Wisconsin, Paul Sally of the University of Chicago School Mathematics Project (UCSMP), and Steve Willoughby of New York University (and NCTM President).

Concerns about Resistance to Reform. Barriers to change were a prominent concern at both meetings (CBMS, 1984; Romberg, 1984b). In the United States, resistance to educational change is complicated by a long tradition of local control of K-12 education. In reconstructing the events of the mid-1980s, it is important to keep in mind that the setting of national standards for school curriculum was seen as a very bold step. NCTM was advised that "even the word *standards* might not be well received, that many people might not be able to separate the notion of national leadership from the specter of national control" (Crosswhite et al., 1989, p. 55). Federal officials were particularly sensitive; according to one NCTM leader, the report of the "Options for the 1990s" conference (Romberg, 1984b) was not allowed to use the word *standards*:

> We were told by the Department of Education that we couldn't use their name or their funds [for a report on standards]. At that point in 1984, the federal government was not about to use the words *standards, national standards*.

In spite of this concern, Jim Gates of NCTM recalled:

> We charged ahead. Bear in mind that we didn't go to the federal government for [funding to develop standards] because of this concern. I believe that the federal government at that point was taking the position that the use of the term *national standards* was politically dangerous.

Gates reported that resistance to the notion of standards was also prominent in some parts of the private sector:

> I remember one of the editors from a publishing house, one of the older people. We asked for feedback, and the response was that publishers could not write textbooks based on these standards because the consumers, school boards, schools, and superintendents would not want to have any part of anything called *standards* from a national group.

By the early 1990s, the notion of standards appeared to be well accepted, and concerns at the state and local levels had largely disappeared.[21] NCTM President Iris Carl indicated that many state governors liked the fact that there was national leadership in developing standards. "They are not being given specifics; they retain control at the state level." States still had the authority to interpret and implement standards as they saw fit.

---

[20]The contributions of Nelson to the Urban Mathematics Collaboratives case study (Webb et al., this volume) and Pollak to the NCSSM Precalculus case study (Kilpatrick et al., this volume) are also notable.

[21]However, concerns about national educational goals returned in the mid-1990s (Ravitch, 1995).

There were also concerns about keeping the mathematical community unified in its support of curriculum change. The Airlie House conference (CBMS, 1984), following the structure of CBMS, included representatives from all the major organizations in the mathematical sciences, most of which had been included in the earlier Pollak meeting on "What is Still Fundamental" (CBMS, 1983). These meetings kept up some involvement with the leadership of the mathematics research community, and representatives of these organizations continued to be apprised of major developments (like the NCTM *Standards*) as they occurred.

The "Options for the 1990s" conference (Romberg, 1984b), on the other hand, reached out more to the education side of the arena. That meeting included participants from school districts, state departments of education, and national educational organizations, including one representative of the publishing industry, Vivian Makhmaltchi of Macmillan. In her paper for the conference, Makhmaltchi (1984) noted that an adversarial relationship could develop between her (representing textbook publishers) and others at the meeting, but that she hoped they all could work together to improve mathematics education. Tom Romberg remembered her comments:

> One key point [at the "Options for the 1990s" meeting] was the discussion about publishers. People were making comments about why publishers don't do something different. Vivian Makhmaltchi made the very strong argument that we live in a supply and demand economy. If you don't create a demand, we don't have the resources to build a different supply. So part of the story is that the *Standards* were developed to create a demand.

Makhmaltchi (1984) reminded the participants that the leaders of the reform effort should not get too far ahead of the teachers in the field. This concern about keeping in touch with teachers, especially with the general membership of NCTM, continued to be an issue in the appointment of the writing teams for the NCTM *Standards*.

The resistance to the notion of standards was significant, but not as strong as some had expected. The leadership of the reform effort in mathematics education persisted, but the specific proposal that NCTM prepared to obtain funding for the development of standards came only in part from the recommendations of the two meetings (CBMS, 1984; Romberg, 1984b) described above. A different call for change arose separately out of the regular work of NCTM committees.

NCTM Committees and the *Standards*

Like many professional organizations, NCTM has a well-established organizational structure. There are committees that deal with instruction, research, professional development, and other educational issues. Members of these committees are appointed by the president of NCTM, and they serve without pay, although they often do a substantial amount of work. An appointment to a committee is generally seen as a tribute to an individual's professional accomplishments.

One important stimulus for the notion of standards came through the committee structure of NCTM. As Past President John Dossey reported it:

> The Research Advisory Committee (RAC) received a request from one of the affiliated groups to censure the Saxon books on the basis of an article that was published in the *Phi Delta Kappan* (Saxon, 1982). RAC members felt that it was inappropriate for professional groups to censure material, especially in the absence of an agreed-upon set of standards. Rather, professional groups should take the leadership in promulgating the beliefs of the profession. We took that idea back to the NCTM Board in the spring of 1983.

Another NCTM leader gave a similar report:

> RAC had a request for information about research on the efficacy of John Saxon's algebra [text]. Our discussion quickly broadened to the general question of evaluating curriculum materials in the absence of standards by which to measure "success."... We were acutely aware that, in asking for the development of standards, we were asking NCTM, by implication, to abandon its long-standing and explicit policy not to endorse specific curriculum efforts.

At about the same time the Instructional Issues Advisory Committee (IIAC) was discussing problems in the textbook selection process. Jim Fey, University of Maryland, was on IIAC at the time:

> There was some concern from several places that textbooks, and therefore curricula, were being driven by non-professional considerations, political log rolling, and so on. One of the issues for our committee was whether we should make a professional statement from NCTM that could be used by mathematics educators in their school systems to select textbooks.

The IIAC had been assigned the task of developing standards for textbook selection in early 1983, before the appearance of *A Nation at Risk* (NCEE, 1983). The NCTM Board had approved funding of $200 for a survey of mathematics supervisors to find out common practices and problems in textbook selection. Fey recalled how the notion of standards got extended:

> Somehow we got onto the idea that what IIAC ought to do was define professional standards in general—not just for selection of textbook material but for curriculum, for teaching, and so on. So we worked out a schematic diagram or concept map [see Figure 1].

The recommendations of the two committees, RAC and IIAC, began to coalesce in the spring of 1983 at a meeting of the NCTM Board of Directors. As John Dossey remembered:

Figure 1.   **Early Plans for Standards**[22]

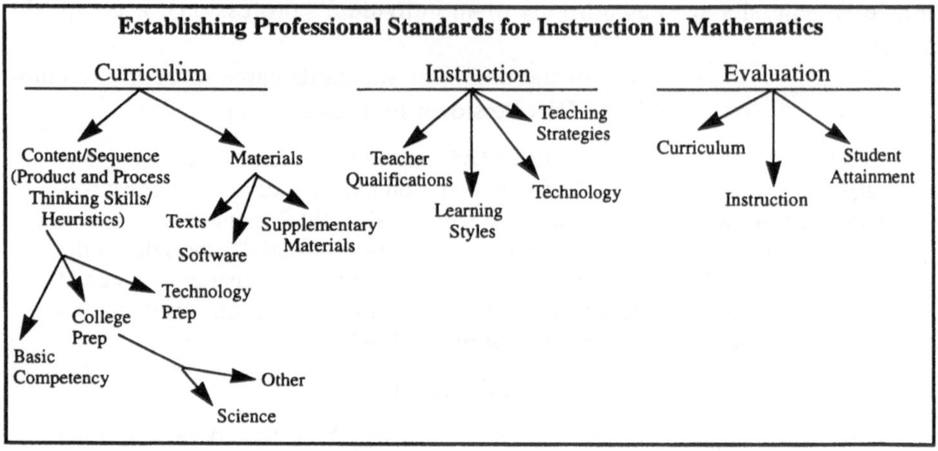

Establishing Professional Standards for Instruction in Mathematics

Curriculum — Instruction — Evaluation

Content/Sequence (Product and Process Thinking Skills/ Heuristics) — Materials — Texts — Software — Supplementary Materials — Technology Prep — College Prep — Basic Competency — Other — Science

Teacher Qualifications — Teaching Strategies — Technology — Learning Styles

Curriculum — Instruction — Student Attainment

It was interesting that not only the Research Advisory Committee was asking the Council to take a proactive stand, but IIAC was arguing a similar line. Some of us had already seen the raw data from the Second International Mathematics Study (SIMS), which wasn't even known yet by the other board members, and data from the National Assessment of Educational Progress (NAEP), too. So a number of issues were coming together. We needed to set some goals to stop this fad and that fad from affecting our curriculum.

Crosswhite (1990) reported another issue that helped motivate the NCTM *Standards*. A committee had been assigned the task of evaluating the impact of *An Agenda for Action* (NCTM, 1980). However, the *Agenda* was "written at a level of generality that proved difficult to translate into criteria for program evaluation. The new standards were intended to define criteria of excellence that would provide more specific guidelines for curriculum and instruction" (Crosswhite, 1990, p. 455). The importance of criteria for evaluating the success of programs and curricula was a central theme in IIAC as well.

Accountability of Textbook Writers. The accountability of textbook writers and others who developed materials for mathematics education was a central issue for IIAC during this period. Committee member Donald Chambers recalled the circumstances:

There was talk about something comparable to the Good Housekeeping Seal of Approval. We would have standards that could be applied to textbooks [and] to tests. The ones that were judged to meet the standards then would be given the

---

[22]A recreation of a hand-written sheet summarizing an early discussion of professional standards by the members of the Instructional Issues Advisory Committee of NCTM. The copy was provided by James Fey of the University of Maryland.

Seal of Approval and the ones that weren't would not. I don't remember how far that idea was taken, but my recollection is that we expected a little more judgment and a little more control than we eventually ended up with.

From the beginning of IIAC discussions, some of the problems of having panels of NCTM members evaluating textbooks were clear. Chambers continued:

> In many cases, the panels might very well include people who write textbooks. We mentally went through the list of all the textbook authors and we said, "These people would all have to disqualify themselves and then who's left? Would anyone feel comfortable—even if they didn't have a personal conflict of interest—passing down a critical review on a highly regarded colleague?"

The problem of conflict between committees of professionals and textbook authors was not just a theoretical issue. A number of NCTM leaders were involved in the California textbook adoption controversy in the mid-1980s, when the state's textbook adoption process resulted in the rejection of all K-8 mathematics textbooks as inadequate[23] (Webb, 1993b). Some of the authors of the rejected textbooks were respected leaders of NCTM, and there continues to be some sympathy among NCTM leaders for the authors and publishers who were inconvenienced and had to absorb the costs of revising their textbooks. There is still no consensus among NCTM leaders on that California textbook selection controversy, and it remains a sensitive point 10 years later.

Textbook writers are often major leaders in NCTM, but there are times when they are subject to critical appraisals. The underlying concern is that textbook authors are too likely to make compromises with publishers, and that publishers are only interested in profits, not quality. A writer of the NCTM *Standards* reported a special admiration for Glenda Lappan, chair of the 5-8 Working Group, because she "steadfastly refused to have any association with publishers so that she would not be influenced." This writer agreed that mathematics education needed the best people to write the textbooks, but lamented that "it is a corrupting process." The conservatism of publishers was also a concern, as a prominent NCTM leader recounted:

> I don't want to go into a lot of personal experiences, but I've had a few that were not terribly pleasing. [For example], when I was trying to suggest that there were other ways to do things, I was told by prominent publishers "That's not the way 'Big Red' does it." And if "Big Red" [the market leader] didn't do it that way, then they weren't going to entertain the notion of doing it that way.

Others do not see publishers as overly conservative, nor the process as corrupting. A major textbook author (who is also an NCTM leader) lamented the

---

[23]One textbook series received relatively favorable reviews, but all the textbooks were later rejected by an advisory commission of the State Board of Education. All publishers had to revise and resubmit their textbooks in order to be approved for adoption in California schools.

rigidity of people who could otherwise be contributing to the field by developing successful materials. Referring to a writer who is no longer active in mathematics education, she noted:

> She was one of the most creative people I have known, but she didn't live in the real world of working with publishers. Her work had to appear just as she presented it or it didn't appear at all. I said to her, "This isn't the way things happen in publishing or anywhere." She just could not see that we had to compromise to provide marketable materials.

One might expect that NCTM's interest in setting criteria for textbooks would be resisted by many publishers, but that was not the case. In an early meeting with publishers, NCTM leaders discussed the tentative plans for writing standards. At that meeting, most publishers were described as "very receptive" to having NCTM provide leadership in this area. Publishers were concerned that different states were developing conflicting guidelines for mathematics; if NCTM could provide direction for all states, the publishers' task of meeting state guidelines would be easier.

Although the relationship between NCTM and most publishers was very cordial, concerns about textbook quality and conflict-of-interest problems continued to trouble some NCTM members. The need to hold publishers accountable was a continuing theme in discussions among some committee members in the early 1980s.

Accountability as a Theme for Standards. The discussions by NCTM committees about developing standards led to a motion at the March 1984 meeting of the NCTM Board of Directors in Houston. At that meeting the Board agreed to appoint a task force[24] to propose a "plan for the development of a comprehensive set of guidelines for the K-12 school mathematics program (Minutes of the March 1984 meeting). Their work led to a grant proposal entitled "Development and Implementation of Professional Standards for Mathematics Education K-12." The proposal makes quite clear the interest in accountability issues:

> In the absence of detailed professional standards, there are only very limited ways that the Council can use its influence to promote excellence and to discourage unacceptable practices in the mathematics education programs of individual schools, districts, or state systems. The goal of the proposed task force is to draw together the best knowledge and experience of the profession into a national statement of standards and to devise effective mechanisms for the implementation and continual review of those standards. (Crosswhite, 1985b, p. 7)

The time line in the proposal (see Table 3) described the plans, including the development of standards for curriculum, instruction, and evaluation. In Year 4

---

[24]James Fey (Chair), Maria De Salvio, Howard Johnson, Steve Leinwand, and G. Edith Robinson.

Table 3.   Proposed Time Line of Activity
The major activities in establishing and implementing professional standards for school mathematics programs can take place on the following time line.

Year 1:	Fall	Appointment of the steering committee and project director; meeting of the steering committee to plan work agendas and to identify working group leaders and members.
	Spring	Preparation of background papers and resource materials for working groups.
	Summer	Working conference to initiate the development of the standards.
Year 2:		Preparation of draft standards documents and circulation for preliminary reaction by working groups.
Year 3:	Fall-Spring	Presentation of draft standards documents at NCTM meetings and to other concerned groups
	Summer	Meeting of the steering committee and the three working group chairs to finalize the standards document.
	Fall	Dissemination of the professional standards statement and planning of implementation efforts.
Year 4 (and beyond):		Continuing NCTM activities to enforce the professional standards.

From the proposal *Development and Implementation of Professional Standards for Mathematics Education K-12: A Project to Improve School Mathematics* (Crosswhite, 1985b, p.11)

the focus was on "continuing NCTM activities to enforce the professional standards" (Crosswhite, 1985b, p. 11). A member of IIAC recalled his feelings about the need for more accountability in mathematics education:

> I remember thinking at the time of *An Agenda for Action* (NCTM, 1980), "Well, here comes another set of recommendations." We're saying all these nice things, but when do we start putting some teeth into them? When do we start pointing the finger and saying that this person or this group is not meeting our standards?

The members of the NCTM Board of Directors tended to be more wary of making judgments about curriculum materials. As John Dossey viewed the work of IIAC:

> That committee came up with something that was much more regulatory, I would say, in terms of approving and disapproving things, which was a little of the style at the time. And it was that committee that formulated, I think, the first tangible recommendation that the NCTM Board should take action in the form of setting guidelines for curriculum.

The proposal, based in part on the report of Jim Fey's NCTM committee, was submitted to the AT&T Foundation on September 3, 1985, and later to several other agencies. The emphasis in the proposal on accountability issues was an important factor in the early discussions of standards. The minutes of the September 1986 NCTM Board of Directors meeting confirm the accountability emphasis by stating that a standard is "a 'specification' by which something can be tested or measured." A slightly different but similar definition appears in the July 1987 draft of the *Curriculum and Evaluation Standards* (NCTM, 1989); that draft states that "a standard specifies a criterion used to judge the quality of the mathematics curriculum" (p. iii). The corresponding statement in the final version is only slightly different, but the shift in emphasis is significant: "A standard is a statement that can be used to judge the quality of a mathematics curriculum" (NCTM, 1989, p. 2).

The result was that standards, which started out as measurable specifications and as criteria to judge quality, were softened to become statements for judging quality. Moreover, the definition of a standard was further refined and explicated in the published document (NCTM, 1989) to emphasize the alternative notions of a standard as a banner used to lead a group or to provide a vision of the future. The emphasis on interpreting the *Curriculum and Evaluation Standards* (NCTM, 1989) as an informed vision of the future has dominated much of the subsequent writing in NCTM journals (see, e.g., Crosswhite et al., 1989; Lappan, 1993).

To summarize the activities of the early 1980s, NCTM committees on instruction (IIAC) and research (RAC) needed criteria to do their work, whether their task was to advise on text selection or to respond to critics of reform (e.g., Saxon, 1982). The committee to evaluate the impact of *An Agenda for Action* (NCTM, 1980) also noted the need for more detailed recommendations on curriculum. Initial work on criteria for textbook selection was expanded to become standards for curriculum, instruction, and evaluation. Over a period of time, the emphasis on standards as accountability criteria shifted to standards as a vision of the ideal. This shift occurred mainly during the writing of the *Curriculum and Evaluation Standards* (NCTM, 1989) in 1987-88; the multiple meanings of the term *standards* were fully explicated only in the 1988 drafts.

Throughout these early stages of planning for the *Standards,* the organizational structure of NCTM had worked effectively. NCTM committees completed the tasks they were assigned and made their views known to the Board of Directors, and the Board appointed a task force to act on the committees' recommendations. The task force, with input from NCTM leaders, prepared a proposal that helped crystallize the thinking and actions of NCTM. That proposal became the foundation for the project that would eventually lead to the publication of the *Curriculum and Evaluation Standards for School Mathematics* (NCTM, 1989) and the standards movement in the US (Atkin, 1994).

The Writers of the NCTM *Standards*

Many of the writers of the NCTM *Standards* were not involved in the meetings and discussion of the 1983 to 1985 period. Most of them were not invited to the meetings of leaders (CBMS, 1984; Romberg, 1984b), and most were not members of the NCTM committees that were discussing the need for standards. From the point of view of the writers, the NCTM *Standards* were a natural step in the continuing efforts of professionals in mathematics education to improve practice in the field. *An Agenda for Action* (NCTM, 1980) had set a new direction for the mathematics curriculum, and the writers saw themselves as building on that initiative, as well as other efforts from within and outside NCTM.

The writers were active professionals in mathematics education, and they were influenced by documents like *An Agenda for Action* (NCTM, 1980) and by their experience in research, development, and teacher education projects. For example, curriculum projects like the "Middle Grades Mathematics Project" had an influence on the NCTM *Standards* through writers like Glenda Lappan (Webb, Schoen, & Whitehurst, 1993), and the initial drafts of the K-4 *Standards* show important similarities to the "Developing Mathematical Processes" curriculum materials, prepared by a team that included Mary Lindquist. At the 9-12 level, a major influence was the 1985 NCTM Yearbook, *The Secondary School Mathematics Curriculum*, edited by writer Chris Hirsch (1985). Mary Lindquist, a writer in the K-4 group, described the origins of the *Standards* this way:

> You can look at the *Standards* as growing in a very natural way out of the concerns of mathematics education. You can trace the *Standards* back to *An Agenda for Action* (NCTM, 1980), back to the NCSM (1978) statement that addressed the back-to-basics movement, and back to the 1975 NACOME report. The *Standards* came mainly from within mathematics education rather than as a reaction to *A Nation at Risk* (NCEE, 1983) or federal policies.

Although Romberg's introduction to the 1989 *Standards* began by noting the role of documents like *A Nation at Risk* (NCEE, 1983) and the NSB (1983) report with their calls for educational reform, the writers tended to see things from a different perspective. As one writer described the origins of the *Standards,* "I don't think it came out of *A Nation at Risk*; we have been saying the same things for years."

Neither the writers nor the leaders put much emphasis on the aspects of international competition that were prominent in the rhetoric of *A Nation at Risk*. However, both groups were concerned about the weaknesses in the US curriculum (McKnight et al., 1987) that were reflected in the data from the Second International Mathematics Study (SIMS). As one NCTM leader put it, "We weren't being motivated by 'world class standards' at that point. But we did have comparative data, especially in terms of the Japanese curriculum, which showed so much more intensity than ours did."

Although the SIMS data were important in the thinking of NCTM leaders, reports of the international study (e.g., McKnight et al., 1987) were not cited in the list of references in the NCTM *Standards* (1989). That omission caused some concern among writers and reviewers, but John Dossey described the reasoning this way:

> I think that if you base your argument for reform on a temporal research result, you're being reactive rather than proactive. The focus was to take the negative, competitive statements out of the document, and make the document a proactive, positive statement. Let's say what we believe and then act on it.

In writing standards, the authors were more concerned about mathematics and the internal forces that were driving change in mathematics education, rather than international competition or political forces from outside of mathematics education.[25] Technology was a major source of these changes. The availability of calculators had made the traditional emphasis on paper-and-pencil computation in the K-8 curriculum an anachronism; calculators also made number sense and estimation more important than ever before. The advent of graphing calculators had the same kind of impact on the algebra curriculum for the writers in the 9-12 group.

The view of the learner was also changing. Kilpatrick (1992) has noted that behaviorist psychology was never fully accepted by researchers in mathematics education, where Brownell and Bruner have always been major influences. The mathematics curriculum, however, had reflected Thorndike's *connectionism* more than Brownell's emphasis on meaning. The plans of the 1960s for new curricula that followed the "cognitive structuralism" (Jenness, 1990, p. 131) of Bruner and others were overturned by a renewed emphasis on basic skills and continued fragmentation of the mathematics curriculum in the 1970s. The popularity of cognitive science in the early 1980s (Romberg & Carpenter, 1986) provided some support for those who viewed the learner as an active processor of information, and the *Standards* are often associated with constructivist approaches to the learning of mathematics (Steffe & Kieren, 1994). One leader reported that he didn't remember "a constructivist approach" being that influential when the *Standards* were planned; however, as the *Standards* developed, the writers were increasingly influenced by changes in conceptions of the learner. Tom Romberg recalled:

> The term that we did not use in writing up the *Standards* (but we certainly talked about) is what might best be called the social constructivist's notion of learning. One of the arguments that people have made is, "Why didn't you call yourself 'social constructivists'?" But that would have put off people who didn't understand that set of notions.

---

[25]The need to update the mathematical content of the curriculum was judged a major influence on the development of the *Standards* by 85% of those who returned our questionnaire, compared to 17% for the issue of economic and global competitiveness. See also Bishop (1990).

Although the NCTM *Standards* were motivated initially by accountability issues, changes in conceptions of mathematics and in theories of learning became important forces as the writers went about their work. These forces influenced all the writers, as did their experiences at the state and local levels.

State and Local Influences. Given their task, writers naturally looked at other statements of curricular goals, including curriculum guidelines from California, Oregon, Wisconsin, and other states. The California *Framework* (California Department of Education, 1985) was mentioned frequently. As one writer noted:

> Certainly the 1985 California *Framework* was one of the documents that was used in helping to formulate the NCTM *Standards*. It was something that everybody in all of the groups was familiar with and looked at for help in thinking about what the *Standards* might contain.

California is the most populous state, and Californians are often concerned that their large numbers within NCTM are resented.[26] A leader from California noted that he was careful not to push too hard since people will reject an idea "just because it comes from California." But he did think that California was a major influence:[27]

> I really think there was a very big influence from the mathematics education community in California on what happened. When I look back, I think that the 1985 California *Framework* really set the stage for a lot of things.

Mathematical power was a phrase from the California *Framework* that was also emphasized in the *Standards*. As John Dossey noted:

> The California *Framework* of 1985 talked about mathematical power. I've tried to trace that back, and that's the first document that I know that really talks about mathematical power—a central theme for the *Standards*. The *Standards* get credit for it, but it was in that California *Framework*.[28]

Other interviews suggest that events in California were not especially influential at the national level for people from other states. For example, the discussion of computation and calculators, which begins the California *Framework* (California Department of Education, 1985, p. 1), makes recommendations that are similar to those in the *Curriculum and Evaluation Standards* (NCTM, 1989, p. 9). However, an NCTM leader felt that similar recommendations were being discussed all over the country, not just in California.

Marilyn Hala, who had served as the State Supervisor of Mathematics in South Dakota before taking a position at NCTM headquarters, said that documents

---

[26]Californians constituted less than 11% of those interviewed for this report.

[27]Questionnaire data confirmed this view; the writers rated the California Framework just behind *An Agenda for Action* in its influence on the *Standards*.

[28]For the earliest use of "mathematical power," we suggest a 1916 paper by W. D. Reeve, cited in Kilpatrick and Stanic (1995).

from the big states (like California or Texas) often set a pattern that is very influential among people in the smaller states:

> If you were to look at the development of curriculum from a state level, you naturally would go to the states that were state-wide textbook selection states because you knew that the materials that your teachers were going to use were influenced by those states. So at a state level, you naturally would go to California, Florida, Michigan, and Connecticut. There were several states that you knew were bellwether states.

The approach to reform in California did have an influence on many writers and on many states. Changes in mathematics education in other countries also were influential.

International Influences. The international community in mathematics education influenced the NCTM *Standards* in several ways. When the writers gathered in Utah in 1987, Anne Zarinnia, an assistant to Romberg and a library and media specialist, provided the writers with a rich set of resources to help stimulate their thinking. These resources included mainly materials written in English, so the number of foreign countries that were represented was small. But the materials did include the Cockcroft (1982) report and other British publications. As Zarinnia put it:

> We just flooded them with stuff, and we asked them what they would like in the way of reading materials. One of the secondary group said that they wanted the School Mathematics Project (SMP) materials from the UK, so we acquired a library of SMP materials.

Tom Romberg reported:

> We tried to organize materials from other countries—England, the Netherlands, Australia. Some of us spent a fair amount of time at the University of Chicago looking at some of the Wirszup materials. The Mathematics Curriculum Teaching Project from Australia had a lot of interesting examples.

At the 9-12 level, the work in England was a significant influence. Chris Hirsch recalled:

> At the time that we were beginning to start work on the *Standards*, there was some interesting work being done at the Shell Centre in England in terms of more qualitative applications of mathematical thinking, for example, the work on the language of functions and graphs. We were also able to examine some of the more recent work of the School Mathematics Project, particularly their more eclectic approach to geometry and topology.

There were also materials by D'Ambrosio (of Brazil) dealing with ethnomathematics, and the writings of Freudenthal (of The Netherlands), whose work on "didactical phenomenology" was thought to be "a little hard for most people" to get through. Only a few of the writers mentioned these works or related research, but the leadership was clearly influenced by them, and saw them as compatible and supportive. Tom Romberg noted:

There was a sense [in the writing groups] that kids ought to experience mathematics—that they're reinventing some of the important ideas. Then teachers negotiate with them the language in terms of signs and symbols that we commonly use.

Romberg's comments indicate the influence of international researchers such as Vergnaud (1982) of France, especially his work on cognitive structures and multiplicative conceptual fields, and de Lange (1987) of the Netherlands, with his curriculum development emphasis on "realistic mathematics education" in the middle school. The presence of researchers on each of the writing teams for the *Standards* ensured that ideas from the international research community in mathematics education would be included in the thinking of each group.

Summary of Origins

During the 1970s, the mathematics education community came of age. Earlier reform efforts had been driven mainly by the mathematicians, but now leadership shifted to NCTM. A knowledge base in areas like mathematics curriculum, learning theories, and teacher education had developed, as reflected in NCTM yearbooks, journals, and other publications.

The "new" mathematics of the 1960s had been part of a "period of triumph for axiomatic methods" (Bass, 1994, p. 921). By the 1980s, however, mathematics had changed and the curriculum needed to change with it. Technology, in the forms of computers and calculators, was a main source of this change. New views of content were accompanied by a new conception of the learner, making pedagogical changes necessary as well. As Bass (1994) pointed out, during the 1960s the mathematics community had sometimes confused logical and psychological bases for organization of curricula. The development of research in mathematics education during the 1970s helped to reduce the confusions of the New Math era (Kilpatrick, 1992) and to support the need for changes in both curriculum and instruction.

Rejection of the "new" mathematics of the 1960s was voiced by the public and some professionals with the slogan "back to basics." The emphasis on "basic skills" tended to fragment and narrow the curriculum in ways that were counter to the main trends in mathematics curriculum and pedagogy. In the NACOME (1975) report, Jim Fey provided a carefully reasoned analysis of the problems that occurred during the reform efforts of the 1960s and the inappropriateness of a narrow emphasis on skills in the 1970s. The NSF-funded status studies (e.g., Stake & Easley, 1978) and other sources of data (e.g., National Assessment of Educational Progress [NAEP], Second International Mathematics Study [SIMS]) provided a basis for describing what was happening in schools with respect to the aspirations of mathematics education. For example, the NAEP data from the 1970s made clear that students were relatively proficient at routine computations, but severely deficient in other areas, particularly problem solving (Carpenter, Corbitt, Kepner, Lindquist, & Reys, 1981). The report of the

SIMS data (McKnight et al., 1987) was effective in identifying weaknesses in the US curriculum that contributed to poor student performance compared to other countries.

As stated in the original proposal (Crosswhite, 1985b) requesting funding for the development of the *Standards*, the problem was "closing the gap between the recommended ideals of professional practice and the reality of mathematics education" (p. 2). The public mood, which was exerting pressure for more traditional instruction, had been countered in part by NCTM's (1980) *Agenda for Action*. The accompanying organizational efforts to implement the *Agenda* gave NCTM a sense of what could be accomplished with a more detailed set of curriculum guidelines. NCTM's knowledge of educational change (Price & Gawronski, 1981) had developed well beyond the educational community's simplistic notions of the 1960s; see McLaughlin (1990) for an analysis of that earlier point of view. NCTM was ready to lay out a broad plan of reform, a plan that could reasonably be described as an effort to raise standards.

Strong leadership from people like Shirley Hill and Joe Crosswhite stimulated NCTM as an organization to take a more activist role in reform efforts. The call for higher standards in mathematics education was originally expressed in terms of monitoring the quality of educational materials, and initially the plan for *Standards* focused on making textbook publishers and test developers more accountable for their actions. Later the discussion of NCTM *Standards* shifted away from accountability issues to the *Standards* as a vision of ideal practice.

Development of the NCTM *Standards*

The initial plan for the NCTM *Standards* came out of the work of the NCTM Instructional Issues Advisory Committee (Crosswhite, 1985b). The plan included the preparation of standards for curriculum, instruction, and evaluation, all of which would have been included in one document. An initial task force report suggested that this combined version of a standards document could be prepared during 1985-86 for a "total estimated cost" of $258,500. As the NCTM Board of Directors planned the development, however, a number of changes were made, including separating standards for teaching from those for curriculum and evaluation.

One of the early debates was over whether to start with standards for curriculum or teaching. Many people thought that teaching should come first because of its importance. One might also speculate that the experience of the 1960s suggested that starting with curriculum (again) would sound too much like the New Math era. As Tom Romberg described the conflict, "I must admit that my own predilection was to start with teaching, because I thought the classroom and what teachers do with kids is probably at the center. But that isn't what NCTM wanted." Glenda Lappan had a similar reaction:

Between the time that John Dossey invited me to be a part of this, and when Tom Romberg brought the group of leaders together, a decision had been made that the first document would only be on curriculum. I don't know how that decision was made. At the time, I argued against it in our first meeting with Tom, but it was clear that the decision had been made.

John Dossey recalled how the NCTM Board decided to focus first on the *Curriculum and Evaluation Standards:*

We were talking about doing both curriculum and evaluation as well as the teaching standards in one document. We realized that it would be too big of a change. We felt that it would be a difficult enough task just to get everybody to say, "We want to try and change our content and our view of content."

The decision to start with a focus on content (and consequently evaluation) was based in part on the fact that most people in mathematics education find it somewhat easier to discuss content than pedagogy. Moreover, the changes that were occurring in technology helped to make the focus on content issues particularly timely. Some NCTM Board members were also concerned about the size of the project; separating out the standards for teaching made the project smaller and easier to fund. In some ways, the decision to focus on content was just "being practical," as one leader put it. Leaders and writers generally agree now that it was the right decision.[29]   As one state supervisor put it:

I think the *Teaching Standards* would not have served as the banner. It was absolutely crucial to have the *Curriculum Standards* be the banner. Very few people really remember that there are also evaluation standards in the document; they sank without a trace. If there'd been a comprehensive document called *Curriculum, Teaching, and Assessment Standards,* teaching and assessment would have dropped off. What would have gotten the attention was the content standards.

NCTM leaders, however, were also aware that it was important to work on changing all parts of the educational system. The experience of the 1960s suggested that you need to "do it all," in the words of one leader. Changing the curriculum would not be sufficient, so the publication of the *Curriculum and Evaluation Standards* (NCTM, 1989) was followed two years later by the *Teaching Standards* (NCTM, 1991) and then the *Assessment Standards* (NCTM, 1995).

Funding for the NCTM *Standards*

The NCTM *Standards* project was planned from the start as a large, multi-year effort. For a project of this size, NCTM leaders would normally have approached

---

[29] A minority view was expressed by a state supervisor, who noted that separating out standards for teaching sent the "wrong message." He thought that "Creatively, it could have been done in one manageable document."

one of the government agencies for support. However, federal policies during the Reagan era were focused on reducing government influence, not extending it, so no federal support was available.

Lacking federal support, NCTM submitted their proposal to several private foundations, requesting funding of $487,000 to develop standards for mathematics education (Crosswhite, 1985b). Funding was requested for writing curriculum and evaluation standards, for extensive review of the draft, for efforts to gain the endorsement of other groups, and for dissemination of the document to members, state officials, legislators, text and test publishers, accrediting associations, and others. Budget plans included the production of 10,000 copies of a 50-page document and 100,000 copies of an executive summary. None of the private foundations[30] was willing to support a project of this size, although the AT&T Foundation did provide a small seed grant of $25,000 (Crosswhite et al., 1989).

When funding was not available from private foundations, the NCTM Board of Directors decided to fund the project internally. During the NCTM Board of Directors meeting of March 1986, Joan Akers presented the report of the Instructional Issues Advisory Committee, which recommended that the development and implementation of professional standards for mathematics education in Grades K-12 be considered "highest priority" and that the curriculum phase of the project should be initiated "without further delay," according to the Board minutes. At that meeting the Board authorized $150,000 for the beginning of the work.[31] The motion passed unanimously, 13-0. Of course, this amount was only the beginning, and did not even cover the costs listed in the proposal for compensating the project leaders and writers. In addition, just the public relations firms eventually cost around $200,000, and other expenses (staff time, a free copy to each member) would drive the total cost of the 1989 *Curriculum and Evaluation Standards* to something over $1,000,000.

After the 1989 *Curriculum and Evaluation Standards* were published and the success of the project became known, the MacArthur Foundation[32] came to NCTM and offered $85,000 to support additional work related to the *Standards*. Other private foundations (e.g., the Exxon Education Foundation) have also provided support for NCTM activities (e.g., see Ferrini-Mundy & Johnson, 1994).

NCTM officials and members took considerable pride in the fact that the 1989 *Curriculum and Evaluation Standards* were "developed and distributed almost entirely with membership funds" (Crosswhite et al., 1989, p. 56). In addition, there were significant advantages in the political arena. For example,

---

[30] AT&T Foundation, the MacArthur Foundation, and Honda were among those contacted.

[31] An NCTM leader noted that internal funding of the project was possible only because of the "dramatic upturn" in NCTM finances that had begun in 1984, following a resurgence in membership.

[32] Peter Gerber of the MacArthur Foundation had been involved in the development of *A Nation at Risk* (NCEE, 1983) as a US Department of Education staff member (Kaestle, 1992).

NCTM President Iris Carl recalled how Governor Romer of Colorado, one of the important advocates of the NCTM *Standards*, was fond of saying, "You have to remember NCTM did it on their own nickel." She, like other leaders, noted that NCTM is "proud of doing the *Standards* independently; it helps to make them acceptable." John Dossey commented further:

> The *Standards* are more powerful because we did it with our own money. Had we done it with others' money, we'd be in the struggle that the other disciplines are going through. Who owns their standards when they are paid for with government money?

Although some members of NCTM reported considerable resentment that other curriculum areas received large amounts of federal funding (up to $3,000,000) to develop their own standards documents ("Guide," 1993), other NCTM leaders were just delighted that NCTM never had to undergo the kind of federal regulation that these other curriculum areas faced. For example, when NCTM leaders decided to divide up the grade levels into K-4, 5-8, and 9-12 for the *Standards*, the decision was based on the leaders' ideas about the most natural way to divide the mathematics curriculum. Now, as an NCTM leader noted, "All the other standards projects are using the same division," following the specifications of the funding agencies. Of course, a division that makes sense for mathematics may not be the best for another curriculum area. However, the pattern set by NCTM has had a major influence on how other curriculum areas were directed to develop standards (Massell, 1994).

Appointing the Commission and the Writers

Two different groups were appointed to develop standards for curriculum and evaluation. The NCTM Commission on Standards for School Mathematics was established first to oversee the project; writing groups were appointed later. As NCTM presidents Crosswhite, Dossey, and Frye (1989) reported:

> The members of the Commission were chosen to represent the NCTM Board, the Mathematical Association of America, the Mathematical Sciences Education Board (MSEB), supervisors, publishers, and the mathematics-education community in general. (Crosswhite et al., 1989, p. 56)

The Commission[33] was to be a "sounding board" and a buffer between the writers and the NCTM Board of Directors. There was some concern among the

---

[33]Members of the Commission included Tom Romberg, Chair, University of Wisconsin; NCTM presidents (and future presidents; see Table 1) Iris Carl, Joe Crosswhite, John Dossey, Shirley Frye, and Shirley Hill; James Gates, NCTM Executive Director; Dale Seymour, a publisher; Lynn A. Steen of the Mathematical Association of America, and the four chairs of the working groups: Paul R. Trafton (K-4), then of National-Louis University; Glenda Lappan (5-8), Michigan State University; Christian Hirsch (9-12), Western Michigan University; and Norman Webb (Evaluation), University of Wisconsin.

leaders that individual members of the Board might try to influence the writers directly, possibly trying to advance their own personal agendas.

As Crosswhite (1990) has observed, the NCTM *Standards* were not intended to be a radical document. The plan was to have standards that were more evolutionary than revolutionary. The appointments to the writing teams reflected this view. The working groups that wrote the *Standards* (NCTM, 1989) were structured to include the different constituencies represented in the organization: teachers, mathematics supervisors, teacher education faculty, and others with expertise in areas like research and technology (Crosswhite et al., 1989). Joe Crosswhite described the Commission's thinking this way:

> In some cases, we wanted individuals working on the inside with us instead of carping at us from the outside. I don't mean anything at all about their qualification for serving at that writing stage. It's just that there were political concerns, geographic concerns, grade level concerns. We certainly wanted people who knew mathematics content [and] who could represent emerging technology. I'm sure the Commission had input but ultimately [NCTM President] John Dossey made the decisions.

Dossey recalled his thinking about how the four writing groups should be structured:

> The composition of groups was five people with representation from universities, classroom teachers, teacher educators, [including] somebody who had been around and had a lot of experience—who could represent not a traditional view, but someone who understood the status quo well, who understood the dangers of change, and who was a worker for change, but who knew that you could not just flip a switch and have it happen.

When asked if there was a conscious effort to have a conservative voice, Dossey responded:

> No, but we wanted someone who understood the problems of change and past efforts, who had a vision at least back to the New Math times and knew why change hadn't occurred, knew what the obstacles were. You need to have some people with "bleeding wounds" who understand change and the problems of change. We also wanted someone who knew research and technology; some people took more than one role.

The effort to include older, experienced NCTM members in the *Standards* project was apparently successful. As of late 1995, 11 out of 34 (6 members of the Commission and at least 5 writers) had retired.[34]

---

[34]Some of these were early retirements.

The members of the writing teams[35] were sometimes aware of their particular role (e.g., the teacher or supervisor in each group), but the role was not part of their assignment. For example, none of the writers who were known for their research reported that they were assigned the role of representing the research community. The leaders of the writing teams did have a special role to fill, as Chris Hirsch, chair of the 9-12 working group, recalled:

> My primary role there was to work with the other chairs in terms of trying to provide a broad framework of thinking that could serve as a starting point for the working groups, and then to serve as both a catalyst and a convener of ideas as working groups developed the *Standards*. Then ultimately [we] served as the editor and final writer of the section of the *Standards* for which we were working group chairs.

Coordinating the working groups was no simple task. The writers came from mathematics departments and colleges of education, from elementary and secondary classrooms, and from positions as district or state supervisors. Turning such disparate groups of people into productive writing groups was a complicated process.

Writing the *Curriculum and Evaluation Standards*

The developmental process used in the preparation of the *Standards* was lengthy. The original four-year plan from the 1985 proposal was included earlier in Table 3, and the actual events (summarized below in Table 4) followed the plan reasonably well (Crosswhite et al., 1989).

During the 1986-87 academic year the Commission and writers were appointed and the leadership met to lay out the tasks. The four chairs of the writing groups met with Romberg in Wisconsin in December 1986. The meeting was organized in response to concerns of working group leaders who felt the need for more detailed planning in preparation for the next summer's writing session. This meeting was viewed by some of the leaders as the origin of the "grain size" issue. As Glenda Lappan recalled the meeting:

> We discussed that this had to be a document that had a small set of statements at each of the levels. As I recall, we set our goal at some number fewer than 15 statements. This was important because it separated, from the very beginning, this document from anything that could be called scope and sequence. This was a document about the big ideas. We already had the shape of the standard, more or less the "grain size" of the standard. For the people in the working groups who came in thinking that we were going to write another molecularized list that would have hundreds of statements in it, it was very helpful for us

---

[35]The Working Groups included Hilde Howden, Mary Lindquist, Ed Rathmell, Thomas Rowan, and Charles Thompson for K-4; Dan Dolan, Joan Hall, Tom Kieren, Judith Mumme, and James Schultz for 5-8; Sue Ann McGraw, Gerald Rising, Hal Schoen, Cathy Seeley, and Bert Waits for 9-12; and Elizabeth Badger, Diane Briars, Tom Cooney, Tej Pandey, and Alba Thompson for Evaluation. Anne Zarinnia served as project assistant.

Table 4.   Major Events in the Development of the *Curriculum and Evaluation Standards*

June-July 1987	Writers met in Park City, Utah, for two weeks and produced an incomplete draft of 90 pages
August 1987	Writers met in Leesburg, Virginia, for two weeks and expanded the draft to 180 pages
October 1987	10,000 copies of draft distributed to NCTM members for review
September 1987-May 1988	Discussion of draft at NCTM regional and national meetings
Spring 1988	Review of sections of the draft by focus groups in MSEB project
June 1988	Writers met in Park City, Utah, for two weeks to revise draft based on feedback of over 2,000 responses
September-October 1988	Final draft of 290 pages was approved by the NCTM Board, edited, and prepared for printing
March 1989	Completed document released at Washington, D.C., press conference

to be committed to the level at which this document was going to be written. We were able to articulate for the working groups what the document had the power to do if we could keep it at this bigger statement level.

Although the "grain size" issue was settled in advance of the meetings of the writers, the issue was not entirely predetermined. As Norm Webb reported:

We knew that having too many standards would dilute their impact, but yet we had to have enough to cover the important topics. I know that the number of standards by a working group changed by combining topics or separating topics and that this was done more by consensus within a working group and negotiation across working groups. Grain size was not a conspiracy of the leadership; it came out of the process where people were put on the top of a mountain with the understanding that they had to come to an agreement. People took consensus seriously.

The way that even the "dictates" of the leadership regarding the number of standards got interpreted by writers suggests how much authority the writers felt they had. Romberg's leadership style was often described as loose or non-directive, and the writers were chosen at least in part for their independence. Thus the lack of direction from the top and the independence of the writers combined to make the task of the group leaders a difficult one. Romberg would rarely take a stand on an issue, preferring to let the groups fight it out, though there were times when people "very much wanted him to take a stand." The chairs of the writing groups were often caught in the middle, and the pressures were intense. One reported feeling "inadequate," not having a direction or a vision of what the *Standards* should be: "I would go away and not be able to sleep at night because I didn't know what to do. It really was a humbling experience."

The writers had their first chance to meet with the leaders at the NCTM annual meeting in Anaheim in April 1987. John Dossey recalled:

> The writing groups got together at Anaheim at the NCTM annual meeting in 1987 for a "launch"—and a speech about the project. Some of the writing group members had not met each other. Then they were inundated with information.

Tom Kieren, a writer in the 5-8 group, recalled that first meeting in Anaheim, as well: "One of the things that struck me was that we were not to be making suggestions that seemed very radical. This really struck me at the first meeting—the narrowness of the task."

The nature of the task was outlined in documents from the leaders in early 1987. Romberg (1987a, 1987b) prepared "some initial ideas" for the Commission and the writers, and the writing group leaders also prepared documents[36] for their own group. The writers had materials to read, but they arrived in Utah with a great deal of uncertainty about their task. After two weeks in Utah in June and July 1987 the writers produced a first draft of 90 pages, mostly made up of the "bullets" for the standards. They added the discussion and elaboration of each standard in the longer second draft (180 pages) that was prepared in Virginia during two weeks in August 1987. Our interviews with the writers provided an indication of how the debates were framed, and diaries and other documents indicated the tensions that the writers felt as they struggled to develop standards in a short period of time.

A log from one of the writers provided details of the first writing session in 1987. After the organizational meeting on Sunday evening, June 28, 1987, the group got down to work on Monday morning:

> June 29 (Monday), 8:30 - 10:00. Met with the large group. Tom [Romberg] talked about the form of the standards [and the] vision of mathematics. The view is that computers have taken away the need to do many procedures. What is important is the understanding of why and when procedures are needed.

---

[36]See, for example, Webb's (1987) "More Detailed Thoughts for Consideration," which was prepared for the Evaluation Working Group.

> Other views of math are:
>   math as a cultural achievement
>   math is a language
> Standards should not be a traditional listing of content areas but would like to
> see them as integrated. (Log 1, p. 2)

Notions of "mathematics and culture" and "mathematics as a language" were part of the first discussion, but the working groups would later re-discover and re-shape these issues in their own way. The day continued with a meeting of the working groups, lunch, cross-group meetings, and working group meetings again. The working group leaders also met at the end of the day. The schedule was similar for Tuesday, when some of the different meanings of the term *standard* began to come out:

> June 30 (Tuesday), 8:30 - 10:00. Met as a large group. ... Some people felt
> strongly about coming up with something new. It appears for them, standards
> are to be a target that is to lead the curriculum. A few have the view that stan-
> dards are criteria for excellence. (Log 1, p. 3)

> July 1 (Wednesday) Some see the standards as a form of communication ... and
> as a banner. (Log 1, p. 5)

The distinction between process and content standards came up, providing a chance to divide up into different groups to discuss each type. Possible statements for standards were suggested and evaluated, usually kindly and thoughtfully. The K-4 group reportedly said that the "need for process standards goes over all grade levels, [but] nobody [else] said anything regarding this" (Log 1, p. 5). Later all the writing groups took some credit for developing the standards on problem solving, communication, reasoning, and mathematical connections that appeared in all levels (K-4, 5-8, and 9-12).

The different groups were beginning to work together and to develop a sense of their own identities as well as their responsibilities. However, the evaluation group found it difficult to proceed in isolation without knowing more about how the content would be described; they split up and met with the K-4, 5-8, and 9-12 groups. This decision led to some difficulties. Some thought the members of the evaluation group were talking too much and disrupting the work (and the new group identity) of at least one of the groups. Also, there were concerns that evaluation was driving the curriculum standards, when it "needs to be the other way around" (Log 1, p. 5).

By the third day, "the groups were feeling some pressure, frustration, and the magnitude of the task" (Log 1, p. 7). Group leaders were conscious of the pressure, and social activities were scheduled to build rapport within the groups and to help reduce the high level of stress. Writers still remember the social events, especially the Alpine slide and the ping-pong tournaments. The log summarized the work to that point:

Day 1 was spent getting to know each other, Day 2 exploring ideas and possibilities, and Day 3 getting down to generating standards.... Groups are wanting to spend more time with each other. (Log 1, p. 8).

By Friday, July 3, 1987, each group had produced a report of their work for the first week. Homework for the weekend was to study what had been produced so far. The outline of the work emphasized the content topics (number sense, fractions and decimals, measurement, geometry, etc.) with only a hint (problem solving, validating, representing) of the general standards that were to appear in all levels. The main content areas summarized in the log are reproduced below as Table 5.

Table 5.   A Summary of the First Week's Writing of *Standards*

K-4	5-8	9-12
number sense	numbers	assumptions with respect to computation
concepts of operations		
methods of computation	computation	
estimating and mental arithmetic		
fractions and decimals	ratio, proportion, percent	
measurement	measurement	
geometry, spatial sense	geometry	2-, 3-dimensional figures interplay of geom and alg
statistics/probability	probability	visual expl of data/stat
patterns and functions	algebra	relations and functions algebraic processing algebraic study of 2, 3 dimensional figures
representing/modeling		
validating		
problem solving/ applications	problem solving	math structure reasoning trigonometry discrete mathematics explorations of calculus

Note: Some writers think these notes are incomplete; for example, problem solving does not appear in the 9-12 list, as other sources suggest it did.

The attention to problem solving was predictable, given the emphasis it had received in *An Agenda for Action* (NCTM, 1980). Most of the mathematical topics were not surprising; the inclusion of "math structure" indicates correctly that at least some members of the 9-12 group were quite comfortable with the emphasis on abstract ideas that had been so prominent in the 1960s. The summary of the 9-12 group's ideas also suggests the role that technology and mathematical connections played from the beginning. There had been considerable discussion of discrete mathematics around this time; since John Dossey's advocacy of discrete mathematics for the secondary school was known, it is not surprising to find the topic included here. As mentioned earlier, the appearance of representing, modeling, and validating in the K-4 group showed the influence of the "Developing Mathematical Processes" curriculum materials.

Comments on the work in progress were not always well received: "I as well as some of the others do not take suggestions or criticism easily," noted one writer. A leader said that a critic was welcome to come to the group meeting but reminded the critic to mention "what was good as well as what could be changed" and to "be sure to bring an example of what the change could be" (Log 1, p. 15). The writers were very aware that one week was already gone and only one week was left. The pressures were intense and the importance of the task was deeply felt; several writers were reduced to tears by criticism of their work.

During the second week, the tensions increased as the time remaining decreased. Old doubts about national standards came up in a discussion early in the second week. One writer felt "the standards were too specific and would confine the curriculum to one curriculum for the nation" (Log 1, p. 17). The same argument that had been addressed in the proposal (Crosswhite, 1985b) and in the documents prepared for the writers (Romberg, 1987a) had re-surfaced. Other writers argued that some specificity was needed if the document was to have any impact at all, and on that note, the work of developing standards continued.

The General or Process Standards. The first four standards for each of the levels (K-4, 5-8, 9-12) deal with problem solving, communication, reasoning, and connections; these are the so-called "process" standards.[37] The special emphasis on these four standards developed over time; only problem solving and reasoning were included at all levels in the draft of July 20, 1987, that was prepared at the end of the first two week session in Utah. Communication did not appear as a standard at any level in the draft of July 20, 1987, but there was a substantial section in the introduction that discussed mathematical literacy. The importance of literacy as a metaphor was noted in the context of the shift from industrial models of schooling to a post-industrial, information age.[38]

---

[37]One writer insists that "process" is not the proper term, since it suggests the idea of a process by content matrix, which was never intended by the *Standards*.

[38]See also Gawronski (1984) and Romberg (1984a) for early discussions of language issues in the information age.

The emphasis on communication was thought to have roots in earlier publications of the College Board (Kilpatrick, 1985) as well, according to John Dossey:

> *Academic Preparation for College* was published in 1983, and there was a companion volume by Jeremy Kilpatrick entitled *Academic Preparation in Mathematics* published in 1985. I think those documents had a real impact on the *Standards* because communication and writing were carefully laid out as goals. All the members of the writing teams got a copy of *Academic Preparation for College*.

The emphasis on communication is one of the important features that distinguish the *Standards* from earlier NCTM documents, according to Dan Dolan, a writer in the 5-8 group. He recalled an early meeting of his group when "we had a bunch of paper up on the wall with 'big ideas'" including communication. Dolan continued:

> Now it's become a hallmark of the *Standards*. If you read the science standards, you find it; if you look at the other standards that are coming out, you find that communication is crossing over into all the curriculum areas.

The notion of a standard on connections was included in the July 20, 1987, draft of the 9-12 group, where the emphasis was on connections within mathematics. The adoption of the connections standard as one of the first four for each level was described by Glenda Lappan:

> The high school group had written a connections standard, and it was more or less connections within mathematics. I remember arguing that the middle school group had decided that what was missing was a standard on connections that would not only do what the high school group had done, but would go beyond that to connections much more broadly construed: connections to other subject areas, and connections to the world of kids. The high school group was somewhat reluctant to give up theirs since they had a well reasoned use of connections in their section. Here we were, the Johnny-Come-Lately, saying, "Not only do we like your idea, but we want to do it differently and we want everybody to do it." I remember that conversation; we had problem solving, communication, and reasoning, and we agreed to add a connection standard as the fourth standard in each one of the sections.

The importance of language in mathematics instruction and the central role of applications, or connections to other disciplines, have been present in earlier recommendations, including *An Agenda for Action* (NCTM, 1980). However, the emphasis in the *Standards* on communication and connections represented an important change, according to Mary Lindquist:

> I still think two of the strongest standards are communication and connections. Not that problem solving and reasoning aren't important, but they seem to be an integral part of mathematics, and we had not given the same attention to connections and communication.

Culture and Standards. The debates over the substance and the wording of the standards were often intense. As an example of one of those debates, consider the case of a "standard" that was suggested by one group but did not garner enough support to survive until the final draft. That proposed "standard" was concerned with the way that history and culture influence mathematics and its teaching.[39] An early version of the standard, entitled "Historical and Cultural Significance," follows:

> In grades 5-8 the mathematics curriculum should foster an historical and cultural awareness of mathematics so that students are able to:
> - explore mathematics in relation to the arts, humanities and sciences
> - appreciate that mathematics is an invention of the human mind
> - appreciate the potential of mathematics as an enjoyable activity
> - appreciate mathematics as a powerful, creative human activity.

The elaboration that was proposed for this "standard" included mathematics and music, history of mathematics, recreational mathematics, numeration systems, and other topics. As some writers look back on it now, the standard would have fit very nicely with the current interest in ethnomathematics, a topic of increasing importance in research on mathematics learning (D'Ambrosio & D'Ambrosio, 1994). At that time, however, the topic was seen as difficult to communicate and not central to the content emphasis of the *Standards*. In the words of Anne Zarinnia:

> The middle school group came up with the standard on culture at the first meeting in Park City. It received short shrift in part because the *Standards* were conceived as focusing entirely on content, and culture was not perceived to be content. The reaction was, "Well, this is too touchy-feely." In fact, the *Standards* were originally couched in a way that would have allowed it to be rejected in those terms. The frustration that I had, and still have, with that rejection is that in fact there is a whole philosophy of mathematics that was developing at that time that looks at mathematics as a cultural creation.

Tom Kieren of the 5-8 group expressed a related view:

> We wanted to show that kids of various heritage had things back in ancient history that connected their ancestors to mathematics. I don't know why that was particularly wonderful. The more interesting thing is what is in the personal culture of each child that is mathematical. Certainly part of that is their history, whether it be racial or ethnic or whatever. We enunciated that kind of ethno-cultural part, but we didn't have anything very strong on the personal-cultural part.

Although the standard on culture didn't survive, it was an example of the kind of argument (like the inclusion of the connections standard) that seemed to be typical of the 5-8 group. As leader Glenda Lappan noted:

---

[39]This version of the "culture" standard was taken from the notes of a writer.

The middle school group was constantly in hot water. We were always pushing for things that nobody else wanted. "Why don't you people go do your work and stop meddling?" That was just sort of the general tenor toward us. But we persevered, and we got some things in the document because of our behaving like middle school children.[40]

Although the culture standard was never included, another attempt at developing a new standard was quite successful. In a discussion of affective issues and ways of thinking about mathematics, Elizabeth Badger suggested that standards on attitudes be changed to mathematical disposition. Noting that attitude scales are administered but that little is done with them, she felt "It would be better to have something that could be observed and that teachers could work with" (Log 1, p. 16). The suggestion was accepted and mathematical disposition, with its emphasis on student confidence, willingness to persevere, interest, and curiosity, became a part of the vocabulary of mathematics education.[41] As Glenda Lappan put it:

> Disposition bubbled up out of the evaluation group, and we talked a lot in all of our working groups about this notion. Starting with disposition allowed us to talk about the need for mathematical experiences that include more than just little problems that you either could do or not do in five minutes. In order to foster the kind of disposition in kids that would make their use of mathematics powerful, we had to engage them in learning mathematics where the ground rules were very different.

The evaluation standards produced other ideas, too; for example, the notion that "you can't separate assessment and instruction" came up early and was included in the August 1987 draft. The idea of *alignment* was discussed at an early stage and included in the September 1987 draft. As Norm Webb recalled:

> I think we came up with the idea of alignment. We didn't want to use the word *validity*. We thought that validity was too restrictive and that it was used in the psychometric world for a specific idea. So we came up with the notion of alignment to mean curriculum validity along with being consistent with how students have been taught. We thought that was new.

The writers were not claiming that all of their ideas were new contributions to the field, and Webb noted that often the writing "was really a reiteration" of earlier work (see also Lambdin, 1993). However, the writers did work hard to make useful recommendations that would deal with major problems in the curriculum, including problems of equity.

Equity Issues. Since the 1989 *Standards* were intended to focus on content recommendations, a number of other issues that are central to the field

---

[40]The 5-8 group did seem to be more controversial and more interesting than the other groups. They were known for passionate brainstorming sessions, for special procedures for limiting debate, and for motivational snacks, including fresh chocolate chip cookies.

[41]See Evaluation Standard 10 (NCTM, 1989, p. 233) for a full description of mathematical disposition.

received less attention. Issues related to equity, for example, were widely recognized as important,[42] but they were not highlighted in the document in the ways that some writers would have preferred. A writer put it this way:

> One big issue that was part of our discussion had to do with equity, and just how strongly we could cast the equity argument. There was great consistency in terms of beliefs about the importance of equity issues, but there was a political reality out there that tempered what we were actually able to get into the document. A lot of the stronger statements were edited out, I believe, to secure the broad range of endorsements. I feel I was fairly outspoken on the issue of tracking, and that was one that they weren't willing to take on.

This writer felt that NCTM leaders were concerned that teachers and the public would not accept recommendations to eliminate tracking, and some writers also had doubts about such a recommendation. As one noted, "I have seen what good students can do together," indicating the view that instruction is not as effective for more advanced students who are grouped heterogeneously.

A state leader in equity issues thought that equity was probably not discussed very much. He used an early draft of the *Curriculum and Evaluation Standards* to revise his state's curriculum guidelines in 1987-88, so he was familiar with the way the *Standards* had developed. He observed, "I don't think people were looking not to be equitable, but I just think that it wasn't part of the conversation. The focus was strictly content."

An NCTM leader confirmed that there was concern that a heavier focus on equity might detract from what would already be a controversial document. Nevertheless, most writers felt they had addressed equity in a significant way, especially in their call for a strong core of mathematics for all students. In their view, the recommendation for a core curriculum that "provides equal access and opportunity to all students" (NCTM, 1989, p. 130) was a strong statement with significant implications for equity in mathematics education.[43]

Equity issues were particularly central to the debates about change in the high schools. As Chris Hirsch reported:

> I think a careful look at the *Standards* would show that, in the case of the high school mathematics curriculum, there were two issues that the *Standards* politically decided not to take a stand on. One was the issue of tracking, and the other was the issue of whether the mathematics studied each year at the high school should be an integrated or unified curriculum, as opposed to a curriculum that was subject-matter oriented each year: algebra, geometry, advanced algebra. That decision was very conscious, in that we felt that we needed to identify in the *Standards* what we believed at that time in history to be the most important mathematics that all students should have the opportunity to study.

---

[42]Questionnaire data indicated that 70% of the leaders and writers saw equity issues as a major influence on the *Standards*, compared to 35% for teacher empowerment and 17% for economic and global competitiveness.

[43]See Johnson (1990) and Secada (1989) for further discussion of equity issues.

And that in itself was advancing thinking on the curriculum quite a ways, because if one looked at the curriculum of the 1970s and 1980s, there was a marked contrast between the mathematics that was in college prep programs and the mathematics that one found in general math, consumer math, remedial courses. We felt it was most important to get out on the table (and over time gain acceptance for) the notion that all kids should be studying different mathematics, rather than getting the *Standards* caught up in a heated debate over how that mathematics could be organized and made available to students—that is, through sequences of courses that may or may not be tracked.

I think without question people who have worked on the *Standards* have now, as they've tried to build curricula that had this common core, addressed the issue of tracking head-on. But the issue of tracking for us seemed to be so political and so sensitive at that time, we feared that the debate would become a debate over tracking rather than a debate over what mathematics all kids should be able to study.[44]

The 9-12 group was able to agree on the notion of a "core curriculum" for all students as the "most fundamental change" proposed by their group. The emphasis on "mathematics for all" was not seen as a complete solution to all the problems of equity, however. As writer Bert Waits noted:

A major problem to all of us was the separating out of mathematics for the college intending versus mathematics for all students. Chris Hirsch ultimately came up with the idea for core mathematics. We all wanted to kill general math, and through our technology, you can get students access to rich mathematics even if their manipulative skills are a little rusty. We all believed that, but then how do you pull that off in an environment that also produces and nurtures your mathematically talented and college-intending students? These are huge issues. I don't think any of us have the answers.

Teachers, parents, students, and society at large are not generally in favor of a radical egalitarianism. Instead, educational privilege is often justified by performance on standardized tests. Mathematics for all is a worthy aim, but parents and teachers fear that advanced study is more difficult when slower learners are present. A core curriculum is a defensible standard, but no one wants schools to provide only a core; beyond-core mathematics needs its standards, too.

Tracking (through use of different curricula for different ability groups) deserved attention from the writers because it can be a practical effort to increase the legitimate concern for individual differences in students. Yet its exclusionary consequences are objectionable, sometimes illegal. Ethical rhetoric of the schools rejects tracking, but it is widely practiced, overtly and covertly. Apparently for both political and substantive reasons, the writers chose to avoid writing the *Standards* in a way that would guide educators on these ethical matters.

---

[44]In 1993-94 the debates in the US Congress over "opportunity-to-learn standards" did draw attention away from curriculum change (Ravitch, 1995).

Technology Issues. Technology, especially the role of the graphing calculator, was always a point of discussion. Since some writers had been chosen in part for their expertise with technology and others were chosen for their more traditional perspective, the stage was set for many disagreements. Moreover, the technology that was becoming available in schools was changing rapidly. Dan Dolan remembered the debates:

> There was always debate on what the public would buy, and on what **we** would buy. For instance, would the public buy the fact that calculators ought to be available all the time? That was debated in the K-4 group and other people hammered away: "Doggone right that's got to be in there!" There was debate right up to the bitter end whether graphing calculators ought to be available by the 11th grade or before. That changed in the final edition. Yes, there were debates, not only about what the public would buy, but even what we thought we could sell within the mathematics community. There had to be compromises at times.

Bert Waits, a writer in the 9-12 group, recalled:

> The most remarkable thing to me was how cautious we were in the 1987 draft. If you go back and look at it, there was only a recommendation that graphing calculators be used in Grades 11 and 12. It turned out that the input during that year showed a strong sentiment for extending this recommendation. It was an assumption that all kids would have access to graphing calculators. To me, that was a milestone.

The tension between the public's interest in "basics" and the needs of the mathematics curriculum was always present. As Bert Waits put it:

> I knew that the basics were dead. I had seen already what the scientific calculator had done to the basic skills that I taught in the 1960s that I don't teach anymore. If you go back and look at a college algebra course taught in 1965, it was shocking what we did in the name of mathematics. The scientific calculator put a lot of that to death.

The writing of the 1987 draft of the *Curriculum and Evaluation Standards* was a tremendously intense experience for the writers. They awaited the results of the year of review with great anticipation.

Feedback from the Field

The plans for the development of the *Standards* always included an opportunity to optimize broad involvement of people in the field. As noted in Table 4, 10,000 copies of the 1987 draft of the *Curriculum and Evaluation Standards* were distributed with an open invitation inviting comments,[45] and over 2000 responses were received. Involving the members in revising the draft of the

---

[45]"WORKING DRAFT — all comment welcomed" was printed at the bottom of every page.

Standards was an important part of the planning. The leadership didn't want to have just another report from one more blue-ribbon panel. As Joe Crosswhite commented:

> I felt that we needed a much broader sense of ownership than that. We had to somehow make the membership at large feel as though they were a part of this process. I think that worked out reasonably well. Rather than just informing the members about their guidelines, I felt that they ought to be part of developing the guidelines. If you can give any group you're working with a sense of owning the document, it's much more powerful.

Although it is common to send drafts of curriculum documents out for review, the extent of the review of the *Standards* was a significant change. The draft was distributed very widely, not just to a select group of leaders. The review process continued the emphasis on the *Standards* as a collaborative effort. Norm Webb commented on the significance of this type of collaboration:

> They had us working in groups rather than as individuals. They put the working groups in a room and said "Here, you have to work it out, together, with consensus." The group work was very important in terms of the quality of the *Standards*. Also, this idea of distributing a draft so widely—I don't know any other models that do that. I think that was really a strength of the project.

Reactions to the draft were mainly positive, but there were important issues that were raised. All of those issues were discussed and dealt with in June 1988 when the writers reconvened in Utah in order to "meld the many comments and suggestions into the final draft" (Crosswhite et al., 1989, p. 57). As Webb recalled, "A lot of people complimented the draft. They said, 'This is me, this is just right on.' And we got a lot of 'This is the work of the Devil,' too."

In addition to the comments received from the field, NCTM solicited comments at all the regional meetings during 1987-88. Materials were prepared for the writers and others who were scheduled as speakers at the regional meetings. These materials provided background information for talks and a common set of visuals for use by the speakers. The materials were designed to bring come consistency to the presentations on the draft of the *Standards*, and to optimize the usefulness of the feedback. As Tom Romberg noted:

> Speakers needed to have something that summarized what we were doing. We needed to focus on goals first, and then how to reach those goals. It had to be done quickly. We pulled out one standard, the 5-8 geometry standard; we picked geometry because we didn't want it to be number. We wanted to emphasize something else. I can't say that we did it with a lot of careful thought. Part of it was associated with the need for reactions.

Further response to the draft was obtained through an MSEB project called "Review by Selected Constituencies outside the Mathematics Teaching Community of the *Curriculum and Evaluation Standards for School Mathematics*" (MSEB, 1988). The focus groups included parents at each of the three grade levels, mathematicians and statisticians. scientists and engineers, school administrators, school board members, state education officials, and employers. The project was chaired by Jim Fey and funded by NSF. There were 18 meetings with the focus groups during April and May 1988. Meetings were led by MSEB members and staff. The focus groups did not include teachers, so the teachers who were encouraged to react to the draft were mainly those who attended NCTM conferences. The review process included teachers who were already active or interested in NCTM, but not the teachers who were most likely to resist the changes implied by the draft of the *Standards*.

Some parts of the *Standards* went through substantial revision based on the year-long review, and other parts didn't change even when they were criticized. For example, parents in general were known to be uneasy with NCTM's policy encouraging the use of calculators in elementary school (NCTM, 1981a). Even though these parental views were well known before the publication of *An Agenda for Action* (NCTM, 1980), the *Agenda* still took a strong position in favor of calculator use. Those parental views had not changed by 1988, when focus groups consisting of parents at different grade levels "expressed real and pressing concerns" (MSEB, 1988, p. 1) about calculators. Even some mathematicians worried that calculators might create "computational cripples" (p. 1). Research studies have not found any empirical support for the hypothesis that "calculators rot the mind" (Suydam & Dessart, 1980), but those studies apparently have not yet reached the public. In any case, consistent with the view that NCTM should lead rather than follow, the *Standards* supported the use of calculators, continuing and extending NCTM policy in this area.

Another important focus group was mathematicians and statisticians. They criticized the *Standards* for being overly ambitious. One suggestion for the probability and statistics standards was that the K-4 standards in this area be moved up to the 5-8 level, and that 5-8 standards be moved to Grades 9-12 (MSEB, 1988). Instead of making that change, the writers tended instead to follow the recommendations of the American Statistical Association/NCTM Joint Committee on the Curriculum in Statistics and Probability. Like other NCTM committees, they were encouraged to respond to the draft, and John Dossey reported that they "had a tremendous impact on the statistics and probability strands." Their influence was especially notable at the 9-12 level.

The writers and other NCTM leaders made good use of many opportunities to obtain feedback on the draft of the *Standards*. As Glenda Lappan reported:

> After the first meeting in Utah, I came back to a summer institute where we had 50 to 75 middle school teachers here on our campus for a three-week institute. I was able to take our fledgling first ideas directly to this group of middle school teachers. From the very beginning we had teachers looking at our work, arguing with us, saying, "Yes, yes. You need to do more with this. What do you mean by this? We need powerful examples; we're not going to be able to

bring this off unless you can give us powerful examples of what you're talking about." The fact that I had ready access to people who were, in some sense, the key players in the documents helped a lot.

Other reviewers and critics from within the mathematics education community also had an impact. The introduction to the July 1987 draft began by criticizing the traditional mathematics classroom as dull and sterile, with an emphasis on paper-and-pencil computation. However, some reviewers thought this approach was too reminiscent of the criticism of teachers[46] that was prominent in *A Nation at Risk* (NCEE, 1983), and the next revision started with a more positive statement of new goals. In another revision, an early draft of the K-4 standard on computation had specified the size of the whole numbers that should be used in paper-and-pencil computation at different grade levels (e.g., three-digit addition and subtraction should be done in Grade 3, but not in Grade 2); this kind of detailed direction raised questions from some reviewers, and was omitted in the final draft. The writers took a different approach in their efforts to reduce the amount of time that students spend doing paper-and-pencil computation with multi-digit whole numbers. As Mary Lindquist described the situation, "I think our biggest controversy was over how much to prescribe. At times, especially in the first draft, we wanted to be more prescriptive than we ended up being in the last draft."

Although the K-4 Standards recommended decreased attention to paper-and-pencil computation, some leaders still found the *Standards* to be quite conservative on this issue. An NCTM leader noted:

> There were some of us who thought the *Standards* were somewhat conservative. For example, in K-4, they have too many standards on computation (estimation, number sense and numeration, concepts of whole number operations, and whole number computation). I still think they were somewhat conservative, but they serve to lead the group onward and that's what the real importance is.

At the secondary level some reviewers criticized the *Standards* by comparing those recommendations unfavorably to the University of Chicago School Mathematics Project (UCSMP). That project, led by Zalman Usiskin and others, was thought by some writers to have taken a more classical, European view of school mathematics. In one writer's words, UCSMP "was committed to a revision of classical math—not a transformation to a changed conception of math." This interpretation of UCSMP's position was shared by several writers. Usiskin's presentation to the Association of State Supervisors of Mathematics (Usiskin, 1988) was probably one source of this view, since Usiskin noted approvingly at that time that many countries teach algebra two years earlier than is typical in the US. Among the writers, however, there was no agreement on the advisability of teaching a formal algebra course in Grades 5-8. As one writer

---

[46]See Nelson (1994) for more on this point.

noted, "I was adamant in opposition, but there were others who came down just as strongly on the other side of the argument." The final version strongly supported algebra for all students, and recommended integrating algebraic experiences throughout the K-8 curriculum (NCTM, 1989, p. 102). However, there is no specific recommendation that formal algebraic ideas appear as early in the curriculum as they do in some other countries.

Outside reviewers apparently had an impact in de-emphasizing standards as accountability criteria. When writers from universities went home in 1987, they found colleagues who were concerned about using standards as criteria for judging curriculum materials. As one writer put it:

> I remember coming back and talking with some of my colleagues and they thought the idea of standards was very authoritarian—they were pretty negative toward it. When we met again at "Fort Xerox" [in Leesburg, Virginia in August 1987], even then we were drifting toward the standards as more of a vision—a less authoritarian perspective.

Reviewers of the draft of the *Standards* had a significant impact on the final document, which was approved by the NCTM Board of Directors at its September 1988 meeting in Reston, Virginia. The final version (NCTM, 1989) was released on March 21, 1989, three years after the NCTM Board of Directors approved the project. The emphasis on the *Standards* as a vision of ideal practice in mathematics education continued to grow during the dissemination phase of the project.

Dissemination of the NCTM *Standards*

The *Curriculum and Evaluation Standards* (NCTM, 1989) were disseminated widely, both inside the mathematics education community and outside to the broader community of educational and political leaders. NCTM took the lead in the dissemination effort and received substantial help from other professional organizations. In commenting on the success of the dissemination effort, Norm Webb observed, "Many of the ideas in the *Standards* are not new. They've been around for a long time and they appear in other reports from NCTM. What was different was the selling job that they did."

The timeline of the original proposal to develop standards (see Table 3) called for "dissemination of the professional standards statement and planning of implementation efforts" (Crosswhite, 1985b, p. 11) in Year 3 of the project. For "Year 4 (and beyond)" the only task listed was "continuing NCTM activities to enforce the professional standards" (Crosswhite, 1985b, p. 11). As the view of the *Standards* changed from an accountability perspective (with "enforcement") to an emphasis on aspiration and vision, the notions of dissemination and implementation also changed.

The original proposal included plans to

seek the endorsement of the final standards document by other concerned professional groups ... and supervise dissemination to the NCTM membership, to designated state education officials, legislators, textbook publishers, test publishers, accrediting associations, and other interested groups. (Crosswhite, 1985b, p. 9)

Clearly, NCTM leaders knew from the beginning that they wanted to reach a wide group of educational leaders; other aspects of the dissemination program only appeared later. The involvement of the field in the 1987-88 review of the draft of the *Curriculum and Evaluation Standards* was in many ways an early "dissemination" program for people in mathematics education, as well as a means of revising the document. The draft was discussed extensively with the membership through presentations at NCTM meetings (regional and annual), and NCTM committees were asked to address issues raised in the *Standards*. State and local organizations of mathematics teachers also made the *Standards* the focus of their meetings. A substantial amount of dissemination had in fact occurred even before the publication of the *Standards* in 1989.

The dissemination effort was quite extensive and expensive. As Judith Sowder, chair of the *Standards* Coordinating Committee for NCTM, indicated:

There were really two prongs to the dissemination. One of them was to disseminate the *Standards* document by giving it away free to the members. This is quite a big membership present.[47] The other prong for dissemination was through the Executive Summary. Within the mathematics education community I don't think the Executive Summary is all that important, but that was what got the *Standards* known outside of mathematics education. The mailing lists were enormous. The NCTM lobbyist took them around personally and handed them to members of Congress. Certainly every dean of sciences, every chair of a mathematics department, every math coordinator, high school principal, and elementary school principal who was on our mailing lists got one. We sent to PTA presidents, school board presidents, and on and on and on. Every mailing list that could possibly be used was used.

The Executive Summary is very brief, but I think it's an impressive looking document, and I think it gives a flavor of what the *Standards* are all about. Many people wanted copies to pass out to the school board, and those who were giving a presentation on the *Standards* to parents or others wanted copies, too. We asked NCTM to make copies available in bulk at cost, and that has been very successful.

One of the reasons for the success of this broad dissemination effort appears to be the use of two experienced public relations firms with strong connections to the education community. Such extensive use of public relations firms was

---

[47]NCTM sells the *Curriculum and Evaluation Standards* for $25. By April 1995 NCTM had distributed over 258,000 copies, including free copies to over 51,000 individual members.

unprecedented at NCTM. Leaders give several different explanations for how the policy for greater use of public relations firms developed.

One explanation suggests that the experience of mathematics education leaders in England was influential. At an annual meeting of NCTM in the late 1980s, a visitor from England discussed over dinner some of the problems that had impeded useful press coverage of the Cockcroft (1982) report. Some members of the NCTM Board of Directors were present and started thinking about how NCTM might have a stronger impact on the press. So when the *Standards* were released in 1989, the press conference in Washington, DC, was run by professionals from the Burson-Marsteller firm. Another interpretation of the change to greater use of public relations specialists suggests that MSEB and the National Research Council had used professionals in dealing with the press, and NCTM followed their lead. A third interpretation is that NCTM leaders were just following the recommendation of one of their own, Shirley Hill, who wrote about the importance of public relations and political action to reform in mathematics education almost a decade earlier (Hill, 1981).

President Shirley Frye recalled how NCTM leaders realized that the *Standards* project would be bigger than anything that NCTM had ever attempted. They knew they needed help in getting their message out; they even brought in another public relations firm to help with training. Frye recalled:

> NCTM leaders spent entire days with the firm learning how to deal with radio and TV reporters, how to answer questions without being defensive, and how to get the message across to the audience. We saw videos of people who had been effective and some who had not, and even watched our own interviews.

NCTM leaders found the training to be helpful and thought that the public relations firms were quite successful in most areas. The "packaging" of the *Standards* involved many different activities. The document itself and the Executive Summary were attractively designed, and the press conference marking the release of the document (as noted earlier) went well. NCTM headquarters developed a Speaker Support Kit, and Gallagher-Widmeyer, another public relations firm, was awarded a "contract for a media/public relations campaign" (see the minutes of the *Standards* Coordinating Committee meeting of December 9-10, 1990). The campaign, financed in part by a grant from the Exxon Education Foundation, involved distributing over 500,000 brochures for parents and school administrators, and included a video (featuring entertainers like jazz musician Wynton Marsalis) describing the NCTM *Standards*. By the end of 1990, the video was reported to have been shown over 6000 times by 121 television stations, reaching an audience in the millions. A representative of Gallagher-Widmeyer traveled with NCTM President Iris Carl to arrange for media events at the regional meetings of NCTM; residents of Iowa, New Mexico, Oregon, and other states would find the NCTM president on their local television news, frequently appearing along with the best of the region's mathematics teachers and footage of local classrooms where the influence of the NCTM *Standards* seemed notable.

The work of the public relations firms was not always acceptable to NCTM leaders; as Shirley Frye recalled, "The draft of the Executive Summary that the firm prepared was not acceptable, so Judy Sowder and Mary Lindquist had to rewrite the document."

This was not the only case in which the public relations firms were unable to communicate NCTM's message in depth. In most areas, however, the experience of the public relations firms was perceived to be useful as NCTM tried to get its message out to the public. NCTM wanted also to help teachers understand the changes recommended in the *Standards*; getting the message out to teachers required a different approach.

The Addenda Project

The original plan for the *Standards* project, as noted earlier, called for the development of professional standards for curriculum, instruction, and evaluation to be published all in one document. That plan was scaled down to include just *Curriculum and Evaluation Standards* in the first document, and the new plan called for that document to be accompanied by "four addenda (one for each work group) containing illustrations, examples, and non-examples of activities and instructional techniques with respect to each standard" *(Curriculum and Evaluation Standards* draft of July 20, 1987). However, as President Shirley Frye noted, there was only so much that could be included in the *Standards* document itself, and teachers would need much more. She appointed a task force[48] chaired by Tom Rowan[49] "to outline an addenda project to provide teaching lessons to exemplify the *Standards*."

The expectation was that such a project would take a year or two, and would produce several booklets. In the end, however, the project produced 22 booklets over a period of 5 years.[50] The booklets included one per grade for K-6, plus 15 others that focused on various content topics (e.g., *Geometry in the Middle Grades*). The Addenda Project was just one of many curriculum development and teacher education efforts that NCTM was involved in during the early 1990s. Grant support, which had been nearly impossible to find in the mid-1980s, was now available for a wide variety of projects.[51] One of the most interesting of the projects was called "Leading Mathematics into the 21st Century."

---

[48]Members included Richard Lodholz, Joan Duea, Christian Hirsch, and Marie Jernigan.

[49]Rowan and Hirsch were writers for the 1989 *Standards*.

[50]Bonnie Litwiller served as project coordinator, and Miriam Leiva, Frances Curcio, and Christian Hirsch were the editors of the three Addenda Series (K-4, 5-8, 9-12).

[51]The minutes of the 1989-90 meetings of the *Standards* Coordinating Committee reported on projects dealing with number sense, geometry, discrete mathematics, calculators, and research in support of reform, as well as other projects that were proposed but never funded.

The 21st Century Project

Since some members of the Association of State Supervisors of Mathematics (ASSM) were involved in the writing of the *Curriculum and Evaluation Standards*, it was natural for them to think about the role that their organization might play in the dissemination effort. Dan Dolan, a writer in the 5-8 group and a state supervisor in Montana, remembered discussions in the summers of 1987 and 1988 about how to "spread the word":

> It seemed as though there was not an organized plan, even on the part of NCTM, for large-scale promotion and dissemination. Some of the state mathematics supervisors thought that this was an ideal project for us. We were the people who were responsible for leadership in mathematics education in each state or province, and the things that we wanted to see happen in our states were consistent with what was being written in the *Standards*. A cooperative proposal was a natural thing to do. We brought together some people to talk about how to formulate the whole thing, and who might lead the parade. It was agreed that ASSM was a logical group to be the lead organization [with] the National Council of Supervisors of Mathematics (NCSM), NCTM, MSEB— because we felt that they had a great stake in what was going on, and the Council of Presidential Awardees in Mathematics (CPAM).[52] The Presidential Awardees were just getting off the ground at that time, and this certainly was a way in which they could exercise some leadership in each of their states.

A proposal called "Leading Mathematics Education into the 21st Century" (Dolan, 1990) was funded by NSF, with ASSM as the grantee. The proposed project included five regional conferences in 1980 (San Jose, Dallas, Atlanta, Chicago, and Hartford) to broaden awareness of the *Curriculum and Evaluation Standards* for 1000 national leaders and to develop networks for dissemination. "Conference participants from each state included representatives from business and industry, state legislators, local and state boards of education, the media, and K-12 teachers, school administrators, university mathematicians and mathematics educators" (Dolan, 1990, p. 2). The five regional conferences included model presentations by NCTM leaders and writers[53] of the *Standards*. Dolan recalled how project resources were allocated:

> First of all, states were allocated a number of teams according to the number of congressional districts; that satisfied NSF in terms of spreading the money evenly. In our directions to state coordinators, we said, "When you choose

---

[52]CPAM is made up of the two teachers in each state who were chosen by NSF each year as the outstanding mathematics teachers. Each awardee received a small grant of $7,500 to support professional activities in the schools.

[53]For example, Sue Ann McGraw, a high school teacher from the 9-12 group, was on sabbatical that year, so she was available to do the workshop for the 9-12 standards at each of the five conferences.

your teams, we'd like to have you pick them in fours. Try to balance your team so that you have somebody from a college or university, preferably a mathematician. We'd definitely like to see you have some people from the business community, from school administration, and teachers." That's the kind of direction we gave.

A state supervisor from the midwest sent her teams to the Chicago meeting for training:

I put together a team of 20 with K-12 teachers, university people, and supervisors. I had a superintendent of a school district, and tried to get somebody from our state legislature and our State School Board, but didn't succeed. We were in Chicago for three days and learned about the *Standards*.

During the following school year, the teams from this state reported conducting conferences in 23 different counties covering all the major populations centers in the state. They appeared to have followed through on the careful planning of the 21st Century project. As Dolan recalled, the project provided significant help for the teams:

We included outlines for talks to various groups, suggesting what to highlight for a parent group, for school administrators, for the business community. We put together a packet that included the *Standards*, as well as other publications. They got the PTA videotape *Math Matters* and a binder including over 300 transparency masters divided into several areas (each section of the *Curriculum Standards*, model workshops, and so forth). Each one of those packets was worth about $150.

Project plans were designed to ensure the fidelity of the information that would be disseminated by many different people, including those from outside of mathematics education. Dolan continued:

In at least four or five states, legislative leaders or staff were part of a state team; in one case we had an education editor from one of the major newspapers in the state. We had the presidents of several state PTAs. At every meeting we had a person from the business community as a speaker; that was a very important part of it. In each case, that business person spoke very forcefully about getting the business community involved. We primed them in terms of the message we wanted them to deliver.

The five initial conferences were replicated in the various states and provinces. The project led to over 50,000 documented contacts within two years, and project reports suggested that more than twice as many contacts occurred as were documented. The overall effort at dissemination appeared to be relatively successful, in the eyes of most leaders. As NCTM President Mary Lindquist reported:

I think that we've been more successful than ever before. In many states a lot happened, but in other states nothing happened. My view now is that the job is never done, but the initial effort was fairly successful. I don't think we ever had a completely coherent plan of what to do when. The PR effort, for example, went in spurts. Now that is not necessarily bad. You have to have new messages and you have to come back fresh. We looked at that seriously. We also depended on other groups to do a lot of the PR work. Sometimes that happened and sometimes it hasn't.

ASSM was one of the groups where a lot happened. The 21st Century project was only one of their activities. Under the leadership of Donald Chambers and Andy Reeves, they also obtained funding from the Carnegie and Exxon Foundations to plan and coordinate additional state efforts in support of reform in mathematics education (ASSM, 1990). They described this project, which included specifying and publishing reform goals for each state, as part of the Year of National Dialogue, one of many projects led by the Mathematical Sciences Education Board (MSEB). We consider next the role of MSEB, especially its efforts to disseminate information to communities outside of mathematics education.

The Role of MSEB

As described earlier, MSEB grew out of recommendations made at the 1983 "New Goals" meeting at Airlie House (CBMS, 1984) and the "Options for the 1990s" conference at Wisconsin (Romberg, 1984b). From the beginning, MSEB worked with NCTM on various aspects of the *Standards* effort. A letter of support from MSEB was attached to the original proposal requesting funding for the *Standards* project (Crosswhite, 1985b). In that letter, dated September 4, 1985, MSEB founding chairman Shirley Hill noted the origin and purposes of the organization:

> The Board was formed at the urging of the mathematics and mathematics education communities ... in order to provide national leadership and guidance on issues affecting the quality of instruction in the mathematical sciences at all levels.

At the time the proposal was submitted, MSEB had not yet had its first meeting, which was scheduled for October 1985. After considerable debate about where MSEB should be housed (a consortium of Big-Ten universities was one option), the group had only recently found a home at the National Research Council.[54] Chairman Hill went on to note in the letter accompanying the proposal (Crosswhite, 1985b) that there was

> a critical need for linkages, communication and cooperation among the many projects, activities, and efforts presently underway or to be initiated in the near future. Thus a coordinating, oversight, and analysis function was seen as a major reason for the Board's existence.

---

[54]The National Research Council is the principal operating agency of the National Academy of Sciences and the National Academy of Engineering.

Linkages among projects and communities, along with communication to the public, were always central to MSEB's priorities.[55] The composition of the 30 to 35 member Board was expected to include members from various communities, including precollege mathematics education, academic mathematics, industry and business, educational policy and administration, and cognate fields and mathematics users, according to Hill's letter. Over the years, the Board has continued to include those groups. The president of NCTM has always been a member, and mathematics teachers and supervisors are well represented in its membership record (MSEB, 1993), along with business executives, school superintendents, and leaders of the National PTA.

An early MSEB "Master Plan" outlined the activities that were planned for the years from 1986 to 1993. During 1986-87 there was a set of meetings on reform issues (e.g., see Hill, 1987, and Steen, 1988), leading up to the publication of *Everybody Counts* (MSEB, 1989), written by Lynn Steen, and the NCTM *Standards* (NCTM, 1989). The year 1990 was planned as "A Year of National Dialogue" to be followed in 1993 by the establishment of a "National Support Structure for Revitalization." Throughout these years there were plans for action programs with business, projects on testing, and other activities dealing with curriculum, teacher preparation, and collegiate mathematics. An MSEB (1993) report includes a summary of selected projects and activities that were completed.

The early years of MSEB were marked by a series of struggles as the organization got established within the bureaucratic structure of the National Research Council (NRC). Marcia Sward, a mathematician with substantial experience as an MAA administrator,[56] became the first Executive Director of MSEB. Starting a new organization and obtaining grant money to sustain it were immense tasks, and Sward's work was made even more complicated by the setting. The NRC was accustomed to appointing bodies that evaluate research in a scientific field and that take positions on research issues, but it was not accustomed to taking positions on educational issues. Lynn Steen described the origins of MSEB this way:

> After the Airlie House meeting (CBMS, 1984) was over, CBMS set up a planning committee for the purpose of implementing the motion to establish a national board (what became MSEB). The committee was chaired by Paul Sally who at that point was just beginning his work on the UCSMP project. Other members were Joe Crosswhite, Henry Pollak, Herb Greenberg, Tom

---

[55]These priorities were listed in Hill's letter as "testing, curriculum goals, standards and criteria for excellence, teacher education, communication and dissemination, the impact of technology on mathematical sciences education, and international comparisons."

[56]Sward was Associate Executive Director of the Mathematical Association of America when Lynn Steen was its president; Sward and Steen provided a strong base of support for the NCTM *Standards* in the collegiate mathematics community.

Romberg, and me. Marcia Sward, the administrative officer of CBMS, served as the key organizer and stimulus for what came to be known as the "Sally Committee."

As the committee developed its ideas and figured out what we wanted to do, we started scratching our heads and asking, "Who on earth could possibly chair such a thing?" We all realized that Shirley Hill, a former president of NCTM and a member of the MAA Board of Governors, was the ideal candidate. Basically, the Sally committee chose her to chair MSEB.

So we recruited Shirley to meet with the committee, and added some others (Tony Ralston, for example) so that this larger group then helped develop the specific proposal for MSEB. There's no question that Shirley's vision and ability to persuade the leadership of the National Research Council was of critical importance. She was able to make a case that the NRC should view MSEB not just as a typical NRC committee in which people come together to reach consensus about issues, but rather that NRC should get behind the idea of standards for mathematics education. Thus the MSEB charter is unusual within the National Research Council in the sense that it talks both about doing studies and working to implement recommendations. That latter phrase does not appear in most other commissions in the NRC.

Frank Press, who was then president of the National Academy of Sciences and chairman of the NRC, was a central figure in getting MSEB established at the NRC. He had been at the Massachusetts Institute of Technology (MIT) earlier in his career, as had Ken Hoffman (of MSEB) and Lynn Steen, and their MIT connection was an important factor in helping MSEB get established. Although MSEB was a part of the NRC, it still had to obtain outside funding, and MSEB staff were always struggling to find support for various activities. Occasionally some confusion developed over the roles of NCTM and MSEB as both tried to obtain funding for their separate projects. John Dossey recalled one example involving the NCTM proposal to obtain funding to develop the *Standards*:

The MacArthur Foundation discussed our proposal, but ended up not funding it because they were unclear on who would be doing the standards for mathematics. In January of 1987, there was a symposium at the National Research Council (NRC) where the results of the international studies were released, and a representative of the MacArthur Foundation was in the audience. The closing remarks at the symposium talked about creating standards, and the MacArthur representative thought MSEB [which was part of NRC] was in competition with NCTM. There was considerable ill-will felt over that. The next summer we hashed this out hard, and it really helped MSEB and NCTM define what their roles were.

One role for MSEB was to set the stage for the release of the NCTM *Standards* by preparing *Everybody Counts* (MSEB, 1989). Since the MSEB publication appeared first, some assumed that its conception had preceded the plans for the *Standards*. However, as Lynn Steen pointed out, the task was "to provide a report to the nation[57] on mathematics education as background to the planned 'standards' being prepared by NCTM."

The task of preparing such a report that could be approved by NRC was a particular challenge, since it was published by three NRC Boards.[58] In addition, the NRC review process includes a wide range of reviewers, and a convincing response is needed for every criticism by every reviewer. The response to the reviews is judged for adequacy by a member of the National Academy of Sciences who acts as a monitor of the process. As usual in such situations, the reviewers often did not agree. Lynn Steen described the situation:

> The reactions were predictable. Those pushing for a more active posture urged that the story line be sharper; those inclined towards conservatism argued for toning down the rhetoric because much of the narrative was not rooted in research that could pass muster with an NRC review.

The debates among the mathematicians were especially vigorous. Many of them thought the draft was too critical of the mathematics community. Other reviewers (from both within and outside mathematics) were strongly in favor of a more activist document, one that would argue strongly for equity as well as for excellence. But getting approval of the strong equity orientation of *Everybody Counts* required substantial amounts of last minute negotiations. NCTM leaders were beginning to worry that delays could undo the plan to have MSEB release *Everybody Counts* in advance of the *Curriculum and Evaluation Standards*. Fortunately, due to heroic efforts by Lynn Steen, Marcia Sward, reviewers, and MSEB staff, the document was ready in time.

Some members of NCTM were still concerned about possible conflicts[59] with MSEB in 1989 when *Everybody Counts* was published. John Dossey noted:

> There were those that worried that MSEB might be trying to upstage NCTM. That was probably just a continuation of earlier discussions about MSEB's role. But I think *Everybody Counts* was a seminal part of the success of the

---

[57]The title of this "report to the nation" is sometimes thought to have been adapted from the Cockcroft (1982) report, *Mathematics Counts*. According to Steen, however, what really happened was that someone in an MSEB meeting used the phrase "everybody counts" in the midst of a brief remark. Mike Atkin of Stanford (who was not aware of Cockcroft's work at the time) interrupted in a half-audible aside with the comment, "That's our title."

[58]Those Boards were MSEB, the Board on Mathematical Sciences (mostly academics from the mathematical sciences), and the Committee on the Mathematical Sciences in the Year 2000 (with many members from academe and industry who came from outside mathematics). The NRC Boards contributed to the stature and credibility of the report, but also to the difficulty of the task.

[59]Other NCTM leaders reported being "rubbed the wrong way when MSEB tried to take credit for the *Standards*," but by this point there appeared to be no major conflicts between the two organizations.

Standards because it spoke to everybody. The *Standards* really speaks more to the people who are involved in curriculum and teaching, but people can understand *Everybody Counts*, no matter what their walk of life.

Through its publication of *Everybody Counts* (MSEB, 1989) and its other projects, MSEB has played a central role in the reform effort.[60] Although there have been times when NCTM and MSEB have had difficulty coordinating their separate agendas, the two organizations have generally worked together well. The bureaucratic regulations that govern NRC, and hence MSEB, have been a source of difficulty from time to time, but those difficulties have usually been of short duration. MSEB publications on curriculum and teacher professionalism (see MSEB, 1993) continued to support NCTM's reform efforts, which turned during 1989-91 to the development of the *Professional Standards for Teaching Mathematics* (NCTM, 1991).

Developing the 1991 *Teaching Standards*

There are many similarities between the development of the *Professional Standards for Teaching Mathematics* (NCTM, 1991) and the *Curriculum and Evaluation Standards* (NCTM, 1989). In this section[61] we discuss the procedures and policies that influenced the development and dissemination of the *Teaching Standards*, often comparing the developmental process to that of the 1989 *Curriculum and Evaluation Standards*.

As indicated earlier, the need for improving the teaching of mathematics was always seen as central to the reform effort, and the initial plan was to include standards for curriculum, instruction, and evaluation all in one document (Crosswhite, 1985b). When that plan was changed to make the development of standards for teaching into a separate project, some leaders thought that the *Teaching Standards* were so important that they should be done first. There were several justifications for putting teaching first. For example, NCTM leaders who remembered the 1960s were quite aware that just changing the curriculum was not sufficient to improve mathematics education. Moreover, the "widespread shortage of qualified mathematics teachers" (NCTM, 1980, p. 24) was a continuing problem, and any move away from traditional instruction would require strong programs for staff development as well as teacher preparation. However, the final decision to begin with content and curriculum received broad support.

The overall plan for developing standards for teaching was quite similar to the pattern established for the *Curriculum and Evaluation Standards* (NCTM,

---

[60]For a listing of major projects, see MSEB (1993).

[61]Data in this section come from interviews with 7 of the 18 writers of the *Teaching Standards* as well as 4 of the 5 members of the Commission on Teaching Standards for School Mathematics. We also included questions on the *Teaching Standards* in our interviews related to the *Curriculum and Evaluation Standards*.

1989). In January 1988 the NCTM Advisory Committee for Professional Development and Status submitted a motion to the NCTM Board of Directors for the appointment of a "Task Force on Professional Standards for Teaching Mathematics"; the motion was approved, and NCTM President Shirley Frye invited James Fey (who had been involved in developing NCTM's 1985 proposal for standards), Glenda Lappan (leader of the 5-8 group for the 1989 *Standards*), and James Sherrill to develop a proposal for Professional Standards for Teaching Mathematics. As one NCTM leader put it, in the wake of the *Curriculum and Evaluation Standards* (NCTM, 1989) "NSF was almost inviting us to come for money," and a grant proposal[62] was submitted to NSF in 1989, with President Shirley Frye serving as the principal investigator (Frye, 1989).

The proposal outlined the major tasks of the project in a manner that was reminiscent of the general plan for the *Curriculum and Evaluation Standards*. The time line began in 1989 with the appointment of a Commission on Professional Teaching Standards. Writing teams were to prepare a draft document in the summer of 1989, and NCTM would hold hearings and gather feedback during 1989-90, a year of discussion and review. Finally, a revision of the draft would be prepared during the summer of 1990, based on the feedback from the field, and NCTM would produce the final document and begin the dissemination phase in 1991.

The plan (Frye, 1989) included standards for teaching (i.e., what a mathematics teacher should be able to do), standards for the professional development of teachers, and standards for the evaluation of teaching. These three areas appeared in the final version[63] of over 200 pages (NCTM, 1991), along with a fourth section on standards for the support and development of mathematics teachers and teaching. The purpose of this last section was to outline the responsibilities of administrators, policymakers, and institutions. The *Teaching Standards* document was released at a national press conference[64] in Washington, DC, in March 1991 (see Carl & Frye, 1991).

Differences from the 1989 *Standards*

Although the overall plans for the two documents had many similarities, there were significant differences in the ways that the *Curriculum and Evaluation*

---

[62]Lappan, who led the work on the proposal, took a "rotator" position at NSF around this time, but still managed to keep her commitment to NCTM to lead the *Teaching Standards* project.

[63]The order was changed slightly; the evaluation of teaching was placed ahead of the section on professional development.

[64]This press conference, like that for the 1989 *Standards*, received only one criticism: The writers of both documents were disappointed that they were not more involved in the press conferences. The writers felt a great deal of ownership of the *Standards*; however, once the documents were presented to the NCTM Board of Directors, they became the property of the Board, not the writers.

Standards and the *Teaching Standards* were developed. First of all, and in contrast to the *Curriculum and Evaluation Standards*, this time the federal government was interested in funding the proposal, and NCTM and NSF agreed to share the costs of the project. The proposal budget for the initial development of the *Teaching Standards* was about $350,000, and about half of that amount was received from NSF.[65]

The project budget for developing the *Teaching Standards* was somewhat less than that of the *Curriculum and Evaluation Standards*, in spite of the fact that NSF was now willing to help fund the project, and NCTM didn't have to take on all of the financial responsibility by itself. The reduction in cost was intentional, and resulted from several factors. As Glenda Lappan reported, there was no way of knowing in 1989 how well the *Curriculum and Evaluation Standards* were going to do:

> It wasn't clear what financial resources NCTM could really provide for this second document, because we hadn't finished the first one before we were actually putting together the makings of the second one. So we were trying to do this as inexpensively as possible. We didn't go off on mountain tops or anything interesting.

Meetings were held mainly at Michigan State University or in the Washington, DC, area, locations that were chosen for convenience rather than for inspirational qualities. Perhaps more important to the work of the group than the location was the way the writing was assigned to the leaders rather than to the members of the working groups.

The proposal listed a commission[66] to oversee the project (just as NCTM had done for the 1989 *Standards*) and three working groups. Each working group had a leader[67] and an "assistant/reactor" who had special responsibility for responding to what the leader wrote. In contrast to the 1989 *Standards*, when writing teams met for three periods of two weeks each to discuss and debate the ideas, the plans for the *Teaching Standards* included three-day meetings of the working groups. There was less time (and less funding) for group meetings, and leaders were assigned more of the responsibility for the writing. The entire working group met to plan the writing and later to respond to the lead writer's

---

[65]When the costs of both the 1989 and 1991 *Standards* are taken together, NCTM officials say $2 million is a reasonable estimate of the total cost.
[66]Members were Glenda Lappan (Chair) of Michigan State University, NCTM President Shirley Frye, President-elect Iris Carl, and Executive Director Jim Gates. Lee Stiff of North Carolina State University served as Board Liaison.
[67]The groups dealt with mathematics teaching, led by Deborah Ball, Michigan State University, assisted by Tom Schroeder, State University of New York at Buffalo; evaluation of mathematics teaching, led by Tom Cooney, University of Georgia, assisted by Donald Chambers, University of Wisconsin; and professional development, led by Susan Friel, University of North Carolina at Chapel Hill, assisted by Nick Branca, San Diego State University.

drafts. Individual members of the groups[68] still made significant contributions, but there was substantially less time for group interaction than in the preparation of the *Curriculum and Evaluation Standards* (NCTM, 1989).

This change was decided upon in part because of the need to reduce costs; it was less expensive to assign the main part of the writing to just the leaders. But the decision was also a consequence of the problems in writing style that hampered the development of the *Curriculum and Evaluation Standards* (NCTM, 1989). As several writers of the 1989 *Standards* reported, "Writing is not one of my strengths," and both writers and leaders indicated that there were difficulties with the quality of the writing in the drafts.

The changes in responsibility for writing, along with the reduced amount of time that the writers spent together, are probably the main reasons for the different impressions one gains from talking to members of the working groups on the two projects. The esprit de corps in the writing groups for the *Curriculum and Evaluation Standards* was especially strong; in the case of the *Teaching Standards*, the sense of group ownership seems somewhat less intense. As one leader noted, "A certain camaraderie developed in the *Curriculum and Evaluation Standards* that was missing in the *Teaching Standards*." Another of the leaders observed:

> There was much more ownership of the *Curriculum Standards* by the writing groups, and they were much more willing to work to get the word out. On the other hand, it is much easier to talk about the *Curriculum Standards* than the *Teaching Standards*, so a sense of ownership is not the only reason.

The change in the structure of the writing groups appeared to be the main source of the change, but the difference in leadership style (noted by several participants) may have been a contributor as well. As a leader put it, "Romberg and Lappan had different leadership styles—Lappan talked about that in the first meeting. She was not going to be a hands-off leader." As noted earlier, Romberg was often willing to remain above the fray of the writers' debates and rarely got involved in arguments. Lappan, on the other hand, was seen as a leader who provided "direction without being dictatorial." Both Romberg and Lappan received many positive comments from the writers who worked with them.

Changes in funding, in leadership and in leadership style, in writing assignments, and of course in the topic under discussion (pedagogy rather than content) contributed to making the *Teaching Standards* different from the 1989 *Standards*. Perhaps the most significant factor of them all is just that the *Curriculum and Evaluation Standards* came first and made a big impression, leaving the *Teaching Standards* always in the "bow wave" created by the arrival

---

[68]Members included Evelyn Bell, Roberta Koss, Steve Krulik, and Jane Schielack in the mathematics teaching group; Marilyn Hala, Tim Kanold, Diane Thiessen, and Sue Poole White in the evaluation group; Bettye Clark, Julie Keener, James Leitzel, Gary Musser, and William Speer in the professional development group.

of the earlier document. The turbulence in the wake of the 1989 *Standards*, however, was a stimulating environment in which to pursue a new vision of teaching. When the writers met to begin preparing the *Teaching Standards*, their mood was one of excited anticipation.

The Writing of the *Teaching Standards*

The writing groups met in the summer of 1989 and began their work. Since only Lappan and Cooney had been involved in writing the 1989 *Standards*, some of the same issues about the meaning of the term *standards* came up again. Some writers wanted to specify standards that would be criteria to judge quality, but most writers (like the writers of the 1989 *Standards*) had by this time moved away from an accountability perspective of standards. More of the new writers were concerned that teachers would find the idea of standards to be too prescriptive, even arrogant, and they worried about how the phrase "professional teaching standards" would be interpreted by teachers. The discussions were intense, but generally good natured and sometimes playful. Brainstorming sessions encouraged consideration of many different ideas about standards, and various concerns were expressed.

One concern was that the teacher unions in large cities would immediately reject the notion of standards, perhaps confusing it with "standardization" (see, e.g., Ball & Schroeder, 1992). This concern was part of the reason that members of Deborah Ball's group decided that they wanted to express the spirit of teaching—an idea that was consistent with telling stories of teaching, rather than stating standards or specifications. The stories of teaching were called vignettes. The use of vignettes to describe teaching led to the development of standards even though some of the writers preferred not to specify standards for teaching. Glenda Lappan recalled:

> We did not want this document to start with anything that looked like an algorithm for good teaching. We wanted it to start with pictures of teaching that would highlight teachers in the process of change and analyzing their own practice. But as Deborah Ball's group began to work on the first part of the document, they said, "Something is beginning to happen. We are beginning to look at the way in which we're writing about vignettes and our conceptualization of this." And standards were just sort of popping out.

Other participants told a similar story of the way that standards emerged from vignettes. Deborah Ball remembered how interesting the process was, both to participate in and to observe:

> We met for a day and we were thinking about what the different vignettes ought to be like. The people in my group thought that it seemed reasonable that we would write vignettes, and we talked a bit about where we would get them[69]

---

[69]Vignettes were based on actual experiences of the writers, and some are "practically verbatim records of actual classrooms," according to writer Tom Schroeder.

and what a vignette would be. At that point, we were imagining the text as being vignettes. In trying to conceptualize how we could have an array of vignettes that would capture differences in teaching for understanding, we backed into having to talk about what dimensions of teaching would be like, or what you would have to think about so that you could have variation among our vignettes. Over night from one day to the next, I remember saying at the end of the day, "Maybe I could try to write down some of the dimensions that seemed to be emerging across the day as we were talking."

The dimensions that came out of the vignettes[70] were the basis for the opening section of the *Teaching Standards* (Standards for Teaching Mathematics), with the first six standards on mathematical tasks, discourse (the teacher's role, the students' role, and tools for enhancing discourse), learning environment, and analysis of teaching. "Structuring the first part of the document around that small set of standards is one of the most creative things in either of the documents," according to Glenda Lappan. She also recalled how the writers worked hard to make the standards for teaching directly related to mathematics:

I remember the conversation when Tom Cooney asked yet again, "How can we make sure that people look at this and have mathematics in mind?" The four of us were sitting at the kitchen table, and we said, "Well, it's obvious. We put mathematics first. The first thing we write about is worthwhile mathematical tasks."

The focus on discourse was controversial at first, but eventually the term came to be one of the hallmarks of the *Teaching Standards*. From Lappan's perspective:

I view these two documents through the eyes of the D-words that the documents gave us. The 1989 document gave us *disposition* as part of what we need to think about, and the second document gave us *discourse*. I think that discourse and disposition are two of the notions that take us further in realizing that what we are talking about is a very different commitment to the mathematics education of children. To some extent, discourse in the second document became another site for worrying about the development of disposition toward mathematics.

Initially the resistance to discourse came from the reluctance to introduce a term that many teachers would find foreign. Tom Schroeder recalled:

It was Deborah Ball who proposed using the term *discourse*, and I believe that she was familiar with a body of research that uses that term in the sense laid out in the *Teaching Standards* (NCTM, 1991, p. 34). There were some extensive discussions in our working group on the pros and cons of the term, including some wisecracks about "discourse, dat course, and dee udder course."

---

[70]One indication of the writers' enthusiasm (and their dedication to the idea of vignettes) was the T-shirts they had made up; they were inscribed "Vignette Visionaries." Our sources did not recall being influenced by the earlier use of vignettes in Kilpatrick (1985).

A state mathematics supervisor was particularly critical in an early review, but eventually was won over, as Deborah Ball reported:

> At first he absolutely hated what we had done. He was one of those negative reviewers. He wrote us this angry, vehement letter about what a stupid word *discourse* was—academic jargon. He disagreed with the framework completely. He went right to the jugular on the whole framework that we did. That was the first summer. We thought a lot about his comments; I didn't completely agree with him, but I really thought hard. I thought hardest about the responses that were vehement because they were the most useful. When we did the revision, which still kept discourse in a more prominent place, he fell in love with them. In his second review, he wrote "not only is this an unbelievable document, but it's scary." What he meant by that was it's scary because it has such a radical vision of teaching. He thought that if people really got it the way that he was interpreting it, they'd be frightened. Now that's one read. But I think that's a good read.

When that state supervisor was asked about how much teachers knew about the two *Standards* documents, he expressed concern about the lack of attention to the *Teaching Standards*:

> I think that the *Professional Teaching Standards* are harder to attain, but more significant by far than the *Curriculum and Evaluation Standards*. I'm telling my friends at NCTM now that we've got to add Addenda programs and videotapes and all those kinds of things to help capture what we mean by tasks and discourse and environment.

Most would agree that the *Teaching Standards* are difficult to attain, and that the *Curriculum and Evaluation Standards* have received more of the attention, partly because they appeared first. There are also some advantages to coming second, however; one might argue that NCTM's experience with the 1989 *Standards* helped the writers of the *Teaching Standards* analyze issues in a deeper way. For example, Ball (1992) noted how the writers tried to balance the competing needs for consensus and change, and the tensions between providing direction and allowing discretion to teachers. The need for balance is crucial, and NCTM leaders are generally pleased with what the writers of the *Teaching Standards* were able to accomplish. This document was also one that they were pleased to disseminate.

The Dissemination of the *Teaching Standards*

The publication of the *Teaching Standards* in 1991 was another major event for NCTM, and the dissemination plans were similar to what was done for the *Curriculum and Evaluation Standards* (NCTM, 1989). The document was distributed free to members[71] and the Executive Summary was circulated widely,

---

[71]By April 1995 NCTM had distributed 121,000 copies of the *Teaching Standards*, including free copies to 58,000 individual members. The cost ($25) was the same as for the 1989 *Standards*, which had 258,000 copies in circulation.

just as for the 1989 *Standards*. NCTM meetings and journals now focused on the *Teaching Standards*, as well as the *Curriculum and Evaluation Standards*. Nevertheless, the dissemination effort did not seem to have the same impact as the 1989 *Standards*. As one state supervisor put it:

> I do not believe that there is anywhere near the level of awareness about the *Teaching Standards* as about the *Curriculum and Evaluation Standards*. The *Teaching Standards* have not been as well disseminated, not been as widely discussed, and have not received the attention that they require. We've not had as many meetings on them, and they have gotten lost in the shadow of the *Curriculum Standards*.

Most leaders do not think the *Teaching Standards* have been lost in the shadows, but many have deep regret that part of the plans to disseminate the *Teaching Standards* were never realized. The attempt to replicate the "Leading Mathematics Education into the 21st Century" project (Dolan, 1990) for the *Teaching Standards* was never completed, to the great disappointment of a number of state supervisors of mathematics, who felt that the 21st Century project had been extremely successful. As one state supervisor noted:

> We decided that it shouldn't be the state supervisors dealing with the dissemination of the *Professional Teaching Standards*; it really needed to be the National Council of Supervisors of Mathematics. The reasoning was that curriculum is more of a state responsibility, and instruction is more for the districts [and the local supervisors]. It really made sense. [Unfortunately, the project] got lost in the shuffle; it fell through the cracks.

The bitterness of some state supervisors was strongly felt, and some preferred not to talk about the circumstances. One NCTM leader noted that all of the work of preparing the proposal was essentially done, but there was just a failure of leadership. The appearance of the *Assessment Standards* (NCTM, 1995) has prompted the Association of State Supervisors of Mathematics to try for another major dissemination effort similar to the 21st Century project, focused this time on assessment. In this case the proposal was submitted in timely fashion, but NSF has so far declined to fund the proposal.[72]

Although the dissemination of the *Teaching Standards* was not carried out to the extent planned, the document was seen by many leaders as an innovative and effective contributor to the reform effort in mathematics education. Its use of language has been identified as a particular strength. Tom Schroeder noted how terms like *discourse* were used in special ways in the document, an example of what he called "*Standards*-speak":

> Each of the *Standards* documents uses terms (some might say jargon, but I won't use that pejorative) that are a bit unfamiliar or unusual, in order to draw attention to issues and re-frame them more broadly than might otherwise be the

---

[72]Some leaders fear that the lack of support from NSF (now as in the mid-1980s) is due in part to the continuing shortage of staff who are knowledgeable about mathematics education.

case. The terms I have in mind include "mathematical disposition" instead of "attitudes toward mathematics;" and "discourse" as opposed to, say, "classroom communication." The term "vignette" is another instance of *Standards*-speak with a purpose.[73]

Although the *Teaching Standards* are admired by many, there have been criticisms of the document. Some in the mathematics community have been concerned that mathematicians did not have sufficient influence on the document; they may not know that many of the writers (e.g., Branca, Lappan, Leitzel[74]) were professors of mathematics. A criticism from an NCTM leader raised a more substantive issue about the kinds of support that teachers need to build curriculum:

> One of my problems with the *Teaching Standards* is that it doesn't talk about how to put a collection of tasks together to make a cohesive unit, a cohesive year, and a cohesive curriculum across years. You have to focus on more than the tasks that kids do.

Equity issues are another area where questions have been raised about both sets of *Standards*. The *Teaching Standards* dealt with the issue in part by highlighting in each major section the 1990 statement by the NCTM Board of Directors on the goal of a comprehensive mathematics education for every child, including specifically minorities and females.

Criticism of the standards movement has grown in recent years, and there are indications that the movement may be coming to a halt. Nevertheless, the *Curriculum and Evaluation Standards* (NCTM, 1989) and the *Teaching Standards* (NCTM, 1991) were judged by most NCTM leaders as a very successful project representing NCTM's finest hour. We consider next the impact of both *Standards* documents (NCTM, 1989, 1991).

NCTM's Impact on Reform

A number of NCTM leaders predicted that the decade of the 1980s would be a time of significant change in mathematics education and several said so in their publications (Crosswhite, 1985a; Hill, 1981; Romberg, 1984a). Nevertheless, all leaders seem to have been a little surprised at the impact of NCTM on educational change in the US. In recalling the writing of the 1989 *Standards*, Glenda Lappan noted:

> I don't think any of us, at that stage of the game, really felt like the *Standards* were going to have the impact that they have had. There was a worry sort of nagging in the back of everybody's mind: We're going to do all this work, and how do we know that it will do anything other than just sit on people's shelves and cause some mild ripple within NCTM?

[73]See also Apple (1992) for a related discussion of the *Standards* as a "slogan system."
[74]Leitzel has been particularly active in the Mathematical Association of America (see Leitzel, 1991).

The broad impact of the NCTM *Standards,* especially the *Curriculum and Evaluation Standards* (NCTM, 1989), has been widely acknowledged within mathematics education, and by authors who come from outside of mathematics education as well (Atkin, 1994; Boyer, 1990; Massell, 1994; Ravitch, 1995). Diane Ravitch, Assistant Secretary for Educational Research and Improvement in the Bush Administration, noted that the timing was fortuitous; the NCTM *Standards* appeared just when President Bush and the US governors were discussing national goals in education. "At the very time that governors and other political leaders wondered about the feasibility of voluntary national standards, there were the NCTM standards as an example for emulation" (Ravitch, 1995, p. 57). Moreover, the NCTM *Standards* were developed by a non-governmental professional organization, so they were seen by the Bush Administration as a "free-market model for educational reform" (p. 28) and hence were politically acceptable, in contrast to the view of standards in the early years of the Reagan Administration. Following the appearance of the NCTM (1989) *Standards*, the US Department of Education made grants to other subject-matter areas for developing standards in 1991 and 1992, and various government programs related to the development of standards were continued by the Clinton Administration (U.S. Department of Education, 1994). As NCTM President Shirley Frye noted, "Most educators are quite grateful that NCTM has been able to put this positive spin on a national direction, arming them with more ability to change their own disciplines than they ever thought they would have."

Other leaders weren't always sure that NCTM had done them a favor. As one noted, the governors seemed to think that it would be easy for other disciplines to develop standards,[75] not recognizing the complexity of the task nor the differences among disciplines (Massell, 1994). The development of standards in all disciplines was a recommendation fro n the National Educational Goals Panel, chaired by Governor Roy Romer of Colorado and Governor Carroll Campbell of South Carolina (Ravitch, 1995). Earlier Romer had become an effective advocate for specification of educational goals at the state level, as well as the use of mandated testing to pressure schools to focus on those goals. He later became noted for his enthusiastic support of the NCTM (1989) *Standards*. Senta Raizen, a leader in science education, recalled, "At every regional meeting held by the National Educational Goals Panel, Romer would wave the NCTM *Standards* and ask how soon we were going to have standards like that in other fields."

The story of how Romer became informed about the work of NCTM is another example of how fortune played a role in the dissemination of the *Standards*. In April 1991, Romer called the NCTM office in Reston, VA, and requested information on the development and impact of the *Curriculum and Evaluation Standards*. Dan Dolan, a member of the 5-8 writing group, was

---

[75]A leader in science education described the directive from the governors in Biblical terms: "Do thou likewise." This "cookie-cutter" approach to reform was seen as unrealistic.

working at the Mathematical Sciences Education Board (MSEB) in Washington, DC, at that time. The NCTM office called Dolan to see if he could go to the Hall of States in downtown Washington, DC, to brief the governor.

By this time Romer was reasonably well-informed about the 1989 *Standards*. Iris Carl, President of NCTM, had been working with the Goals Panel and Romer had attended the April 1991 National Summit on Assessment.[76] Mary Harley Kruter, who had coordinated this assessment conference for MSEB, went with Dolan to meet with Romer and one of his staff. Romer was particularly interested in the word *standards*. Would standards become objectives for teachers in the classroom? How would they be measured? Dolan pointed out that standards were to be criteria by which states, publishers, local districts, and others could measure the quality of their program; they were not intended to be behavioral objectives. Dolan recalled the discussion this way:

> As we opened the book [*Curriculum and Evaluation Standards*] and reviewed the standards themselves, he became more aware of the broad base of the document rather then the prescriptive nature which he thought the standards might take on.

Governor Romer expressed his particular interest in assessment, noting that the Goals panel was keen on developing a national standardized test that would provide scores for individual students. According to Dolan:

> He seemed convinced that a standardized test could be developed and given to all students to measure how well they had achieved. In several of his public discussions, he had compared assessment of students to the testing he had gone through during pilot training. He pointedly asked again, "Why can't we devise a test for all students to find out how well they do at mathematics in the same way that we [determine] how well a person is progressing toward being a pilot?[77]

At this point Dolan used the governor's example to show the weaknesses of traditional assessment:

> I pointed out that the assessment of a pilot needed to measure the degree to which the person was successful in flying the plane. Although there was a great deal of class work involved, the real measure of the person's ability to fly was determined by having the person sit in the pilot seat and fly the plane. I asked him if he would feel comfortable getting in an airliner piloted by a person who had passed every one of the paper-and-pencil tests but never sat in the pilot's seat.

---

[76]President Bush and Secretary of Education Lamar Alexander also attended this MSEB conference.

[77]The military origins of the US testing establishment are well documented.

Dolan argued that no simple test was adequate, and felt that the governor was persuaded. At least he did not hear Romer pushing for a national test after that meeting. The meeting continued past the appointed time, with regular visits from one of the governor's aides, who kept reminding him that he was late for his next appointment. But the governor wanted more information, and the meeting continued, to the satisfaction of Dolan and Kruter.

There are other examples of NCTM's influence on the development of education policy, especially through the work of its presidents; Iris Carl, for example, was appointed by President Bush and Secretary Alexander[78] to serve on several commissions. She also served as an advisor on educational policy to several states, the level of government that has typically been most involved in setting policy related to curriculum.

Influence on State Documents

The *Standards* clearly had an influence on national educational policy, and the impact at the state level was also substantial. In a study of changes in state curriculum frameworks in mathematics and science, Blank and Pechman (1995) noted that 41 states reported changes in mathematics frameworks that were related to the NCTM *Standards*. In their analysis of key elements of state frameworks, they observed that all recent mathematics frameworks showed the *Standards'* influence in their language, the typical (K-4, 5-8, 9-12) grade-level organization, and their attention to process as well as content. As Blank and Pechman (1995) pointed out, there was a "high degree of consistency" with the emphases of the NCTM *Standards*. However, sections of frameworks that specified ideas in more detail were more likely to show differences with the *Standards*, and these differences were more apparent at Grades 9-12.

Some states (Webb, 1993b) had their reform efforts well underway by the time the NCTM *Standards* appeared in 1989. In those cases, the *Standards* were used to support and confirm the direction that the state had taken. As an NCTM leader put it, "The NCTM *Standards* document was going along parallel to our state framework; both supported the same things. Publication of the *Standards* helped support what was going on in my state."

In other states, the NCTM *Curriculum and Evaluation Standards* were often used as a model to develop state guidelines. A member of a state committee that was preparing curriculum guidelines in 1987-88 described how their committee approached its task:

> We got the 1987 draft copy of the *Standards* as background reading for the actual committee meetings. Although I can't say that the final product is in a one-to-one relationship with the *Standards*, I can say that it definitely impacted the direction of the statement that we put together.

---

[78]Tom Romberg and other NCTM leaders have also served as advisers to Alexander, both when he was a governor and when he became Secretary of Education.

A state supervisor of mathematics from a different state reported a similar use of the *Standards*:

> When I took the job here at the state, one of the first projects was to revise our curriculum guide. The *Standards* were still in draft form, but we used the draft to write our curriculum guide. We thought, "This is a good document; why should we reinvent the wheel?"

Based on the data from the Blank and Pechman (1995) report, it seems likely that many states worked in this way to develop guidelines and frameworks for curriculum. The impact of the NCTM *Standards* on state policy documents is relatively clear and easy to document. The influence of the *Standards* on other communities[79] has also been substantial.

Influence on Other Communities

Collegiate mathematics and other professional organizations are among the groups that have been influenced by NCTM's work on standards. The main organizations concerned with mathematics education at the post-secondary level are the Mathematical Association of America (MAA) and the American Mathematical Association of Two-Year Colleges (AMATYC[80]). The MAA has a long record as a major influence on the collegiate mathematics curriculum, including calculus reform.

The movement to reform the teaching of calculus during the 1980s was taking place in parallel with the development of the NCTM *Curriculum and Evaluation Standards*. Lynn Steen, past president of the MAA and a member of the NCTM Commission for the 1989 *Standards*, recalled the origins of the calculus reform effort:

> The other issue in the air at the time, which is interesting because it is partly independent and partly not, is the motivation that led to the calculus reform movement. All of the discussion about calculus was going on exactly at the same time as NCTM was planning the *Standards*. Superficially, calculus reform was disjoint from the *Standards* Two different communities had simultaneously come to the same conclusion: The methods they were using weren't working and they needed to do something about it.

> It is very clear that there was practically no intersection among the people working on these two reform movements. The origin of the calculus reform movement grew out of the increasing emphasis on discrete mathematics, which in turn came out of the swelling enrollments in computer science. This had little or nothing to do with high school mathematics, or with the constructivist

---

[79]Others have reported data on some of these communities; for example, see Romberg and Webb (1993) for comments on publishers.

[80]AMATYC (1994) began developing standards in the summer of 1993, and has recently released standards for intellectual development, content, and pedagogy in mathematics.

movement in education, or any of the other issues that were on the minds of NCTM leaders. Yet they emerged simultaneously in a parallel fashion, doing things that were intellectually very consistent.

I think it is a very interesting question as to why the timing of those movements just happened to coincide when they clearly had very different driving forces. If it hadn't been for the impact of rising enrollments in computer science and the falling numbers of mathematics majors, the mathematics community might have tolerated the 40-60% failure/drop-out rate in calculus for two or three more decades.

The many factors that combined to support educational reform during the 1980s were an interesting aspect of this case study. Later we speculate about the common "driving forces" that were involved. As the example of calculus reform at the college level shows, reform went beyond the K-12 level in mathematics,[81] and beyond mathematics to other disciplines.

Since the release of the NCTM *Standards*, standards have been developed in other content areas. They have been funded by professional teaching organizations, federal grants, and other organizations with interests in education. Most have followed the NCTM format and philosophy. The work on standards in science education is discussed by Atkin and colleagues (this volume) in their study of Project 2061. The similarities between NCTM's work and standards in science are substantial; there are science standards for teaching, professional development, assessment, content (for grade levels K-4, 5-8, 9-12), education programs, and education systems. Standards for history, published in 1995, have been criticized for excesses in multiculturalism; the document contains standards for US and world history organized around specific chronological eras. The status of standards for English is unclear; the federal government stopped funding this project in 1994. Other curriculum areas that have developed standards include geography, arts, economics, and physical education, the only area that seems to have proceeded with internal funding the same way that NCTM did.

The influence of the NCTM *Standards* on other groups in mathematics education,[82] other professional organizations, and other school subjects has been substantial. As one state supervisor noted, "The politicians kept saying, 'These folks in mathematics education have their act together.' In 1990 I heard that phrase at least once a week. It almost became embarrassing."

---

[81]Leonelli and Schwendeman (1994) have also modified and adapted the NCTM *Curriculum and Evaluation Standards* to meet the special needs of adult learners.

[82]For example, NCTM is connected to numerous affiliated groups, over 250 in 1994. These organizations of mathematics teachers and leaders may be national in scope, like the National Council of Supervisors of Mathematics, or state or regional groups that represent the teachers in a particular area.

At the policy level, the *Standards* clearly expressed a powerful idea. The influence of NCTM *Standards* on administrators and teachers, however, is not as clear.

Influence on Teachers

The 1993 National Survey of Science and Mathematics Education (Weiss, Matti, & Smith, 1994) reported data from a national probability sample of administrators and teachers in the US. Although the survey had many purposes, one of its main research questions involved the extent to which teachers support reform notions embodied in the NCTM *Standards*. We summarize some of the data from that study here, along with data of our own obtained in interviews with state supervisors of mathematics and with NCTM leaders.

Table 6 summarizes mathematics teachers' familiarity with the NCTM *Standards*, including both the *Curriculum and Evaluation Standards* (NCTM, 1989) and the *Teaching Standards* (NCTM, 1991). As the data indicate, awareness of the *Standards* increased as one moves up the grade levels from elementary (18% for Grades 1-4) to high school (56% for Grades 9-12). Since 59% of the mathematics teachers in Grades 9-12 reported that they had a major in mathematics or mathematics education, one might speculate that most of the teachers whose primary interest is mathematics have some knowledge of the *Standards*. However, the *Teaching Standards* are not as well known as the *Curriculum and Evaluation Standards*. Of those teachers who said they were well aware of a *Standards* document, roughly half agreed with the statement that they were "prepared to explain" the document to their colleagues.

When one considers the teaching practices that are associated with the NCTM *Standards,* the data tell a different story; when it comes to pedagogy, the elementary teachers are more in line with the *Standards*. The percentage of mathematics teachers that indicate support for "an emphasis on solving real problems" and "cooperative learning groups" declines as one moves from Grades 1-4 to 9-12. For example, 59% of teachers in Grades 1-4 indicated that cooperative learning groups should definitely be a part of mathematics instruction, but the corresponding figure for teachers in Grades 9-12 was 27%. For Grades 5-8, the percentage was 41% (Weiss et al., 1994, p. 25).

The data from our interviews with state supervisors of mathematics were in general agreement with the data from the survey, although there were some differences. For example, state supervisors tended to estimate that a higher percentage of teachers would know about the *Standards*. When asked about teachers' awareness of the *Standards*, one state supervisor responded, "I would have said 95% if I hadn't been out visiting classrooms, but it's probably more like 50 or 60%. I'm still amazed at the number of teachers out there that don't know the *Standards* exist."

Table 6.   **Mathematics Teachers' Familiarity With the NCTM** *Standards*

	Percentage of Teachers		
	Grades 1-4	Grades 5-8	Grades 9-12
Curriculum and Evaluation Standards			
Well aware of the NCTM *Standards*	18	28	56
Heard of the NCTM Standards, but don't know much about them	39	41	33
Not aware of the NCTM *Standards*	30	22	8
Not sure	13	9	3
Professional Teaching Standards			
Well aware of the *Teaching Standards*	12	19	40
Heard of the *Teaching Standards*, but don't know much about them	38	48	44
Not aware of the *Teaching Standards*	38	25	13
Not sure	13	8	3

Adapted from Weiss, Matti, and Smith (1994, p. 22).

Another state supervisor gave higher estimates than the survey found. He suggested that for his state, "If we're talking about the secondary math teachers, probably 95% of those people are aware of the *Standards*. If you talk about the K-6 elementary teachers, maybe half. It's significantly less than that for the *Teaching Standards*."

In a survey of the 50 state supervisors of mathematics conducted for this case study, we asked the supervisors to estimate the average level of change related to the *Standards* that had occurred in mathematics instruction in their state. On a scale of 1 to 4, where 1 indicated no impact and 4 indicated a major impact, the average of the estimates went from 3.2 for Grades K-4 to 2.4 for Grades 9-12. Although we make no claim of precision for these estimates, we note that they are in agreement with the trends identified in the survey data from Weiss et al. (1994).

Although the state supervisors of mathematics were generally optimistic about the direction of change, they were also realistic about the difficulties of obtaining significant change in mathematics classrooms. As one put it, "Let's not fool ourselves—the *Standards* have had an impact, but not a significant impact." Another state supervisor observed:

I don't think that the changes that are advocated in the *Standards* have occurred in that many classrooms. I would be very happy if 25% of the classrooms have changed. I don't say that in a negative way; I think that's something we could be proud of.

As this state supervisor implies, it is easy to become so enthusiastic about the impact of the NCTM *Standards* in general that one loses sight of what is happening in classrooms. Change comes slowly, or not at all, for many teachers, even those in reasonably progressive school settings. Some of the reasons for the slow progress of change in classrooms were reflected in the reports of visits to schools, which appear below.

The Standing of the *Standards* in the Schools

In this study of the development and influence of the NCTM *Standards* (NCTM, 1989, 1991), we thought it important to observe and reflect upon current appearance of the NCTM advocacy in schools. We drew upon field research conducted by several teams,[83] and made visits ourselves to half a dozen schools. We prepared brief synopses of our visits to three of the schools to include in this report, the first for an elementary school on the West Coast, the second for a K-8 school in Chicago, and the third for a secondary school on the East Coast. A sample of three, or six or even twenty-six, is not representative of the nation's schools.[84] However, from these three cases we can increase our understanding of the complexity of situations in which the *Standards* are pertinent, whether or not acknowledged at the scene. This use of case studies is conventional as summarized in various sources (e.g., Stake, 1994).

Our report has focused on the origins and influence of two NCTM *Standards* documents (NCTM, 1989, 1991) and NCTM's role in stimulating systemic change. Now, however, we change our focus from systemic change to change in individual schools. We wanted to observe how the forces that drive systemic change[85] are reflected in schools, in teachers, and in students. The issues for these observations differed somewhat from those we set forth earlier for the study of the development of the *Standards*. Among our questions here, the following were prominent:

1. As perceived by people in the schools, are the *Standards* an instrument of accountability, drawing school operations more closely toward the pursuit of goals identified by state mandates, media pressure, and parent consternation?

---

[83]Particularly helpful were the projects headed by Ferrini-Mundy (Ferrini-Mundy & Johnson, 1994), Secada (Byrd, Foster, Peressini, & Secada, 1994), and Silver (Stein, Grover, & Silver, 1991), as well as the related work of Ball (1990), Cohen, McLaughlin, and Talbert (1993), Garet and Mills (1995), and Talbert and Perry (1994).

[84]For a proper sample of the nation's schools and teachers, we look to the recent national survey described by Weiss, Matti, and Smith (1994).

[85]See Fullan (1993) or the discussion of macro and micro forces by Raizen (1996).

2. Is there within these schools an evolution of the teaching of mathematics in directions advocated by NCTM, and have the *Standards* played a direct role in that evolution?

3. For teachers and schools influenced by the *Standards*, is there an accompanying reconceptualization of mathematics as well as a reconceptualization, particularly a constructivist reconceptualization, of mathematics education?

4. As greater use is made of computers and other technology in all teaching at the school, is student progress toward the goal of deeper mathematical understanding served or deflected?

In addition, following the observations of Ferrini-Mundy and Johnson (1994), Fullan (1993), Hutchinson and Huberman (1993), and others, we looked for evidence of those forces that influence school change at the local level: school leadership, the collaboration among teachers, professional development efforts, the influence of outside specialists, and similar issues. Although we value studies of change that focus on individual teachers (e.g., Parker, 1993; Romagnano, 1994), we have chosen to report on the experience of change at the school level.

Our first report is from Lion's Mane Elementary,[86] a school in the west. We visited this school in both fall and spring, and conducted interviews with leaders over two years. The report, prepared by Robert Stake in his own voice, is based mainly on his fall 1993 visit.

Lion's Mane Elementary School

The first impression you get approaching the Lion's Mane School is how closely it is nested into the neighborhood. You drive a short lane between small houses, circling toward the front door only a few yards from others. A spacious playground behind the complex is out of view, hidden by houses and a stand of hundred foot cedar trees. The building itself is ultra-rectangular, newly refurbished in silver gray and glass façading for nine clusters of squared-up classrooms surrounding a central library and staffroom. Squares within squares, nests within nests.

The school has been recognized as a place where reform in mathematics teaching has taken hold. Lion's Mane teachers are in their third year of staff development with Columbine District consultant, Amy Alexander. Alexander had worked in leadership training programs and a project to help schools align with the State Framework and Model Curriculum Guide. She was a strong advocate of NCTM *Standards*. Lion's Mane was one of the two elementary schools in Columbine identified by the Assistant Superintendent to lead local reform of mathematics teaching. It was part of the Superintendent's campaign

---

[86]In this section of our report we use pseudonyms for schools and individuals.

called "Columbine 2000" to prepare students for "a changing world," partly by emphasizing thinking, personal expression, problem solving, self-directed learning, and working with others—key aims also for NCTM.

To the 28 teachers currently at Lion's Mane, this meant radical change, voluntary change, in what happens in math class, particularly posing, for interactive solution, practical tasks using ingredients such as I saw on Elaine Eddy's whiteboard:

> Billy has nine dimes to spend. He needs 25¢ for the bus. List of items [he could buy]: Eraser 10¢   Ruler 29¢   Pencil 25¢   Pen 39¢   Book Cover 20¢

When I saw them, the 23 second, third, and fourth graders in Eddy's multiage classroom were writing out reasons for their choice of purchases. Eddy sat at the tables with them, at the moment thinking through an interpretation that Carolyn posed. Three students stood, paragraphs in hand, waiting their turn, listening to the exchange between Carolyn and her teacher.

Gretchen Golding, the principal, was proud of what Lion's Mane teachers were doing with interactive mathematics, but mentioned her irritation at one teacher who had "undercut us" at Parent Night, complaining about a lack of attention to traditional skills. Golding recognized that it called for major change in self-perception for many teachers, had encouraged each to "partner" with another for planning (a partnering several told me they liked), and had provided half day blocks for such planning. Golding noted the expense [district money] of this staff development with Amy Alexander, regretting that it left not enough for staff development on computers, social studies, or for her most immediate advocacy, multi-age classrooms, where students were increasingly involved in intellectual discourse and attitude development.

Creative pedagogy and interactive mathematics have long been part of innovative schools like Lion's Mane. *Architecture and Children* had been a Lion's Mane project while the school was undergoing renovation, with many practical problems posed by the builders. The *Healthy Children Project* was another effort at reconceptualization. I tried to ascertain which of those innovations, especially the whole language approach, might have provided early stimulation or underpinning to taking up the NCTM *Standards*. I did not come up with any good examples. Principal Golding said that it took "the mastery of Amy [Alexander] to bring us to a higher level of teaching."

I asked third grade teacher Heather Holt if in these last ten years she had started to move toward a problem-solving pedagogy even before the consultancy began. She said that of course they used to do story problems and that she had become increasingly aware of the importance of concrete learning situations, integrating writing and mathematics, a whole language emphasis, and cooperative learning, but that it was not until Amy Alexander demonstrated how groups of children involved in mathematical investigations can solve real and complex problems that she and fellow teachers radically changed their teaching. Some primary teachers at first felt they already had been doing it, having followed

Math Their Way, but now, for the first time, they were releasing pupils to seek *their own* interpretations of and solutions to situationalized problems.

Holt showed me a recently completed geometry exercise: forming new geometric shapes by cutting a rectangle on the diagonal and then describing these new shapes in writing. On the first day, the descriptions were bare and brief, indicating to Holt a need for vocabulary of geometric shapes. In spite of what she felt to be a regression into didactics, she provided some terminology (obtuse triangle, quadrilateral, etc.) and had the students do the exercise a second time, with greatly improved description.

In reflecting on her current mathematics teaching, Holt was pleased "how fully engaged and comfortable the children are" in the activities. But she felt "way behind," knowing that she was not providing the computational exercises that many parents wanted and that she herself felt were needed. "Parents want their kids good at basic math. They don't want their kids to become reliant on calculators. They support what we are doing but they are not 100% satisfied." In sharp terms, she indicated that consultant Alexander opposed any return to didactics and drill, that Alexander favored using all the time on interactive problem solving. But to get more visible productivity, Holt had just shifted student tables from clusters back into rows. The pressures from both philosophies were heavy upon her.

To many Lion's Mane teachers, Alexander's strategy was "all or nothing." From their remarks, I guessed that Alexander's strategy was: "You have to work 'round the clock to convince a tradition-fixed populace, including kids, and yourself, that in the long run, the criteria are not algorithmic readiness and multiple choice testing; the criterion is *understanding mathematics*. The experience of problem solving is the means to that end. That experience should be fixed in student recollection as definition of what mathematics learning is."

When I presented my synopsis to her, Alexander objected to the oversimplifications, in part saying:

> Algorithmic thinking, not proficiency with standard US algorithms, is our goal. We do want students to work with numbers confidently and competently. They can't be mathematically powerful unless they have facility with numbers. It is how we teach number concepts, not whether or not we teach them, that is the issue. I would argue that frequent practice with mental computation and estimation is far more essential to mathematical power than is frequent practice with paper-and-pencil algorithms. Certainly I would agree that performance-based assessment, not multiple choice tests, is what is needed.

I watched Alexander conduct a day-long staff development session at district headquarters. Some forty specially-selected elementary teachers from Columbine elementary schools—whose other teachers were not yet participating in the staff development—had read an article by Blais (1988): "Constructivism—a Theoretical Revolution for Algebra." Volunteering passages, they probed its relevance to elementary school teaching: "Knowledge is something that learners *must* construct for and by themselves. There is no alternative" (p. 627). Gathered around seven tables, the teachers next performed the same exercise Heather Holt had shown me earlier: cutting rectangles on a diagonal, pasting the parts to form new shapes, and looking for patterns. For more than half an hour, the concentration was intense and the discussion vigorous. Alexander indicated that good experiences are extemporaneous, and teachers have to be ready to decide what is mathematically relevant.

Alexander used that exercise to illustrate "an assessment task that could be used prior to beginning a unit on geometry." It was followed by involvement with numerous "geometry tasks in a MENU structure" for the bulk of the day. Finally, the NCTM *Standards* on K-4 and 5-8 Geometry were examined, as Alexander said, "to determine whether or not the tasks had involved teachers with essential mathematical ideas."

Repetition of exercises was something of an issue. One way pupils have of taking advantage of the teachers is to say, "Oh, we did that last year." Traditionally, most teaching is highly repetitious, but it has been easy for traditionalist teachers to show that "the children still haven't mastered it." Alexander was able to persuade most workshop participants that the same interactive exercise often could be used for increasingly advanced conceptualizations—but many teachers were not comfortable with their own ability to identify those advancements.

Curriculum Issues. Back at Lion's Mane, second grade teacher Kris Knickerbocker told me that Alexander occasionally cited the NCTM *Standards* but did not offer them formally as a conceptual map. Assistant Principal Jana Johnson, with a doctoral dissertation in mathematics education and with her own meticulous advocacy of the *Standards,* had distributed a five-page summary. The teachers did not perceive the *Standards* as scope and sequence statement, as inventory, as road map to guide their own thinking of "What next?" and "Where eventually?" Alexander and Johnson appeared to appreciate the resistance of NCTM authors to provide an inventory of essential tasks.

Many at Lion's Mane felt some discomfort about this shortfall in disciplinary specification. They knew the strands, they had a collection of activities to keep the same students busy for multiple years, but they lacked assurance that they were teaching the mathematics that needed to be taught. Process was essential but what about content? Geometry was essential but were tetrahedrons and tessellations? Statistics was essential but were sampling and the standard deviation? And if content at these lower levels *were* so prioritized, then wouldn't there need to be prioritized subdivisions of each of these? The teachers were not convinced that disciplinary content is best left subordinate to process. The *Standards* only partially satisfied what they saw as the need for topical checklists.

So on two counts, Lion's Mane teachers in their third year of exciting, satisfying teaching were disquieted They lacked the sense that they could make the right decisions (i.e., they lacked criteria for taking command of the curriculum).

With regard to content, especially computation, they were not convinced that Lion's Mane students were getting ready for the real world of the middle school and on the right road to college preparatory testing. Somehow they lacked a contract, a legitimation of the experiential education that they apparently had practiced now for going-on three years.

The dilemma had sharpened the previous summer. The mathematics teachers at Hay Meadow Middle School, themselves beneficiaries of Alexander's consultancy, were faced with incoming seventh graders from two participating and four non-participating elementary schools. So about a third of their new students had been solving problems interactively for two years, while the remainder had received more traditional preparation. The teachers wanted to follow a problem-solving pedagogy, but did not want a third of their classes "old pros" and two thirds "novitiates." They, just as the teachers at Lion's Mane, were apprehensive about the sallies, "Oh, we did that before!"

The Hay Meadow mathematics teachers decided that all incoming students should take a test to assist in assignment to classes. The Lion's Mane staff found the proposed test an inadequate reflection of what they had been teaching and perhaps not even a good test of what they had not, and refused to administer the test. They were despondent. Here they had been notably successful in providing a mathematical experience consistent with urgings of the world's leading mathematics educators, consistent with *Standards* held up to a television audience by President George Bush, and their work appeared not only not to be celebrated but held to rejected standards by their colleagues in the nearby middle school. Lion's Mane faced a fundamental problem of reform in teaching. Sooner or later, their children will face a larger world that has not endorsed or understood the reform.

Actually the quandary subsided at Hay Meadow before it peaked at Lion's Mane. The middle school principal did not support grouping newly arriving seventh graders on the basis of previous experience and a move toward team-teaching took attention away from the grouping issue. In the fall, mathematics chair Larry Liggitt observed that the seventh grade teachers were not finding it problematic to have a mix of students engaged in a mix of exercises, both interactive and skill-oriented. As at Lion's Mane, the Hay Meadow teachers expressed appreciation for consultant Amy Alexander's guidance into a different pedagogy and were relieved when she allowed that they should mix the new with the old as they saw fit. Still using textbooks outmoded and badly worn, Liggitt felt that the right new text would anchor a comprehensive pedagogy. He said he continued to feel that the intended entry testing could be useful for fitting instruction to the needs of individual children.

The issue of conflicting expectations popped up repeatedly. Second grade teacher Kris Knickerbocker noted that when she and her husband bought a house in nearby Mission Valley, her own children had had to leave Lion's Mane. "Our second grader was embarrassed by not knowing what multiplication is, and the sixth grader with fractions and ratios." Indeed, she said that even before that, she had been doubtful, but in spite of it all, now—thanks to the consultant and her new experience—she was "at peace." You "pick up the child where they are and advance them as far as you can." That is a fine philosophy but not one that works long for an individual teacher surrounded by teachers with a different philosophy, who treat students as if certain topics should be learned by all students at specific grade levels.

In teaching children to seek options and patterns interactively, Knickerbocker saw that "the kids don't get bored doing the same basic things over and over," even if they sometimes say, "Oh, we did that last year." She saw the assignments as worthy and the response as gratifying. Other parents, of course, had far less opportunity to study the paradigm shift, and Knickerbocker encouraged a greater use of the newsletter to parents.

Assistant Principal Johnson had periodically distributed to parents a newsletter, an update on the mathematics teaching at Lion's Mane. A few had approached her in tears, partly because they missed the drill sheets brought home to post on the refrigerator and partly because they did not know how to help their children with their projects. Recognizing that parents should be teaching mathematics at home, Johnson provided exercises that the whole family could work on, exercises very much like the tasks being worked on at school. A few parents remained dubious, and so informed the district office. Their queries were answered, encouraging the parents to talk with Johnson. And for whatever reason, none took that further step.

In Maddie Morganthau's Grade 4-5 room, a teacher intern, Nell Naustrow, was reminding the class of previous days' work setting up a "super rectangle." Some recognized it as a 25x25 multiplication table. In groups of four, they had cut out strips one inch wide by lengths running from one inch to twenty five inches. They would lay the strips across the top of the grid, count the number of squares, lift the bottom right corner, and write down the number on the grid; then continue with 2x2, 2x3, 2x4, etc. They noted that some numbers did not appear at all within the table and soon, without Naustrow's telling, were referring to them as "prime numbers."

Until they got their hands on the materials, many were inattentive. Thad and Bryan scanned the room for worthwhile distraction; Carmen made herself a cup of hot chocolate; Lisa, here for the first day, seemed awestruck by the hubbub. One child said, "This was too easy. We did multiplication tables in the third grade." But they had not dealt then with the concept of "prime," or perhaps even with the concept of "multiples." Morganthau's comment after class: "Some of these kids [it was a mainstreamed class] have some problems, but they do okay. What I am not enthusiastic about is the interactive, hands-on activities that leave us without materials for individual student portfolios."

Leadership Issues. Clearly, for our research, the consultancy at Columbine was a major interest. Had the vision of the *Standards* been effectively advanced by Amy Alexander and the participating teachers? The alignment between the consultancy and the *Standards* was clear. What Alexander endorsed and illustrated as pedagogical process and task content was certainly compatible with NCTM advocacies. But of course, even with the elaboration in the *Standards* volume, much detail remains to be developed by both consultant and teachers. The NCTM *Standards* is an open framework for teaching. Even major strategies for using the *Standards* will differ.

Alexander more or less provided a single strategy. According to what I found, Alexander did not help teachers savor alternative strategies, alternative experts (though later she showed me a diverse list of resources she used). She drew an array of interpretations from workshop participants, but (according to the teachers) did not encourage confrontation, usually settling the arguments before they began. Another expert might have said (in fact, Jana Johnson did say) that perhaps 20% of the time in most classrooms, depending on the teacher, should be spent on computation, on terminology, on "drill sheets for the refrigerator," on enculturation in the traditions of mathematics teaching. The teachers said Alexander saw computation as necessary and repetition as healthy, but they needed to be contained within contemplative situations. Sessions devoted to drill opened the door to backsliding. One teacher claimed to quote Alexander: "Twenty percent would not do the job because soon it would be 50%."

Another consultant might have worked more directly on having the teacher make some strategic decisions, working toward the contract-indicated day, after three years or five, when the teachers would no longer have a consultant, when the management of the curriculum would be more clearly in the hands of the faculty, and in fact, in the relatively unsupported hands of each classroom teacher. Alexander saw these preparations as premature. The magnitude of the change to an interactive and interpretive problem-solving pedagogy was so great that each minute of staff development time was needed for that. Each participant would be working toward special moments of insight. During the just-previous summer, sixth grade teacher Oliver Osterman had found a way to store and file his strand activity materials so that movement through the year made sense to him. So even though Alexander might have suggested such a file, Osterman worked it out, and probably was the better for being the inventor.

Still, almost sole reliance on a single consultancy view for five years seemed a risk for the District. There were escape clauses in the contract, but the expectation was that one voice was better than many, that one pristine voice is better than the disharmony of counter-advocacies. And a good voice it was, recorded contextually in a book she was writing. One could recognize in her manuscript and in the *Standards* the joint invitation for problem solving and the invention of options among teachers as well as among students—of course such will occur whether there is invitation or not—but the workshop strategy chosen by Alexander at Lion's Mane did not encourage teacher choice of classroom strategy.

The participants appeared not yet ready for the issue of whether or not the content of mathematics is changed by the change in pedagogy, whether a constructed mathematics converges on the historically delivered mathematics. Consultant and teachers treated content as remaining essentially the same, that mathematics was mathematics, in spite of an advocacy of constructed knowledge in which vastly different experiences now led to mathematics learning. Clearly students now were getting more geometry and statistics.

But was the idea of addition changing at Lion's Mane? How to add and subtract, the process, did not apparently change but did the idea change? In working out the problem of spending Billy's nine dimes, the children added and subtracted in an other-minded way, thinking of what they needed to know rather than what you do to get sums and remainders. And spending lots of time on strategy and justification, without calling it that. It was not obvious that they were coming to behold the idea of mathematics differently, although coming to experience mathematics in a differently contextualized way. Clearly, they were not amassing so great a pool of "facts," with algorithms to be browsed as practicality arose. "I wonder if this needs casting out nines." For mathematical learning, a pool of experiences would be remembered, boxes within boxes, to be browsed as need arose. A constructed mathematics cannot escape its ontology. Lion's Mane knowledge of mathematics would not be context-free: Some glint of Billy's dimes would long remain.

Chicago Westlake, K-8

Westlake School[87] had been referred to, somewhat mischievously, as a Christmas Tree school. That's a school that participates in an unusual number of special programs and gets itself written up in feature stories.[88] One teacher told me that she had heard that Westlake participates in several hundred special programs. What she may have been referring to was the extra long list of items in the Westlake School Improvement Plan, but the perception was there. The comment was apocryphal, but there *was* a lot going on at Westlake. Some of the special programs were staff development activities such as that provided by the Chicago Teachers Academy for Mathematics and Science. The Academy was the main link between Westlake School and the NCTM *Standards*. In this report, we looked carefully at the work of the Academy in the school. Just as it costs money to make money, it takes time and effort to learn how to use class-

---

[87]This report, also by Robert Stake, is based on an evaluation study (Stake et al., 1994).

[88]All of Chicago's schools draw upon assistance programs sponsored by the District, foundations, the business community, universities, the teachers' union, etc. Westlake drew upon well more than the average.

room time better.[89] The issue of how special programs, including staff development programs, cost and benefit a school became an issue for this study.

The Teachers Academy was a free-standing 1989 creation of the U.S. Department of Energy under the leadership of Nobel physicist Leon Lederman. In its workshops and school improvement programs, the NCTM *Standards* were rarely mentioned but Academy agendas closely approximated NCTM aspirations. For the two years previous to this study, Academy assistance at Westlake had been substantial. Academy staffer Steve Walsh had worked long hours with the teachers on-site, subbing, even team-teaching a bilingual first grade class with one of the school's more-or-less traditionalist teachers, Marta Saavedra. Walsh's work continued to be spoken of in high regard and, lately, with disappointment that "he didn't come around much any more." Mathematics teaching was a major responsibility in every classroom. Improvement was a schoolwide aim but the means was left to the individual teacher. A third of Westlake's 21 "current" teachers had involved themselves regularly in Academy workshops. But another third had had almost no involvement with the Academy at all.

For staff development, in addition to the Academy's Teaching Integrated Mathematics and Science (TIMS) for the upper elementary grades and MathTools for Grades K-3, Westlake teachers were involved with the Algebra Project (primarily for Grades 6 and 7) and two primary school mathematics activities, Karen Fuson's from Northwestern University and Jerry Foster's field experience for trainees from DePaul University.[90] Most teachers were not simultaneously working with more than one, and half the teachers worked with none, but mathematics assistance was visible throughout the building.

An array of staff development activities as extensive as this was available at only a small proportion of Chicago schools, according to Sarah Spurlack of the University of Chicago Center for School Improvement. The Christmas Tree schools were the ones that seemed to try everything, creating opportunities as well as seizing them. Principal Kathryn Lindstrom had worked vigorously to be involved with many school improvement projects, recognizing that most required commitments. In her four years at Westlake, she had committed the school to the Comer Process and to the District's CANAL staff development, feeling that these and many other school reform efforts were based on sound thinking and capable of serving the Westlake teachers and children.

Student Achievement. Strong teaching and weak teaching were observable at Westlake, and strong and weak student performance. Primary school learning was particularly noteworthy. Spanish-language teaching in most

---

[89]Most Westlake teachers already knew how to teach better. They taught the way they did more by custom, pressures, and constraints than by choice. Tradition is more the resolution of competing interests than a reflection of competence. Each teacher knew there was more that could be learned about teaching.

[90]Somehow, none of Westlake's teachers had participated in the previous summer's staff development in mathematics provided by the Chicago Public Schools.

bilingual classrooms reflected professional competence, as was the case in most English-language rooms. Efforts in 6th and 7th grade preparation for later algebra teaching were slowly moving toward the conceptualizations of Robert Moses' Algebra Project. More generally, a good bit of the mathematics teaching was moving toward the NCTM *Standards*. Challenging activities were visible—more common than they had been several years back, according to principal and teachers, and perhaps a bit more common than researchers were observing in similar schools across the city.

The signs of good schoolwork were these: Students were using various desired forms of communication and symbolic representation to express the ideas of mathematics. They were examining tasks in an analytic fashion, noting what was critical and what was irrelevant to problem solution. They were working cooperatively. They were writing their classroom exercises, calculations, their homework assignments, and other matters in their own blank-page notebooks, thereby taking somewhat greater responsibility for composing the tasks of their own learning. It was not apparent that these behaviors would result in better test scores—the tests are not oriented to this kind of learning. But according to the NCTM *Standards*, as well as staff development specialists visiting Westlake, these are desirable habits and mindsets for mathematical thinking in the classroom. These signs did not typify mathematics performance of the students across the school. In some rooms, few signs were visible. In rooms taught by teachers participating in Academy staff development, more signs of the *Standards* were apparent.

Announcement of Illinois Goal Assessment Program (state mandated) testing appeared on the huge school calendar in the Westlake hallway and, on the Friday before that week, Counselor Evelyn Vokurka was readying test materials to be used. But I heard almost nothing about these tests from teachers, parents, children and administrators.[91] With admiration, I concluded that professional knowledge more than mandated testing was driving instruction at Westlake.

Teaching. In Patrick Willen's[92] eighth-grade classroom, only 14 students were present when I arrived at math time; the rest were across the hall at the computer lab. Gradually others returned. At the outset, the students were casually sitting or standing, chatting, gesticulating. Willen, a husky man, was correcting a scattering of homework papers—the few handed in, he grumbled. Then he said, "Wow!" and went over to one of the boys and pointed at an answer. I couldn't tell whether he was pleased or distressed. They chatted a few moments.

---

[91]Part of the "silence" might be attributed to the fact that Westlake is a bilingual school and I am a monolingual researcher.
[92]Mr. Willen's class was the bilingual eighth grade class, perhaps because of the physical and social maturity and intellectual immaturity of the students, the most difficult-to-teach classroom at Westlake. Willen taught it in English.

Willen asked Trino, Oscar, Sylvia, and Paul to go to the board, each to work one of these:

$$- |16| + |-68| = \qquad - |16| + |+68| = \qquad - |-9| = \qquad -2.1 + 8.6 =$$

Willen announced this was a lesson on absolute values, as represented by the problems on the board. When asked, Trino declared that the first answer was -52. Willen went to Trino's board, worked at his problem silently, then wrote the answer, 52. He frowned and shouted at Trino and others, urging them not to treat an absolute value $|x|$ as a value in parentheses (x).

After checking that the second one was, in fact, a different problem, he worked Oscar's, assigning again the value, 52. He alluded to but did not directly make the point that $|-68|$ and $|+68|$ are to be treated the same. As answers to the third problem, he chalked both -9 and +9, asking "Who says it is -9?" "Who says it is +9?" "How many vote for positive?" "Negative?" He got perhaps four or five votes from the 14 adolescent boys and girls in the room.

One student, Philip, already enroute to becoming a huge man, tracked most of what Willen was doing, frequently commenting aloud, often correctly—and regularly being told to be quiet. Three or four boys, possibly all Hispanic, volunteered an occasional answer, but regularly lost track of what Willen was doing. Most of the boys kept an eye on him—warily, I thought. Most of the girls ignored him, as he did them.

Speaking next of line segments and rays, Willen drew attention to the board showing

and asked the students to identify the parts of the line, which they did. Then he added another ray at point K, like this, passing through point T.

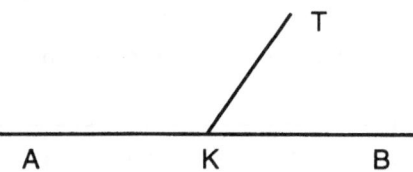

He asked, "How many degrees in angle TKB?" Several students offered guesses: 80 degrees, 90 degrees, 30 degrees. Exasperated with the 90 degree answer, Willen drew a right angle on the board and brusquely said, "*That's* 90 degrees. So what is that?" jabbing at angle TKB. Someone said 80 degrees. "I don't want your guesses. How would you find out?" He produced a demonstration protractor, handed it to Artur and asked him to measure the angle. Artur placed the unfamiliar protractor correctly but did not know how to read the scale. Willen pointed and quickly counted by tens from KB up to KT: "80 degrees. See how I did it, Artur?" Interpreting this as release, Artur sat down. Willen went on to distinguish between negative and positive angles, for perhaps fifteen minutes demonstrating more than one way of getting negative degrees for angle TKB, sometimes asking a boy for what seemed to me to be more reasoning than he could deliver.

For the remainder of the period, they examined pages 160 and 161 in Usiskin's textbook, *Transition Mathematics* (a book recommended by the Teachers Academy), taking up the subject of commutative and associative properties in addition. A student read a paragraph and Willen explained it, often writing an equality on the board of the form $(a + b) + c = (a + c) + b$. An able reader, Aggie, read a paragraph on earning and spending and Willen made the point that the *order* of earning and spending events does not change the result. The negative value of each expenditure was emphasized. With seven minutes remaining on the clock, the students spontaneously began putting their books away, moving about, talking. A few slipped into coats, Chicago Bulls coats prominent. Willen announced the homework; then, with comments, changed it—perhaps a clever thing to do if one wants students to think about what the assignment amounts to. Following his lead, Sylvia wrote the assignment on the blackboard for others to copy. Later in the afternoon, just before school was out, I saw several of these students working at the problems assigned.

Carmen David was the teacher of the other eighth-grade class, also taught in English. On the day prior to the day I visited her room, Mrs. David had divided the class into five groups, with each to find the metric area of the bulletin board, the classroom floor, and the blackboard. This day, she asked Angel to describe what they did yesterday. He said that they used meter sticks and a formula, L x W = A. When asked, "What terms did you use?" he answered, "two dimensions." Mrs. David, a no-nonsense teacher with long experience teaching in the Philippines, revealed the answer she wanted by stating it, then pressed upon them the importance of speaking of the *unit* of area as "square meters, not meters square."

Each of the five groups was to make a formal table with data headings as follows:

Object	Length	Width	Area

Mrs. David reminded them (or perhaps me) of previous activities: measuring the diameter and circumference of bubbles without ruler or scissors; producing a square by folding paper; developing spreadsheet data about US presidents and their parties, then showing fractional parts with pie charts.

She had divided the students into five groups to work on problems in the *Transition Mathematics* book (e.g., "Find the area of a square that is 1.5 inches on a side.") After a while, Mrs. David drew them back into a single group, asking for oral descriptions of what they did. "I want to see every hand up." Then, "Tell me what you did, Donny." Donny had trouble articulating how one squares a number. As his voice faded, she pressed him with "I didn't hear it!" When another youngster offered 3.25 as the answer, she said, "3.25 what? We need a *unit*. We don't say, 'One million live in Yokohama.' What do we say? We say, 'One million people'." Later she asked them to calculate the number of square inches in a square yard.

At the end of class, I asked Jamie if today's class was typical; he said yes. I asked what Mrs. David insisted on and he responded: "Working. And that's good. We can learn a lot in this room." Wondering if he was acquainted with the term, I asked if there was "cooperative learning" in this class. He was as cautious as I about our ability to communicate with each other. After an effort to be sure I knew what he was talking about, he said, "It is one of the best ways to learn. If you don't know it, someone else does. You get different answers and you have to figure out why." "Don't you get discouraged sometimes?" "No, no one gets discouraged *here* just because they don't know something."

From talking to Mrs. David and others in the school, I found confirmation that she did work her students hard, got them to assume responsibility, and that they did learn a lot. She provided a certain wholeness to the experience of learning mathematics. I got the sense that, unlike many teachers, she carried through on her threats and the students "did not mess with her." She had participated in professional development, integrating mathematics and science, and seemed to understand it, seemed to understand what the NCTM *Standards* were all about. As indicated above, she gave students lots of opportunity to solve problems, to collaborate, to deal with the mathematics concretely, and to express themselves.

School and Setting. Westlake School was a neighborhood school, built as the city of Chicago was building, surrounded now by weathered homes and small businesses, more than a few signs of prosperity. A half dozen buses poured African-American and Latino students from more crowded areas but most of the children walked, many younger ones accompanied by a mother or grandmother. About 400 children in all. The playground was anchored by an old church and a Mexican restaurant. The first day I visited the school the streets were icy. A grandfather was broadcasting salt crystals to the sidewalk, "helped" by a three year old struggling to reach into his bucket.

Inside the school were slogans: "El pueblo unito siempre triunfara." "What is best for children." "We believe in our children." "Apprehende Hoy Trianfuna Mañana." "Be a Winner. Attendance Counts." Upbeat. Occasionally something like "Thou Shalt Not Whine." Happy. But on one Tuesday I wrote in my notebook:

It's not a happy place today. It's back to school after Pulaski Day, concluding a long four-day weekend. And somewhere in the school, there is a major discipline problem to resolve. And in the basement, condolences are mounting for a first grader's father shot dead last night.

Everyone needs breaks. In schools like Westlake, children usually have to take breaks on schedule. In most rooms, individuals have a hard time getting permission to leave the classroom. Trips to the bathroom and the drinking fountain need to be delayed until the teacher escorts everyone. In Westlake halls they waited, lined up in two lines, often boys against girls, sometimes with both feet inside a 12x12 floor tile, as required, until all had had a turn. Strict enforcement occurred mostly when the children were volatile, which was not all the time. Knowledge that rules are sometimes enforced and sometimes not was reinforced again and again, an important lesson. Opal Jimenez was one of the teachers pleading for more uniform enforcement.

Staff Development. Rosalina Coronado and Kathryn Lindstrom had given thought to the continuing professional development of each member of the Westlake faculty. They considered the school's affiliation with the Teachers Academy, and with Karen Fuson, Comer, and other teaching assistance projects as providing some of the best possible support for upgrading teaching. While they encouraged individual teachers to consider their careers in the long term and perhaps to enroll in higher education degree programs, they concurred with Nell Cobb of the Chicago Algebra Project who placed a higher premium on professional development to deal with each teacher's *immediate* responsibilities.

After brief observations and discussion with the Assistant Principal, I made individual estimates of the qualification of Westlake teachers to teach mathematics. The Westlake people prided themselves on the maturity of the staff but 8 of these 21 teachers, as teachers of mathematics, were beginning teachers. Based on both conversation and observation, I rated 6 of the 21 as highly capable of teaching traditional elementary school mathematics, another 3 as of middling capability. The potential of some of the beginners was high, but considerable staff development was needed. And perhaps not surprisingly, these six highly capable, traditional teachers were among the most active participants in Academy workshops. As a group, they knew the content and vision of the NCTM *Standards* for mathematics. Yet, in my eyes, for various reasons, they were quite a ways from fully implementing that vision in their own classrooms.

Of the three experienced teachers I rated as low in teacher capability, all three were participating in some Academy activities but none had developed the beginnings of that pedagogy. Each had had over two years of access to Academy help. Only two of the regular teachers at Westlake and the regular substitute had not taken any advantage of Teachers Academy assistance, and another six teachers only a small bit. Clearly, participation could be much better. Lots of room for improvement existed across the board. It was apparent to me that participation was encouraged by the principal and her assistant, as well as by the Academy liaison teacher and the Academy trainer, but I saw nothing to indicate that pressure was put on the teachers to get more involved. The reasons several teachers did not were some combination of: seeing little merit[93] in the issues raised by the NCTM *Standards*, dreading the work required to acquire the new perspective, and resisting the appearance of losing control of instruction when youngsters are working together noisily or individually on a diversity of tasks. The Teachers Academy had not been able to sell perhaps half of this generally qualified faculty on the merit of change.

According to Steve Walsh, Lindstrom recognized the considerable difference among her faculty members in readiness to teach mathematics. She wanted them to have options, to work with specialists each had confidence in. She recognized that Anna (from the Fuson project) could give more help with an understanding of how children learn and Sylvia (Academy Mathematics Coordinator) more help with the understanding of what should be taught. Several Westlake teachers told me they appreciated the adaptability of the Academy to the uniqueness of classrooms and teachers and the reluctance of Academy people to make them feel guilty about what they did not know.

Since the beginning of the school year, Melinda Delzotti had been Westlake's in-school instructional facilitator. An instructional facilitator often works at integrating the curriculum, helping teachers coordinate lesson plans, obtain instructional materials, doing "whatever" to help teachers—at times, a staff development specialist. But with many staff development specialists visiting the school and holding workshops elsewhere and with a number of veteran teachers, Delzotti concentrated on the language arts, her "strongest area," with a few of the newer teachers. "You can't help teachers who aren't interested," she said, referring particularly to a veteran teacher who was popular with kids but using her own basic skills syllabus, adding, "Really no way to help her." Delzotti had no regularly assigned teaching load and did not have the know-how in mathematics and science. "My weakness," she said, but she did not take advantage of Teachers Academy workshops.

It might seem that the school's instructional facilitator would be the ideal person to provide liaison with the Academy. Lindstrom had committed Westlake to providing such a liaison person but assigned the responsibility to Debbie Hendrix. Although Hendrix was a kindergarten teacher, she had a good comprehension of K-8 school mathematics and understood complex organizational problems. The problem was that she found her personal situation outside school leaving very little time to do anything beyond her own teaching. She did keep faculty members informed of Academy workshops and other staff development activities. Most of the other teachers also found it difficult to commit themselves to professional development. Academy workshops encouraged partnering and collaboration with other teachers but the tradition, the environment,

---

[93] These teachers did not appear to see a need for a changed conception of mathematics; they were satisfied to teach rules and procedures.

the schedules worked against it. There was money for substitute teachers, but most of them only worsened the problem.

Laura Epp was the regular substitute. She came to school every day, not knowing which room she would be teaching. She was a treasure: humane, principled, intelligent, patient, vigorous—and bilingual. But even with Laura, a day with a substitute is likely to be no better than half a day on the instructional track. Many of the substitutes were no more than wardens and not good ones at that. In my second week, one retiree showed up to substitute as much in need of social service as anyone in the neighborhood. The problem of teacher absence and the substitutes' poor command of teacher responsibility was huge. It was not uncommon for effectiveness to drop to a fraction of what the regular teacher could do. On many days, Westlake needed several substitutes. All Chicago schools knew these problems.

The Burden. Delzotti commented on the number of help-programs working with the school. "Such a burden," she said. "Each program needs committees, and a teacher in charge of the effort here, and hospitality there. Each comes with a gift but each extracts a price, such as workshops, liaison duties, ..."

Everyone in Chicago knows that the schools could be better. And surely they are trying to be. But it is not clear to the careful observer that they can carry out their responsibilities *and* work at being better. The professional people at Westlake are serious about their responsibilities. They do not put as high a priority on being better. It is possible they could do a better job of educating the children if they did not have such a heavy burden. If anyone thought that decentralization would help alleviate the burden, they were wrong. Citywide, some Local School Councils are still not engaged in running the school, but, according to research coordinator Tony Bryk, most are. They draw enormously on the Principal's time and perhaps that time is well spent. But it is not spent in making the responsibility of the school more realistic.

The load these schools can bear is not being assessed adequately. Unrealistically, the State, the mathematics reform movement, the assistance organizations have been increasing the load. They are not helping school people decide, given the resources available, what responsibilities to diminish. Part of the burden is those long lists of state, national, community, and professional goals and standards, idealistic views of what ought to be. Voltaire's words were "The best is the enemy of the good." Some schools like Westlake have been getting better gradually but far, far too slowly. Long term failure is almost certain, partly because the outsiders, as well as the Westlake Council, have too grand a view of what they should be. Perhaps the vision of democracy-in-action to effect school reform is not possible until communities are setting the goals as well as the programs.

If a Christmas Tree school is glitzy, superficial, and self-promotional, Westlake was not one. Even during a difficult spring, it was a hard-working, self-reflecting school. The Teachers Academy shared blame with many for giving Westlake teachers too much to do—but it pushed with a gentle hand. It helped guide and endorse the teachers who were attracted by the vision of the NCTM *Standards*.

Pinewood High: We Are Moving[94]

"We are not where we want to be, but we are moving!" says Dr. Simmons, the district mathematics supervisor. Her statement reflects the essence of the reform effort in mathematics at Pinewood High, a large urban/suburban high school of 2500 students, the vast majority of whom are African-American. Reform moves slowly in this school, with its forceful administrators and cautious teachers; well-dressed students also move slowly through the long, narrow halls as they make their way from one classroom to the next.

Pinewood has a substantial record as an exemplary school. It includes magnet programs that were originally established as part of a desegregation effort, and it offers advanced course work in mathematics for college credit. The school has been doing well in many areas—they take great pride in their success in arts education, for example—but there has also been concern in recent years that mathematics achievement scores were too low.

In the opinion of Dr. Simmons, the mathematics supervisor, the district has been suffering from an elitist point of view towards mathematics. The supervisor, an energetic, resourceful African-American woman, felt that low test scores in mathematics were too easily accepted, and that teachers and others in the district conveyed a feeling that *these* students just weren't able to do well in mathematics. So when the district appointed a task force about five years ago to review the mathematics program for the more than 100,000 students in the district, Dr. Simmons went to work. Although it was not always easy, the end result was that the school board and the superintendent went on record as supporting the *Standards* as the vision of the future for their mathematics program.

Meanwhile, back at Pinewood High, Donna Bell had been appointed the new coordinator for the mathematics department. She, too, was working to strengthen the mathematics program, in her case at the building level. Her review of the final exams that her teachers planned to administer led her to believe that some teachers weren't expecting enough of the students, and their finals had to be revised and strengthened to get her approval. Did she have the right to do that, some teachers asked? "Yes," said the principal, and his support for the efforts to strengthen the mathematics program have never wavered.

The School. Pinewood High sits on a heavily trafficked street where gas stations and strip malls provide the only decoration. The school grounds are relatively spacious; the school sits on a large open lot with grass and trees, and there is plenty of parking for both students and teachers. Ms. Bell says that most students come from low to middle-income families, and the school has a socially

---

[94]A similar version of this report, prepared by Doug McLeod, will appear in reports from the project of Ferrini-Mundy and Johnson (1994).

conservative atmosphere. Students are well dressed, refer to adults as "sir" and "ma'am," and follow the injunction by one of the doors that says, "No hats inside the building, please." The building itself is well maintained, although the drinking fountains are unusable at the moment—too much lead in the water for safety.

The teachers in the school are mainly white, with about 30% of the mathematics department of African-American heritage. Like the mathematics departments in most secondary schools, perhaps half of the teachers majored in mathematics in college. The rest have come to mathematics through retraining programs, sometimes with a renewed zest for teaching. The teachers, like the students, seem relatively conservative, and cautious about change. They worry that the students will not get enough practice on basic skills, and they are concerned about saying anything that might offend the school administrators.

The Program. The reform effort started several years ago, but the most active changes at the present are the new precalculus program in Grade 12 and the "algebra and geometry for all" program in Grades 9 and 10. The Grade 12 program is a pilot, part of a national effort to strengthen the teaching of mathematics at that level, and has many similarities to the vision of the NCTM (1989) *Standards*. Students are expected to work on investigations of significant problems that last for days, even weeks. Appropriate technology is available for the investigations, and the teacher serves as a guide, not a lecturer, for the students.

The program for Grades 9 and 10, also part of a national project, started with an effort to improve the Grade 8 program in the middle schools, and continued with algebra "for most students" in Grade 9 two years ago. That was a difficult year, reported Dr. Simmons; many teachers were not happy about the changes. The algebra classes now included students who would previously have been enrolled in General Math, not Algebra. Some teachers who were used to drilling students in General Math or a traditional algebra course were struggling with new approaches to the content. Dr. Simmons reported that "We had to say to teachers, 'Folks, this is not optional'." And most difficult of all for many teachers, there was no textbook. The cry from the teachers was "When are we going to get a book?" But there was no single text that was suitable, said Dr. Simmons, so the school provided multiple resources for the teachers to use. "Teachers have to be decision-makers," according to Dr. Simmons. The end of the year was a decision point, but the teachers agreed to continue the reform effort for one more year—"a real victory," felt Dr. Simmons—before doing a serious evaluation of the change. After getting through the first year, the teachers thought they had a better idea of what they were doing in Algebra, and wanted to give it one more try.

This past year the teachers agreed to expand the reform effort to tenth-grade geometry. The school did provide a text for geometry, but the text is reform-minded, putting less emphasis on two-column proofs and more emphasis on geometric discoveries. The teachers were relieved to have a text to use, and the administrators were pleased to find a text that showed some similarity to the vision of geometry teaching that is presented in the NCTM (1989) *Standards*.

The Classrooms. Classrooms in both programs (Grades 9-10 and Grade 12) show evidence of change. Students sit in groups around tables or at clustered desks. Motivational posters, student art work using tessellations, and recommendations for small-group work ("use your two-foot voice," "ask for reasons," "respect ideas") decorate the walls. Teachers indicated that they were often apprehensive about putting the students in small groups; administrative evaluations of teachers had always appreciated quiet classrooms in the past, and some teachers just didn't like all that noise, either. But the students seemed to enjoy group work, and reported that they now know they are supposed to figure things out in the group, if they can. If they can't, then it is all right to ask the teacher.

Teachers have had some support as they have tried to initiate new ways of teaching mathematics. A two-week institute the previous summer was a positive experience for the geometry teachers. They heard the author of their text talk about how he got started teaching this new kind of geometry course, and realized that he, at least, didn't expect them to do everything perfectly the first year. They learned about manipulative materials that could be useful, and strategies for classroom management of small groups. And they got paid for their time, too.

But perhaps the most important support for the teachers comes from their work in "divisions"—especially the one for algebra teachers, and also the division for geometry teachers. Division meetings provide an opportunity for teachers to share materials and teaching tips. As teacher Frank Schmidt put it, "Working in those groups has really helped; they are always throwing ideas back and forth." The teachers also do some visiting in each others' classrooms. The change in atmosphere and the mutual support were noted by several teachers as an important impetus for sustaining the reform and for extending it to those teachers who have been reluctant to participate.

The Assessments. The reform was motivated initially, in large part, by concerns about achievement as measured by traditional test scores, and assessment concerns continue as an important factor in the Pinewood reform effort. An early change, now nearly complete, was to move the required state test of basic skills—a "fifth-grade" test in terms of content—out of the high school and down to the middle grades. Some time is still being spent at Pinewood High on this test, but students are now expected to finish that state requirement before arrival. The district also uses a locally-generated criterion-referenced test for their algebra and geometry courses. Making sure that those tests are aligned with a changing curriculum is a formidable task, requiring coordination with the district test office, and last year the tests weren't ready as early as the teachers wanted, said Dr. Simmons. But she hopes that the alignment of the testing program with the reform effort will make assessment a lever that supports, rather than inhibits, change.

Although it is early in the change effort, several people think there are data on how successful the change has been so far. They don't, however, necessarily agree on whether the limited data now available show improved or weakened performance by the students. Mr. Schmidt, for example, thought that the placement tests for the advanced classes showed a decline from previous years, but Ms. Bell was generally pleased with the effects of the reform effort. Both would agree that it is too early to tell, but both expect to see positive changes in test scores as the reform effort goes forward.

The Concerns. The reform effort is clearly visible in classrooms at Pinewood High, as described above, but there are concerns, too. The administrators worry about problems with funding, about teachers resistant to change, about keeping the assessment program aligned with the new curriculum. The teachers worry about keeping up with the work involved in teaching algebra out of several different sources, rather than from a single text, especially when the copy machine is a long walk away and "half the time doesn't work anyway." And a few of the best students worry that they might be held back by small group work—that they could get their work done faster if they were working alone. The most common concern expressed by teachers involved increasing student skill levels. In Grade 12, for example, the materials are designed for reasonably advanced students. At Pinewood, however, those advanced students were in the Advanced Placement courses; the new Grade 12 program was directed toward the "C and D" students. So the teachers spent considerable time trying to make sure that students improved their skills, adapting the curriculum as they felt necessary.

In Grades 9 and 10, the algebra and geometry teachers also showed a concern for basic skills. Although the materials that were available to those teachers emphasized problem solving, reasoning, and connections in ways similar to what the *Standards* recommend, considerable amounts of class time were spent on low-level skills. Although the students were working in small groups, the tasks that the groups addressed were often very similar to those worked on individually before the reform effort started. What is the appropriate balance between teaching skills and problem solving? The NCTM (1989) *Standards* suggest that the traditional classroom placed too much emphasis on skills, but these teachers were cautious about change in this area. Administrators talk very progressively about higher order thinking skills, rather than basic skills, but what will the assessment program concentrate on? These teachers were betting that student performance on basic skills would continue to be an important factor in how the school and teachers are evaluated.

Pinewood Themes. Several themes characterize the reform effort at Pinewood High. Equity is certainly among the most important. Since the district is about two-thirds minority students, and the school predominantly African-American, equity concerns are an important part of the agenda of the administrators. "Algebra for all" and now geometry, as well, are one rallying cry for this district. Combating elitism is part of what the reformers want to do, but the more general issue is changing the cultural belief in the community, and eventually the entire country, that doing mathematics is largely a matter of ability rather than a matter of effort and opportunity. The kinds of change required to make "algebra and geometry for all" a reality are a significant challenge for every school, and especially for schools with large minority populations.

Leadership is a second theme. The reform certainly began as a top-down change, an effort out of the district office, with support from the leaders at the school site. Subsequently there has been substantial effort invested in supporting reform from the bottom up: support for inservice training, time for teacher collaboration, funding for graphing calculators in classrooms, and re-directing some district resources toward mathematics. The impact of these efforts on some teachers has been substantial. Teachers who formerly were content to stay within the walls of their own classrooms are now out going to workshops and conferences, bringing back information for other teachers who are interested in learning about reform. However, financial support from the district office has been crucial in bringing about this change. As Mr. Schmidt said, the chance to go to conferences just "opened up" his whole world, giving him the desire to be a leader after years of self-imposed isolation.

Another theme is accountability. The district is concerned about low test scores, the school administrators worry about low grades, and the teachers are afraid of being held accountable by the administration for problems that they cannot control. For example, some teachers worry that they will be "called on the carpet" if they give too many low grades. Since the criterion-referenced tests in algebra and geometry are given by all the teachers, the testing program within the school provides useful information, according to Ms. Bell, the coordinator for the mathematics department. She "can hold teachers a bit accountable" through her analysis of students' test scores, noting which classes "are doing better." Both she and the teachers realize that alignment of the tests with the curriculum is a central issue that requires significant effort by both the administrators and by teacher committees. Neither teachers nor administrators questioned the basic premise that student performance on district tests was an important indicator of the effectiveness of the reform effort.

Which themes will need more attention in the years ahead? One is the change in the conception of mathematics that characterizes the reform documents, like the NCTM (1989) *Standards*, but is rarely found in classrooms. The vision of mathematics that characterizes Pinewood High so far is quite traditional—an emphasis on procedures with a contemporary overlay of calculator-based technology. There are some exceptions, of course, but in most classrooms a traditional view of mathematics is quite evident. The changes in pedagogy, especially the small-group work, provide some of the context that may eventually support the notion that mathematics is more than rules and right answers.

Other pedagogical changes appear to be on the horizon. Some of the school administrators have made a "constructivist" pedagogy one of the major goals for the program, and greater change in this area can be expected. But such change is difficult for teachers, especially for those teachers whose own mathematics background and experience provide no model for a constructivist approach to learning and teaching. Some teachers are making great progress in assessment, using journals and projects as part of their grading system. And many teachers have taken advantage of the opportunities for professional development. But teacher change is slow, and the task of changing an entire school is truly daunting. In fact, the reform effort as a whole is daunting, a monumental task that seems insurmountable when you look at it broadly. But at Pinewood High, they are moving. Moving forward. Moving toward reform. And toward a better mathematical future for their students.

Change Themes: The Case of the NCTM *Standards*

Information gathered during our visits to schools provided a useful counterpoint to our interviews with leaders and writers of the NCTM (1989, 1991) *Standards*. The vision of the future that is presented by NCTM seems much more clear when imagined from the office of an NCTM leader than it does from the back of a classroom. The commonalities of the change process in schools are sometimes difficult to detect and less easily bounded, especially when one tries to capture them in sentences and paragraphs. Nevertheless, there are important themes that arrested our attention repeatedly in our work, and we present some of them here.[95]

We start with two foundational themes, the changing conceptions of mathematics that helped to drive this reform effort, and the changing conceptions of the learner, a set of ideas that spread in the mathematics education community at the same time as did the awareness of the NCTM *Standards*. We also discuss the burden of change as it is carried by teachers and schools, and the notion of *grain size*—an emphasis on the bigger ideas in reform rather than the specifics—which has been so important in the movement toward standards in the US. The importance of developing teacher communities as a way of supporting change attracted our attention, especially as it came up in our visits to schools as well as in the reports of others (e.g., Byrd, Foster, Peressini, & Secada, 1994; Talbert & Perry, 1994). Finally, the many different interpretations of what the developers of the NCTM *Standards* intended, and the dangers in some of the interpretations, seem to us to indicate points that need attention from leaders of NCTM and others who care about the improvement of mathematics education. We describe these misinterpretations using terms that appeared in the NACOME (1975) report—the notion of false dichotomies or false choices, and their relation to the "swinging pendulum" metaphor for educational change.

---

[95]For alternate views, see Bosse (1995) or Massell (1993).

Changing Conceptions of Mathematics

The difficulty of changing students' and teachers' conceptions of mathematics is widely recognized by leaders in the field. In most districts, even those where change has been substantial, the reform effort has not had much of an impact on teachers' and students' beliefs about what constitutes mathematics. As writer Alba Thompson put it, the 1989 *Standards* have had a tremendous impact if you consider how often they are mentioned in schools, textbooks, and the corridors of educational power centers. But if you look in classrooms:

> You can still see that the message in the *Standards* has not made much of a dent in the very robust orientation to ritualize and routinize everything that is mathematical. The *Standards* say that there should be an integration of algebra and geometry and that multiple representations ought to be incorporated. Well, now the drill and practice is on changing from a table to a graph to an equation, so there are fifty exercises doing this. But these kids have no clue as to what a function is.

Pinewood High School shows much of this traditional orientation toward mathematics as routine. Few other schools are successful in changing their "implemented" curriculum to reflect a new conception of mathematics (see Ferrini-Mundy and Johnson, 1994). Such changes will be difficult at all grade levels. At the elementary school level, it is very difficult to move away from the public perception of school mathematics as "arithmetic plus"; parents, teachers, and some mathematicians as well still conceive of elementary mathematics as computation, with a few other things like geometry added in (MSEB, 1988). NCTM has been working to overthrow this idea at least since *An Agenda for Action* (NCTM, 1980). However, even in very progressive schools (like Lion's Mane Elementary), some teachers continued to portray mathematics as primarily computation.

The impact of technology is a major reason for change in conceptions of mathematics. The role of the computer was of particular importance to many leaders of NCTM. As Tom Romberg noted:

> Changing technology has vastly expanded the uses of mathematics. It's all the capability of the computer to build mathematical models. It shifts the emphasis from paper-and-pencil calculation and shop-keeper arithmetic to building models. It really forces us to rethink how much time we spend on various topics. Is it really important that we teach kids how to add fractions with unlike denominators, spending two months every year for four years or so?

Similarly, John Dossey commented on "the emerging power of computers to deal with problems. Across the country you saw the explosion of journals in discrete mathematics. In statistics, what was really powerful was exploratory data analysis—that whole movement."

The need to update the mathematical content of the curriculum[96] was the factor most often identified (in our questionnaire data from leaders) as a major influence on the development of the *Standards*. Moreover, changes in the philosophy of mathematics (Kitcher, 1984; Ernest, 1991; Hersh, 1995) have influenced certain leaders. But the impact of these changes on people in the field was limited. As one leader noted:

> I'm not sure that changes in philosophy had much to do with either calculus reform or the NCTM *Standards*. No, mathematicians ignore philosophy of mathematics. That's true in many other fields, too. The people that are in the discipline focus on the discipline, not on the philosophy of the discipline.

A writer of the *Teaching Standards* made a related point about mathematics as the science of patterns:

> I don't see looking at patterns as anything new. To me that's what mathematics has been for generations. The difference to me is you can use machines to look at patterns and do things that a human couldn't do. That's the big difference.

Opportunities to advance this new conceptualization of mathematics[97] are related to increasing acceptance of a constructivist epistemology (Steffe & Kieren, 1994) that encourages the teacher to de-emphasize the mathematics collectively defined in authoritative works and to legitimate the mathematical thinking of the individual child, not just as thinking that is "on the road" to mathematics but as mathematics itself. The issue was apparent in the way a highly competent, *Standards*-oriented consultant at Lion's Mane Elementary continued to see students and teachers as working toward a non-traditional but discipline-authorized mathematics. In this way we see a connection between the changing conceptions of mathematics and the change in researchers' views of the learner.

Changing Conceptions of the Learner

Many ideas that dominated thinking in education from the early 1970s to the mid 1980s found their way into the NCTM (1989, 1991) *Standards*. In particular, changes in theories of learning had an impact on how the *Standards* were written and interpreted. In an effort to better understand the development of the *Standards*, we consider the shift in influence in mathematics education from behaviorist psychology[98] to cognitive science to constructivist approaches to learning.

---

[96] Some reformers have made changes that are not in accord with the recommendations in the NCTM *Standards*; Haimo (1995), for example, notes that proof is "barely mentioned" in some projects.

[97] For a full discussion of changes in mathematics as a discipline, see Steen (1990).

[98] For a more detailed discussion of these psychological influences on mathematics education, see Kilpatrick (1992), Romberg and Carpenter (1986), and Steffe and Kieren (1994).

Most descriptions of constructivist approaches to learning encompass the idea that individuals construct their own reality through actions and reflections on their actions. As Tom Romberg noted, constructivism was central in the development of the *Standards*:

> It was very clear that we needed to think about psychological development in ways that were quite different from the behavioral tenets that were behind most education in the United States. Constructivism goes back to Dewey in the early part of this century and certainly was influenced by the things that Piaget and others were preaching in the early 1960s.

Mathematics educators were influenced by constructivist ideas found in developmental psychology, along with other researchers and educators in areas like language arts and science education. During the 1970s and 1980s constructivist ideas were important to the work of many reading researchers who were working towards formulating guiding principles for the betterment of teaching and learning. The work of Anderson, Hiebert, Scott, and Wilkinson (1984), *On Becoming a Nation of Readers* (BNR[99]), was seen by one reading expert as an effort to "get across powerful ideas" about how readers are "actively constructing meaning." Another reading researcher (who was a consultant on BNR) stressed what he saw as the importance of constructivism to the thinking at the time:

> I view the math *Standards* as the instantiation of a constructivist view of learning within mathematics, parallel to the constructivist instantiation of curriculum in English/language arts and a similar view that's beginning to emerge in science education. It's not an isolated, insular kind of operation, but rather just a natural outcome of intellectual forces that were operating in the mid-1970s through the early 1990s. When I look at the math *Standards*, I see all the elements of child centeredness and constructivism that hearken back to Dewey and progressive education.

Chris Hirsch, a writer of the 1989 *Standards*, also noted the influence of Dewey's ideas on the *Standards*, as well as on current curriculum development efforts. He added:

> I think a reading of the *Curriculum and Evaluation Standards* and the *Professional Teaching Standards* will show that both documents were heavily influenced by contemporary thinking on students' building meaning and constructing their own knowledge. There are now curriculum efforts that are attempting to organize school mathematics programs around that constructivist point of view where students reinvent important mathematics.

Not every writer was a proponent of a constructivist approach, and one NCTM leader suggested that the popularity of constructivism only came along

---

[99]Some leaders in mathematics education saw BNR as an early influence on the 1989 *Standards*.

after the *Standards* project was well underway. However, an emphasis on a constructivist approach to teaching has become associated with the *Standards*. New
views of the learner were thought of as part of Dewey's influence, as was the
new (or actually old) emphasis on connections between different school subjects. As a reading researcher put it, "Those are old, progressive education ideas
that have been rediscovered in the last 20 years or so. They were in the air."

How did it happen that these ideas were "in the air" at the same time?
Romberg (1984a, 1987b) argued that the need for communication and connections came from the shift to the "information age" and away from the "machine
age." This shift required a change from a curriculum where concepts were frag-
mented (to be absorbed more easily) to a more holistic view, as well as a view of
the learner as a problem solver who builds up knowledge and understanding.
The requirements of the information age provide one explanation for how similar forces impinged on different curriculum areas at the same time. Changes in
conceptions of mathematics (as well as the learner) developed at around the
same time, requiring changes in content and pedagogy in mathematics. Similar
changes are occurring in other subjects. One consequence of trying to change
all these dimensions of schooling, and changing several curriculum areas at
once, is that we have increased the burden of change that teachers have to carry.

The Burden of Change

As the discussion of Westlake School in Chicago indicated, every innovative
and assistance program that comes into a school to improve teaching has costs
attached to it: Teachers need to attend workshops, try out or develop new materials, organize the classroom differently, report to evaluators, respond to interviewers, engage in more elaborate assessments, reassure concerned parents;[100]
the list goes on and on. What is being done to keep such burdens reasonable?
When are new projects energizing, and when enervating?

Most teachers work hard in an atmosphere of unrealistic expectations. The
demands on urban teachers in a school like Westlake are particularly severe,
partly because the children's safety and social plight are always a concern.
When a district-wide change (like decentralization in Chicago) occurs along
with special staff development efforts in mathematics and science (as well as
other projects), policymakers need to consider ways to limit the increasing burdens on teachers.

In Lion's Mane Elementary School, most of the teachers were energized by
the efforts to change their mathematics program. But even there, the demands
were heavy. A talented teacher of Grades 5-6 was ready to give it all up and go
back to using a traditional textbook, saying "I was tired. I just wanted something easier." The struggle to put new activities on the table every day was time
consuming, even though the principal was supportive. The burden of innovative

---

[100]See Dillon (1993) for an example of how parents can influence reform efforts.

teaching is a heavy one, even for the most talented teachers in relatively advantageous settings.

Another example of the burden at Lion's Mane Elementary was the extensive staff development workshops that were part of the reform effort. Although the workshops were a central part of the effort to help Lion's Mane teachers change their practice, they were also a part of the burden. All of the teachers went to the workshops at the same time, so the school was being run entirely by substitute teachers. Principal Golding noted the difficulties:

> My kids do well with their teachers. Their behavior is contained and they're successful because of their relationship with the classroom teacher. When that person leaves and a new person comes in, they don't seem to be able to adjust. They write petitions to get rid of the substitute. They come and complain to me all day. They're out of line when they have substitutes.

A teacher at the secondary level (Pinewood) who was teaching an innovative 12th grade mathematics course also felt the burden of change:

> Everybody thinks they are swamped every year, but this is my 21st year and I've never been swamped like this. I am still trying to get it all together. Just the introduction of technology [graphing calculators] has changed completely the way we teach; it makes books we thought were bad *really* bad now.

The administrators at Pinewood recognized that many teachers were uncomfortable in developing courses based on multiple resources, rather than on a single text, but not all of them felt that teachers were justified in considering course development an overload. As one administrator put it:

> You hear the cry, "When are we going to get a book for Algebra?" But we can't find one book that does it all like the *Standards* recommend. The teachers have to be decision makers. We gave them three primary resources that were most reflective of the *Standards* [and asked them to choose their own instructional materials].

Schools that are working on changing the mathematics program are usually working on changing other aspects of school, too. They may not be referred to as "Christmas Tree" schools (as was Westlake). However, as urged by specialists in the study of institutional change (Fullan & Stiegelbauer, 1991; McLaughlin, 1991), they are usually attempting simultaneous change in multiple areas. How much of a burden is reasonable?[101] How much of a burden is reform based on the NCTM *Standards*, a reform effort that is characterized by complexity and a lack of specificity?

---

[101]Leinwand (1994) suggested that changing more than 10% of the teacher's work in a year is unreasonable, but changing much less is unprofessional.

Specificity or Generality of Change Advocacy: Grain Size

Several of our sources commented on the importance of having the right "grain size" in curriculum documents. As we reported earlier, NCTM leaders took the position that standards were different from behavioral objectives. They did not conceive of standards as a long list of things to cover. A state supervisor of mathematics put it this way:

> One of the brilliant characteristics of the *Curriculum Standards* is that the grain size is big enough that the bullets aren't damaging; they cannot be treated like behavioral objectives. Brilliant move! They're not sweepingly general; they can't be rejected because they're too general. But they're not at the small grain size where you'd have 150 per grade span, [the grain size] that people are used to. So the group that put the *Standards* together hit it just right.

By choosing a large grain size for the *Curriculum and Evaluation Standards,* the developers had promoted their cause in several ways. They had made it clear that they were not setting down a scope-and-sequence chart for a national curriculum in mathematics. They had indirectly undermined those in government and elsewhere who wanted a narrow set of accountability criteria by which to judge schools, teachers, and children. They had especially undermined those who were pushing a back-to-basics agenda. And they had remained true to their vision of teachers as active professionals who would construct their own "curriculum and evaluation standards," their own interpretation of what the mathematics curriculum should be.

Of course, this openness to interpretation caused difficulty in communicating in a simple way the instructional goals of NCTM (and added to the burden that the reform placed on teachers). An elementary school administrator reported on a regional meeting about the NCTM *Standards* in her state:

> Four people from my district [attended this meeting] on the *Standards* and [were] involved in these [pedagogical] activities. But they never connected the activities to the *Standards*. What is the *Standards*, what does it say? It's an activity that we do? I know that the activities are supposed to illustrate the *Standards*, but it doesn't happen and it always make me feel really uncomfortable.

So the larger grain size of the *Curriculum and Evaluation Standards* gets a good reaction from some who want to move the mathematics education community toward a new vision of mathematics teaching. But foregoing the smaller grain size leaves a great deal of room for misunderstandings and misinterpretations of NCTM ideas about reform.[102] The district mathematics supervisor is the person at the local level that NCTM leaders were counting on for help in inter-

---

[102]An interesting contrast to the NCTM *Standards* on the grain size issue is the reform effort in Australia (Ellerton & Clements, 1994) and in England (Black, 1994). In both cases, one could argue that the use of a smaller grain size caused great difficulty.

preting the *Standards*, but budget reductions in education over recent years have resulted in significantly fewer supervisors. Also, NCTM leaders had hoped to provide greater direction for the reform effort through a large monitoring project (Schoen, Gawronski, & Porter, 1989; Gawronski, Cooney, & Dougherty, 1991). The work of these task forces[103] did result in a proposal to NSF; however, the proposal was not funded, and the monitoring of the *Standards* project that NCTM did carry out was supported by the Exxon Education Foundation (Ferrini-Mundy & Johnson, 1994).

Teacher Communities

The importance of teacher communities in educational reform is receiving increased attention from researchers (Byrd, Foster, Peressini, & Secada, 1994; Talbert & Perry, 1994). The evidence suggests that teachers need time to work together and opportunities to support each other in order for educational change to be sustained in a school. Some school administrators recognize this need and have arranged innovative ways for teachers to spend time working together. An elementary school principal addressed the issue in the following way:

> The state allocates money for "staffing" kindergarten through the third grade. We have choices to make at the building. Rather than reducing class size and adding another class, our choice was to hire two half-time teachers, one science teacher and one in health. Those teachers [each half-time] have their own curriculum; they replace two teachers at a time and those teachers are released to plan or observe or go to other buildings. We've been doing that for about three years.

Principals who use these kinds of innovative strategies can provide significant opportunities for teacher collaboration in the elementary school. It is also possible to arrange time for teacher collaboration at the secondary level. The following excerpt is from a summary of our visit to Pinewood High School:

> When asked what has been the major change within their school and where they get the most support, the teachers usually answered "division groups"—a group of teachers who all teach a particular class, like the Algebra 1 division group. Four years ago the mathematics coordinator decided to institute common subject exams. To coordinate this, division groups were formed so that teachers who were teaching the same subjects could meet to produce the exam. These division groups have developed into a support system for many teachers. The teachers visit each others' classrooms and give mini-inservices within their groups. The change in atmosphere and the mutual support were noted by several teachers as an important impetus for sustaining the reform and for extending it to those teachers who have been reluctant to participate.

Teachers who are isolated and fail to articulate their teaching with the work of other teachers are not likely to be ideal advocates of a curriculum featuring

---

[103]MSEB had intended to follow up on the work of the first of these task forces with a monitoring project; that project failed to occur, so NCTM appointed a second task force.

mathematics as communication and mathematical connections. Leaders call for a community of practitioners, but the conditions in many schools deny teachers the opportunity for a true community. The improvement of all teaching, certainly mathematics teaching, depends on a restructuring of teachers' work so that they have opportunities for substantive communication with colleagues.

Dichotomies and Dilemmas

The NACOME (1975) report and later Crosswhite (1990) directed attention to "false dichotomies" that helped to undermine the efforts at reform in mathematics education in the 1960s. In their enthusiasm for new insights and practices, and also to illustrate change, innovators take unnecessarily extreme positions, rejecting useful ideas more because they are associated with the old ways than because they do not fit with the new. Overstatement on both sides leads to a perception that the positions are dichotomous when in fact the differences are matters of degree.

Two examples of false dichotomies in mathematics education regard the centrality of algorithms and reliance on textbooks for course scope and sequence. Classroom instruction shifts from concentrating on algorithms to avoiding discussion of traditional algorithms, and back again. Classrooms where the textbook defined a bounded curriculum are transformed into classrooms where no textbook is used, sometimes where no scope and sequence are apparent. Severe swings of the pendulum undercut the confidence of teachers, parents, and the learners themselves.

Some false dichotomies have developed from misinterpretations of the *Standards*. As Alba Thompson, a writer of the *Standards* indicated, some converts have trouble maintaining a balanced view:

> The 1989 *Standards* has lists of topics to be emphasized and de-emphasized for the different ... grade levels, but anything that says "de-emphasize" gets interpreted as "Take it out. Delete it. Omit it." There again is the perpetual pendulum-reaction kind of thing. Instead, it should be a way of alerting us that these aspects of the curriculum have to be brought back into balance. The *Standards* attack the two-column proof format [used in high school geometry], but that doesn't mean that now proof is going to disappear completely. Some people are very superficial in the way they interpret these things. They throw out the baby with the bath water.

Another instance of overreaction to the *Standards* was noted by an NCTM staff member. Some schools report that they no longer use a textbook in certain courses; the lack of a text was presented as evidence that a school was implementing the *Standards*. Given long-standing close ties between NCTM and textbook publishers, this interpretation was especially painful to some NCTM leaders.

Developing one's own curriculum materials is difficult for many teachers; in our visits to schools, many of the teachers—even those who are enthusiastic about reform—begrudge the time and effort that it takes to gather appropriate instructional activities from various sources. What kind of curriculum is made available when teachers don't have the support of a text? A former teacher and state supervisor of mathematics told what she saw happening in elementary classrooms:

> Most teachers who throw away their textbook in elementary school go through all their files and try and put together their own curriculum through a series of activities. Maybe they do all the geometry activities for October and maybe November will be a month to focus on something else. Typically, this is how teachers do it. That process doesn't look at a mathematics program across grade levels. It only looks at what happens in a given year. It's probably better for the kids than what's in typical textbooks with an emphasis on memorizing and regurgitating, but it is far from ideal.

Schools and teachers are not likely to realize the vision of the NCTM *Standards* by getting rid of textbooks. There are no simple solutions, and no easy way to resolve the many instructional dilemmas faced by teachers (Romagnano, 1994). The dilemmas that Romagnano describes involve many of the practical problems of teaching: When should the teacher tell students what to do, and when should the teacher just ask another question? How should teachers encourage students to work on mathematically rich projects when the students find them boring? These are examples of the important questions that pose dilemmas for teachers.

Standards and Goals

What are the NCTM *Standards*? Before 1989 the term *standards* would have conveyed to most people some kind of criteria for measuring satisfactory performance. If one had attained a standard, one would have made measurable progress toward an educational goal. Typically, a goal would be more general, a standard more specific. The NCTM *Standards,* however, have more of the characteristics of goals than what used to be called standards. We suggest that NCTM leaders finessed the definitions, provided some excellent goal statements, and then reified them by giving them the special name *standards*. NCTM took advantage of the power that the word *standards* had with politicians and the public in the 1980s.

What are the crucial forces that influenced the development of the NCTM *Standards*? There was no Sputnik in the 1980s to prompt federal officials to invest in educational improvement, although *A Nation at Risk* (NCEE, 1983) did its best to catalyze public and private agencies. We find more convincing the argument that the broad reform effort coalesced as a by-product of the shift from an Industrial to an Information Society, as argued by Romberg (1987b) and others. Although the mathematics education community has been cautioned against making economic arguments for educational reform (Bishop, 1990), fundamental economic and cultural forces, especially the need for "mathematically literate workers" (NCTM, 1989, p. 3), appeared to generate support for the NCTM initiative. These forces are intimately connected to changes in technology and the need for new goals in mathematics education.

Although changes in technology and the economy were important driving forces, the call for *standards* was heard by a receptive audience, and that receptiveness seems to come from a different source. We maintain that social trends from the 1960s through the 1980s have caused the general population to perceive societal decline in areas like the stability of families and the quality of public institutions, as well as in the economy. These general concerns about an overall decline in the quality of life made the public respond positively to calls for a return to high standards, even when the standards of the past were never especially high, and when the new goals of the curriculum (e.g., reduced emphasis on paper-and-pencil computation) were contrary to public expectations.

How did NCTM *Standards* become the centerpiece of a broad reform movement in education? We suggest that the primary goal of most NCTM members—helping more students understand more mathematics—is in some ways far removed from issues related to economic change or societal decline. However, NCTM leaders saw the call for standards as a tool that they could use to reach their own goals for improving mathematics education. Some leaders were originally not interested in specifying standards because they were not convinced that an accountability emphasis was the best way to promote change. With the switch in emphasis to standards as a vision of the future, the mathematics education community became much more united on the importance and usefulness of standards.

A unified membership, strong leadership, and a useful organizational structure made it possible for NCTM to have a significant impact on national reform, even though reform in other disciplines was not their goal. A succession of NCTM presidents—not a single charismatic individual—kept the organization focused on the goals expressed in the *Standards* documents, and that focus attracted a broad range of reformers. Partly this broad impact was due to serendipitous political circumstances, but it also developed out of NCTM's willingness to invest in public relations and media. The messages conveyed by the media, however, may be too superficial to portray the depth of change that the *Standards* will require. For example, small-group instruction is a useful strategy, and many people (e.g., see the earlier discussion of Pinewood) associate that strategy with the recommendations of the NCTM *Standards*. However, a teacher who is using small groups is not necessarily teaching for understanding in the ways that NCTM has recommended. The media are good at disseminating the simple messages (e.g., small groups are useful), but not at dealing with more complex issues. Some NCTM leaders are worried that the association between the NCTM *Standards* and these kinds of superficial messages may lead to public disappointment and eventual rejection of the *Standards* when it becomes clear that such simple messages are inadequate. The public may perceive that the *Standards* have failed when in fact they have never been tried.

What is the standing of the *Standards*? NCTM as an organization is clearly concerned; in 1995 a new task force on Professional Outreach (known familiarly as the "backlash" committee) was trying to address forces that oppose the *Standards*. However, reform in schools continues, and there are notable signs of progress, including some progress in the schools that we visited for this project. But change occurs slowly, and the public shows little inclination to provide the sustained effort and financial support that significant change would require.

When the current interest in national goals and standards disappears, an event that seems imminent in 1995 (Ravitch, 1995), the effort to improve mathematics education will be far from reaching even its intermediate goals. Could NCTM spark another reform movement? Definitely. Based on its past efforts, NCTM can embark on another reform effort whenever the organization can find a politically popular movement to adapt to its own purposes. However, the goal of helping students develop a deeper understanding of mathematics will still be out there, beyond the *Standards*, past Project 2061. A wag has already suggested a name for NCTM's next initiative: Goals 3000.

At the current rate of change, the year 3000 does set an appropriate time frame for the tasks that we face. NCTM has made little progress toward promoting a new conception of mathematics; too few teachers and parents see understanding mathematics as the goal, and too many are still satisfied to focus on rules and procedures without understanding. We did see some pedagogical changes in a few classrooms; for example, there was much more emphasis on small-group discussion and writing as a mode of learning than ever before. However, it is still a rare classroom that takes these opportunities for mathematical discourse and uses them in ways recommended by the *Teaching Standards*. The new technologies that are available provide new tools for discourse that the writers could not have anticipated, but technology has not driven reform very far in the classrooms that we have observed. Significant improvement will require deeper changes in curriculum, assessment, instruction, and especially teacher education.

In the area of curriculum, we found substantial change in state curriculum guidelines, but evidence of change in other areas (e.g., textbooks) was limited; NCTM's attempts to obtain NSF funding to monitor the impact of the *Standards* (including change in textbooks) were unsuccessful. There was evidence of at least superficial change by textbook publishers (Romberg & Webb, 1993), and a set of new curriculum projects (funded mainly by NSF) that use new approaches to school mathematics are now available. A concern about some of these new projects is that their emphasis on the applications of mathematics, and their lack of attention to traditional topics (like proof), will cause them to be rejected by mathematicians. If such projects do not reflect the balance between pure and applied mathematics that is found in the *Curriculum and Evaluation Standards*, perhaps they deserve to be rejected. In the area of assessment, it is too early to see any effect of the newest *Standards* (NCTM, 1995); however, that document summarizes changes that are currently underway. For instruction, a major concern is the frequent misinterpretation of what a constructivist approach to teaching might mean. The simple message that "Good teachers don't tell; they let students discover" is one example of such a misinterpretation; a consequence is that some people tend to assume that the teachers' knowledge of mathematics is not as important as it was for a more traditional curriculum. In fact, just the reverse is true; if we are to make progress in realizing the vision of the *Standards*, teachers will need a much stronger foundation in mathematics than ever before. Finally, any long-term solution to the problems in mathematics education requires improvement in teacher education. Too many mathematics departments still "deliver" instruction in large lecture halls, unaware of alternatives or unable to obtain funding for them. Too frequently there is no one in the department who has a serious interest in preparing mathematics teachers. In colleges of education, there also may be no specialist in mathematics education; a state supervisor estimated that only 20% of the future elementary school teachers in his state would have a methods teacher with expertise in mathematics.

The problems of mathematics education are deeply embedded in the culture. Changing the public's conceptions of mathematics and the learner will take generations, and those changes are unlikely to occur without significant change in teacher education, an area where the reform effort is just getting underway. The process of changing the public's view of mathematics education requires a change in the culture, a change that NCTM started at that 1989 press conference to release the *Curriculum and Evaluation Standards* and has continued in the years since that time. Each NCTM president has contributed to the effort, adding support from the nation's governors and from politicians and presidents from both political parties. The reform of mathematics education has just gotten started, but it is a start worth sustaining.

References

American Mathematical Association of Two-Year Colleges. (1994). *Standards for introductory college mathematics*. Memphis: Author.

Anderson, R. C., Hiebert, E. H., Scott, J. A., & Wilkinson, I. A. G. (1984). *Becoming a nation of readers: The report of the Commission on Reading*. Washington, DC: National Institute of Education.

Apple, M. W. (1992). Do the *Standards* go far enough? Power, policy, and practice in mathematics education. *Journal for Research in Mathematics Education*, 23, 412-431.

Association of State Supervisors of Mathematics. (1990). *The coordinated implementation of national and state-by-state reform in school mathematics*. Tallahassee: Florida Education Center.

Atkin, J. M. (1994). Developing world-class education standards: Some conceptual and political dilemmas. In N. Cobb (Ed.), *The future of education: Perspectives on national standards in America* (pp. 61-84). New York: College Entrance Examination Board.

Ball, D. L. (1990). Reflections and deflections of policy: The case of Carol Turner. *Education Evaluation and Policy Analysis*, 12, 263-275.

Ball, D. L. (1992). Implementing the NCTM Standards: Hopes and hurdles. In C. M. Firestone & C. H. Clark (Eds.), *Telecommunications as a tool for educational reform: Implementing the NCTM Standards* (pp. 33-49). Washington, DC: Aspen Institute.

Ball, D. L., & Schroeder, T. L. (1992). Improving teaching, not standardizing it. *Mathematics Teacher*, 85, 67-72.

Bass, H. (1994). Education reform from a national perspective: The mathematics community's investment and future. *Notices of the American Mathematical Society*, 41, 921-926.

Bishop, A. J. (1990). Mathematical power to the people. *Harvard Educational Review*, 60, 357-369.

Black, P. J. (1994). Performance assessment and accountability: The experience in England and Wales. *Educational Evaluation and Policy Analysis*, 16, 191-203.

Blais, D. M. (1988). Constructivism—a theoretical revolution for algebra. *Mathematics Teacher*, 81, 624-631.

Blank, R. K., & Pechman, E. M. (1995). *State curriculum frameworks in mathematics and science: Results from a 50-state survey*. Washington, DC: Council of Chief State School Officers.

Bosse, M. J. (1995). The NCTM *Standards* in light of the New Math movement: A warning! *Journal of Mathematical Behavior*, 14, 171-201.

Boyer, E. L. (1990). Reflections on the new reform in mathematics education. *School Science and Mathematics*, 90, 561-566.

Byrd, L., Foster, S., Peressini, D., & Secada, W. G. (1994, April). *Teachers' collective action for the enhancement of school mathematics.* Paper presented at the annual meeting of the American Educational Research Association, New Orleans.

California Department of Education. (1985). *Mathematics framework for California public schools: Kindergarten through Grade Twelve.* Sacramento: Author.

Carl, I., & Frye, S. (1991). The NCTM's *Standards*: New dimensions in leadership. *Mathematics Teacher*, 84, 580-585.

Carpenter, T. P., Corbitt, M. K., Kepner, H. S., Lindquist, M. M., & Reys, R. E. (1981). *Results from the second mathematics assessment of the National Assessment of Educational Progress.* Reston, VA: National Council of Teachers of Mathematics.

Cockcroft, W. H. (1982). *Mathematics counts: Report of the committee of inquiry into the teaching of mathematics.* London: Her Majesty's Stationery Office.

Cohen, D. K., McLaughlin, M. W., & Talbert, J. E. (Eds.). (1993). *Teaching for understanding: Challenges for policy and practice.* San Francisco: Jossey-Bass.

College Entrance Examination Board. (1959). *Program for college preparatory mathematics.* New York: Author.

Conference Board of the Mathematical Sciences. (1983). *The mathematical sciences curriculum K-12: What is still fundamental and what is not.* Washington, DC: Author.

Conference Board of the Mathematical Sciences. (1984). *New goals for mathematical sciences education.* Washington, DC: Author.

Crosswhite, F. J. (1985a). *An Agenda for Action:* Continuing commitments and mid-course corrections. *Mathematics Teacher*, 78, 574-580.

Crosswhite, F. J. (1985b). *Development and implementation of professional standards for mathematics education K-12: A project to improve school mathematics.* Grant proposal to the AT&T Foundation. Reston, VA: National Council of Teachers of Mathematics.

Crosswhite, F. J. (1990). National Standards: A new dimension in professional leadership. *School Science and Mathematics*, 90, 454-466.

Crosswhite, F. J., Dossey, J. A., & Frye, S. M. (1989). NCTM Standards for school mathematics: Visions for implementation. *Arithmetic Teacher*, 37, 55-60.

D'Ambrosio, U., & D'Ambrosio, B. (1994). An international perspective on research through the *JRME*. *Journal for Research in Mathematics Education*, 25, 685-696.

de Lange, J. (1987). *Mathematics, insight, and meaning.* Utrecht: Rijksuniversiteit Utrecht.

Dillon, D. R. (1993). The wider social context of innovation in mathematics education. In T. Wood, P. Cobb, E. Yackel, & D. Dillon (Eds.), *Rethinking elementary school mathematics: Insights and issues* (Journal for Research in Mathematics Education Monograph No. 6, pp. 71-96). Reston, VA: National Council of Teachers of Mathematics.

Dolan, D. (Ed.). (1990). *Leading mathematics education into the 21st century: A preliminary report.* Helena, MT: Office of Public Instruction.

Dossey, J. A. (1990). The political realities for mathematics education. In I. Wirszup & R. Streit (Eds.), *Developments in school mathematics around the world, Volume 2* (pp. 151-158). Chicago: University of Chicago School Mathematics Project.

Ellerton, N. F., & Clements, M. A. (1994). *The national curriculum debacle.* West Perth, Australia: Meridian Press.

Ernest, P. (1991). *The philosophy of mathematics education.* New York: Falmer Press.

Ferrini-Mundy, J., & Johnson, L. (1994). Recognizing and recording reform in mathematics: New questions, many answers. *Mathematics Teacher, 87,* 190-193.

Fey, J. T. (1979). Mathematics teaching today: Perspectives from three national surveys. *Mathematics Teacher, 72,* 490-504.

Frye, S. (1989). *Professional standards for teaching mathematics.* Grant proposal to the National Science Foundation. Reston, VA: National Council of Teachers of Mathematics.

Fullan, M. (1993). *Change forces: Probing the depths of educational reform.* London: Falmer Press.

Fullan, M., & Stiegelbauer, S. (1991). *The new meaning of educational change.* New York: Teachers College Press.

Garet, M. S., & Mills, V. L. (1995). Changes in teaching practices: The effects of the *Curriculum and Evaluation Standards. Mathematics Teacher, 88,* 380-389.

Gawronski, J. (1984). The problems of change from the school administrator's perspective. In T. A. Romberg & D. M. Stewart (Eds.), *School mathematics: Options for the 1990s. Proceedings of the conference* (pp. 155-160). Madison: Wisconsin Center for Education Research.

Gawronski, J., Cooney, T., & Dougherty, B. (1991). *Proposal to monitor the effects of the Standards.* Unpublished report of a Task Force. Reston, VA: National Council of Teachers of Mathematics.

Guide to national efforts to set subject-matter standards. (1993, June 16). *Education Week, 12,* 16-17.

Haimo, D. T. (1995). Experimentation and conjecture are not enough. *American Mathematical Monthly, 102,* 102-112.

Hersh, R. (1995). Fresh breezes in the philosophy of mathematics. *American Mathematical Monthly, 102,* 589-594.

Hill, S. A. (1980). An agenda for action: President's address, 58th annual meeting. *Mathematics Teacher*, 73, 473-480.

Hill, S. A. (1981). The "Agenda for Action" as a potential agent for change in the mathematics curriculum. In J. Price & J. D. Gawronski (Eds.), *Changing school mathematics: A responsive process* (pp. 3-10). Reston, VA: National Council of Teachers of Mathematics.

Hill, S. A. (1987). New perspectives on the education of teachers. In *The teacher of mathematics: Issues for today and tomorrow* (pp. 5-14). Washington, DC: Mathematical Sciences Education Board.

Hirsch, C. R. (Ed.). (1985). *The secondary school mathematics curriculum: 1985 Yearbook*. Reston, VA: National Council of Teachers of Mathematics.

Hutchinson, J., & Huberman, M. (1993). *Knowledge dissemination and use in science and mathematics education: A literature review*. Washington, DC: National Science Foundation.

Jenness, D. (1990). *Making sense of social studies*. New York: Macmillan.

Johnson, H. C. (1990). How can the *Curriculum and Evaluation Standards for School Mathematics* be realized for all students? *School Science and Mathematics*, 90, 527-543.

Kaestle, C. F. (1992). *Everybody's been to fourth grade: An oral history of federal R&D in education*. Madison: Wisconsin Center for Education Research.

Kilpatrick, J. (1985). *Academic preparation in mathematics*. College Entrance Examination Board: New York.

Kilpatrick, J. (1992). A history of research in mathematics education. In D. A. Grouws (Ed.), *Handbook of research on mathematics teaching and learning* (pp. 3-38). New York: Macmillan.

Kilpatrick, J., & Stanic, G. M. A. (1995). Paths to the present. In I. M. Carl (Ed.), *Seventy-five years of progress: Prospects for school mathematics* (pp. 3-17). Reston, VA: National Council of Teachers of Mathematics.

Kitcher, P. (1984). *The nature of mathematical knowledge*. New York: Oxford University Press.

Lambdin, D. V. (1993). The NCTM's 1989 Evaluation Standards: Recycled ideas whose time has come? In N. L. Webb (Ed.), *Assessment in the mathematics classroom: 1993 Yearbook* (pp. 7- 16). Reston, VA: National Council of Teachers of Mathematics.

Lappan, G. (1993). What do we have and where do we go from here? *Arithmetic Teacher*, 40, 524-526.

Leinwand, S. (1994). Four teacher-friendly postulates for thriving in a sea of change. *Mathematics Teacher*, 87, 392-393.

Leitzel, J. R. C. (Ed.). (1991). *A call for change: Recommendations for the mathematical preparation of teachers of mathematics*. Washington, DC: Mathematical Association of America.

Leonelli, E., & Schwendeman, R. (Eds.). (1994). *The Massachusetts adult basic education math standards.* Malden, MA: The Massachusetts ABE Math Team.

Lund, L., & Wild, C. (1993). *Ten years after A Nation at Risk.* New York: The Conference Board.

Makhmaltchi, V. (1984). The problem of change from the publisher's perspective. In T. A. Romberg & D. M. Stewart (Eds.), *School mathematics: Options for the 1990s. Proceedings of the conference* (pp. 137-140). Madison: Wisconsin Center for Education Research.

Massell, D. (1993). *National Council of Teachers of Mathematics, with references to the "New Mathematics": Case study.* A report prepared for the National Education Goals Panel. New Brunswick, NJ: Rutgers University, Consortium for Policy Research In Education.

Massell, D. (1994). National curriculum content standards: The challenges for subject matter associations. In N. Cobb (Ed.), *The future of education: Perspectives on national standards in America* (pp. 239-257). New York: College Entrance Examination Board.

Mathematical Sciences Education Board. (1988). *Review by selected constituencies outside the mathematics teaching community of the Curriculum and Evaluation Standards for School Mathematics.* Washington, DC: National Research Council.

Mathematical Sciences Education Board. (1989). *Everybody counts.* Washington, DC: National Academy Press.

Mathematical Sciences Education Board. (1993). *A strategic plan.* Washington, DC: National Research Council.

McKnight, C. C., Crosswhite, F. J., Dossey, J. A., Kifer, E., Swafford, J. O., Travers, K. J., & Cooney, T. J. (1987). *The underachieving curriculum: Assessing US school mathematics from an international perspective.* Champaign, IL: Stipes.

McLaughlin, M. W. (1990). The Rand Change Agent Study revisited: Macro perspectives and micro realities. *Educational Researcher, 19* (9), 11-16.

McLaughlin, M. W. (1991). Enabling professional development: What have we learned? In A. Lieberman & L. Miller (Eds.), *Staff development for education in the 90s: New demands, new realities, new perspectives* (pp. 61-82). New York: Teachers College Press.

National Advisory Committee on Mathematical Education. (1975). *Overview and analysis of school mathematics grades K-12.* Washington, DC: Conference Board of the Mathematical Sciences.

National Commission on Excellence in Education. (1983). *A nation at risk: The imperative for educational reform.* Washington, DC: US Government Printing Office.

National Council of Supervisors of Mathematics. (1978). Position statement on basic skills. *Mathematics Teacher, 71,* 147-152.

National Council of Teachers of Mathematics. (1980). *An Agenda for action: Recommendations for school mathematics of the 1980s.* Reston, VA: Author.

National Council of Teachers of Mathematics. (1981a). *Priorities in school mathematics.* Reston, VA: Author.

National Council of Teachers of Mathematics. (1981b). *Statement on a federal role in mathematics for the 1980s.* Reston, VA: Author.

National Council of Teachers of Mathematics. (1989). *Curriculum and evaluation standards for school mathematics.* Reston, VA: Author.

National Council of Teachers of Mathematics. (1991). *Professional standards for teaching mathematics.* Reston, VA: Author.

National Council of Teachers of Mathematics. (1995). *Assessment standards for school mathematics.* Reston, VA: Author.

National Science Board. (1983). *Educating Americans for the 21st Century.* Washington, DC: US Government Printing Office.

Nelson, B. S. (1994). Mathematics and community. In N. L. Webb & T. A. Romberg (Eds.), *Reforming mathematics education in America's cities: The Urban Mathematics Collaborative Project* (pp. 8-23). New York: Teachers College Press.

Osborne, A. R., & Crosswhite, F. J. (1970). Forces and issues related to curriculum and instruction, 7-12. In P. S. Jones (Ed.), *A history of mathematics education in the United States and Canada* (pp. 155-297). Reston, VA: National Council of Teachers of Mathematics.

Parker, R. E. (1993). *Mathematical power: Lessons from a classroom.* Portsmouth, NH: Heinemann.

Porter, A. C., Smithson, J., & Osthoff, E. (1994). Standard setting as a strategy for upgrading high school mathematics and science. In R. E. Elmore & S. H. Fuhrman (Eds.), *The governance of curriculum* (pp. 138-166). Alexandria, VA: Association for Supervision and Curriculum Development.

Price, G. G., & Carpenter, T. P. (1978). Review of *On further examination: Report of the advisory panel on the Scholastic Aptitude Test score decline. Journal for Research in Mathematics Education,* 9, 155-160.

Price, J., & Gawronski, J. D. (Eds.). (1981). *Changing school mathematics: A responsive process.* Reston, VA: National Council of Teachers of Mathematics.

Raizen, S.A. 1996. The general context for reform. In *Bold Ventures: U.S. Innovations in Science and Mathematics Education, Vol. 1: Patterns Among Eight Innovations,* ed. S.A. Raizen and E. D. Britton. Boston: Kluwer Academic Publishers.

Ravitch, D. (1995). *National standards in American education: A citizen's guide.* Washington, DC: Brookings Institution.

Romagnano, L. (1994). *Wrestling with change: The dilemmas of teaching real mathematics.* Portsmouth, NH: Heinemann.

Romberg, T. A. (1984a, August). *Curricular reform in school mathematics: Past difficulties, future possibilities.* Paper prepared for the Fifth International Congress on Mathematical Education, Adelaide, South Australia.

Romberg, T. A. (1984b). *School mathematics: Options for the 1990s.* Washington, DC: U.S. Department of Education.

Romberg, T. A. (1987a). *The NCTM Commission on Standards for School Mathematics: Some initial ideas.* Madison: Wisconsin Center for Education Research.

Romberg, T. A. (1987b). *Standards: Goals, knowledge, work and technology.* Madison: Wisconsin Center for Education Research.

Romberg, T. A., & Carpenter, T. P. (1986). Research on teaching and learning mathematics: Two disciplines of scientific inquiry. In M. C. Wittrock (Ed.), *Handbook of research on teaching* (pp. 850-873). New York: Macmillan.

Romberg, T. A., & Stewart, D. M. (Eds.). (1984). *School mathematics: Options for the 1990s. Proceedings of the conference.* Madison: Wisconsin Center for Education Research.

Romberg, T. A., & Webb, N. L. (1993). The role of the National Council of Teachers of Mathematics in the current reform movement in school mathematics in the United States of America. In *Science and mathematics education in the United States: Eight innovations* (pp. 143-182). Paris: Organisation for Economic Co-operation and Development.

Saxon, J. (1982). Incremental development: A breakthrough in mathematics. *Phi Delta Kappan*, 63, 482-484.

Schoen, H., Gawronski, J., & Porter, A. (1989). *Final report of the NCTM Task Force on monitoring the effects of the Standards.* Unpublished report. Reston, VA: National Council of Teachers of Mathematics.

Secada, W. G. (1989). Agenda setting, enlightened self-interest, and equity in mathematics education. *Peabody Journal of Education*, 66, 22-56.

Stake, R. E. (1994). Case studies. In N. K. Denzin & Y. S. Lincoln (Eds.), *Handbook of qualitative research* (pp. 236-247). Thousand Oaks, CA: Sage.

Stake, R. E., Cole, C., Sloane, F., Migotsky, C., Flores, C., Merchant, B., Miron, M., & Medley, C. (1994). *The burden: Teacher professional development in Chicago school reform.* Urbana, IL: University of Illinois.

Stake, R. E., & Easley, J. A. (1978). *Case studies in science education.* Urbana, IL: University of Illinois.

Steen, L. A. (1988). Forces for change in the mathematics curriculum. *Wisconsin Teacher of Mathematics*, 34, 3-7.

Steen, L. A. (Ed.). (1990). *On the shoulders of giants: New approaches to numeracy.* Washington, DC: National Academy Press.

Steffe, L. P., & Kieren, T. (1994). Radical constructivism and mathematics education. *Journal for Research in Mathematics Education*, 25, 711-733.

Stein, M. K., Grover, B. W., & Silver, E. A. (1991). Changing instructional practice: A conceptual framework for capturing the details. In R. G. Underhill (Ed.), *Proceedings of the Thirteenth Annual Meeting of the North American Chapter of the International Group for the Psychology of Mathematics Education* (Vol. 1, pp. 36-42). Blacksburg: Virginia Tech.

Suydam, M. N., & Dessart, D. J. (1980). Skill learning. In R. J. Shumway (Ed.), *Research in mathematics education* (pp. 207-243). Reston, VA: National Council of Teachers of Mathematics.

Talbert, J. E., & Perry, R. (1994, April). *How department communities mediate mathematics and science education reforms.* Paper presented at the annual meeting of the American Educational Research Association, New Orleans.

U.S. Department of Education. (1994). *High standards for all students.* Washington, DC: U.S. Government Printing Office.

Usiskin, Z. (1988). *National standards, states' rights, and local autonomy.* Presentation to the annual meeting of the Association of State Supervisors of Mathematics, Boston.

Vergnaud, G. (1982). A classification of cognitive tasks and operations of thought involved in addition and subtraction problems. In T. P. Carpenter, J. M. Moser, & T. A. Romberg (Eds.), *Addition and subtraction: A cognitive perspective* (pp. 39-59). Hillsdale, NJ: Erlbaum.

Webb, N. L. (1987). *More detailed thoughts for consideration.* Madison: Wisconsin Center for Education Research.

Webb, N. L. (Ed.). (1993a). *Assessment in the mathematics classroom: 1993 Yearbook.* Reston, VA: National Council of Teachers of Mathematics.

Webb, N. L. (1993b). Mathematics education reform in California. In *Science and Mathematics Education in the United States: Eight Innovations* (pp. 117-142). Paris: Organisation for Economic Co-operation and Development.

Webb, N. L., Schoen, H., & Whitehurst, S. D. (1993). *Dissemination of nine precollege mathematics instructional materials projects funded by the National Science Foundation, 1981-91.* Madison: Wisconsin Center for Education Research.

Weiss, I. R., Matti, M. C., & Smith, P. S. (1994). *Report of the 1993 national survey of science and mathematics education.* Chapel Hill, NC: Horizon Research.

Willoughby, S. S. (1984). President's report: Mathematics education 1984: Orwell or well? *Mathematics Teacher, 77,* 575-582.

Appendix: Case Study Team

Douglas B. McLeod has been professor of mathematics at San Diego State University since 1972. He also served from 1979 to 1981 as a program manager at the National Science Foundation and as a professor of mathematics and education at Washington State University from 1986 to 1993. His Ph.D. is from the Department of Mathematics at the University of Wisconsin. Previous research projects have focused on mathematics teaching and learning, with a special emphasis on affective issues in mathematics education. Within the National Council of Teachers of Mathematics, he has served on several committees, including 10 years on the editorial board of the *Journal for Research in Mathematics Education*. He has been a member of NCTM since 1969, but he was not involved in the development of the NCTM *Standards*.

Robert E. Stake is professor of education and director of the Center for Instructional Research and Curriculum Evaluation at the University of Illinois. Since 1963, he has been a specialist in the evaluation of educational programs. Among the evaluative studies he has directed are works in science and mathematics in elementary and secondary schools, model programs and conventional teaching of the arts in schools, development of teaching with sensitivity to gender equity, education of teachers for the deaf and for youth in transition from school to work settings, environmental education and special programs for gifted students, and the reform of urban education. Among his writings are *Quieting Reform*, a book on Charles Murray's evaluation of Cities-in-Schools; *Custom and Cherishing*, a book with Liora Bresler and Linda Mabry on teaching the arts in ordinary elementary school classrooms in America; and *The Art of Case Study Research*, a book on research methods.

Bonnie P. Schappelle is a research specialist at the Center for Research in Mathematics and Science Education, San Diego State University. She has a master's degree in mathematics from San Diego State University and has taught mathematics at the secondary and community college levels. She is currently the assistant to the editor of NCTM's *Journal for Research in Mathematics Education*.

Melissa Mellissinos is a doctoral student in mathematics education at the Center for Research in Mathematics and Science Education, San Diego State University. She has a master's degree in mathematics from the University of California at San Diego and has taught mathematics at the college level. She has also held an administrative position in institutional research at Los Angeles Trade-Technical College.

Mark J. Gierl is a doctoral student in measurement and evaluation in the Department of Educational Psychology at the University of Illinois at Urbana-Champaign. He received both his bachelor of arts degree in psychology and his master of education degree in school psychology from the University of Alberta, Canada.

Chapter 2

Teaching and Learning Cross-Country Mathematics
A Story of Innovation in Precalculus

Jeremy Kilpatrick
Lynn Hancock
Denise Spangler Mewborn
Lynn Stallings

University of Georgia

Contents

2

Teaching and Learning Cross-Country Mathematics
A Story of Innovation in Precalculus

Preface

In many countries, teachers are expected to work from a national syllabus or examination framework to develop their own courses. They often have little to guide them other than their own professional preparation plus a set of standard textbooks. Mathematics textbooks often are largely used as sources of problems rather than as determiners of lesson structure. In the United States, in contrast, mathematics teachers use not only detailed teachers' versions of textbooks to guide their work but also elaborate packages of what are called "ancillary materials"—worksheets, software, sample tests, videotapes, calculator booklets, laboratory materials—that help lift the burden of curriculum innovation from the teachers' shoulders. That burden has been largely assumed in the United States by commercial textbook publishers, who lavish resources on the preparation of curriculum packages, and by curriculum development projects funded by the national government or charitable foundations. Although collegiate mathematics teachers usually feel quite free to develop their own courses and the materials for them, precollege teachers do so only in rare cases. The case study reported in this volume concentrates on one of those exceptional cases in which a group of high school teachers initiated curriculum reform on their own, albeit eventually with substantial assistance.

This case study was, at one level, a study of an innovative precalculus course. But what is a course? Is a course at my school the same as one with the same title at yours? Is it the same in my classroom as in yours next door? Although we often resort to the term "the course" in this report, we really found ourselves studying various versions or embodiments of a course over time and across classrooms. We also found ourselves studying the teachers of that course as they engaged in change.

Our study was designed and conducted as what Bob Stake (Stake 1986 and Stake 1995) would term a "responsive-naturalistic" research study. It was *responsive* with regard to our efforts to let the teachers we worked with help us identify the important themes and issues. Going into the study, we knew something about the textbook and other materials that had been produced in connection with the course, but we knew almost nothing of the people who had worked on it. In retrospect, our preconceptions of how the course developed and was then adapted at various schools seem almost laughably naive. Much of that naiveté undoubtedly remains, but we tried to respond intelligently to the many teachers who took the trouble to instruct us otherwise, sharing their perception and understanding of the changes they had made. Our study was *naturalistic* in that we looked as best we could at the everyday work of teachers teaching pre-calculus mathematics and of students learning it. We also looked at teachers learning from each other and reflecting on their practice.

Our case study was also historical. Many of the events we discuss in this report took place well before we went into the field. We were fortunate to obtain access to considerable documentation, but nonetheless we often had to rely heavily on people's memories. In most cases, we could easily reconcile conflicting accounts of what happened, but a few puzzles remained for the historians in us to work out.

We have tried our best to capture what we came to see as a unique and intriguing story of change. Lynn Hancock and Jeremy Kilpatrick vividly remember returning from our first visit to the North Carolina School of Science and Mathematics (NCSSM) with a stack of interview tapes and boxes of photocopied documents. We drove out of Durham, tired yet exhilarated, knowing that the main job before us was to do some kind of justice to what we had seen and sensed we would be seeing. We knew we had a story. The problem ever since has been finding a way to tell it.

While conducting this case study, Jeremy Kilpatrick was Regents Professor of Mathematics Education at the University of Georgia, and Lynn Hancock, Denise Mewborn, and Lynn Stallings were his doctoral students there. None of us had done a study of this sort before. Jeremy had supervised doctoral students, including these, who had done or were doing case study research, but had little firsthand knowledge of all the research techniques involved. Lynn Stallings had taken two courses on qualitative research methodology and was assisting in a qualitative evaluation of the 1993 annual meeting of the American Educational Research Association when she joined the research team. Denise and Lynn Hancock had taken qualitative research courses as well. All three students used observation and interview techniques in their doctoral research.

Our on-site data gathering began in May 1993 and continued until February 1995. We visited NCSSM once in 1993 and once in 1994 to interview teachers and observe classes. We also observed the summer workshops at NCSSM in 1994 and the winter conferences in 1994 and 1995. We observed classes and interviewed teachers at Woodward Academy twice in 1994, at the Webb School twice in 1994, at the Westminster Schools once in 1993 and once in 1994, and at Eisenhower High School once in 1994. From the summer of 1993 until the spring of 1994, we were assisted in our observations and analyses by Karen Brooks, who was also one of Jeremy's doctoral students. We are grateful to Karen for field work at Westminster, the initial analysis of school surveys, and the initial textbook analysis. Between site visits, we obtained further data from sources at the schools and elsewhere by telephone, electronic mail, postal service mail, fax, and occasional in-person interviews.

Modern technology—photocopy machines, tape recorders, computers—eases the collection of vast quantities of data. Analysis is another story. We used the HyperRESEARCH computer program (Hesse-Biber et al. 1993) to do the initial analyses of our field notes but ultimately resorted to discussion and old-fashioned hand compilation of printed documents (supplemented by keyword searches of electronic versions) for our final analyses. To help with the task of analysis, each of us became a specialist on at least one site: Lynn Hancock and Jeremy on NCSSM, Denise on Woodward, Lynn Stallings on Webb, Karen (and later Lynn Hancock) on Westminster, and Jeremy on Eisenhower.

From March 1993 to June 1995, we typically met each week when the University of Georgia was in session to plan data collection activities, discuss our analyses, and review our findings. Preliminary versions of our results were presented at a research colloquium at the Department of Mathematics Education, University of Georgia, in January 1995; a meeting of the working group on implementation of reform at the National Center for Research in Mathematical Sciences Education in Madison, Wisconsin, in January 1995; a research presession at the annual meeting of the National Council of Teachers of Mathematics in Boston in April 1995; and an international conference on regional collaboration in mathematics education in Melbourne, Australia, in April 1995.

We deeply appreciate the extensive contributions of time, information, and insight by those colleagues with whom we worked on this study, particularly those teachers who generously allowed us to observe their classes or whom we interviewed. At NCSSM, they were, from the then-current faculty: Gloria Barrett, Kevin Bartkovich, Helen Compton, Peggy Craft, Dot Doyle, John Goebel, Julie Graves, Tracey Harting, Jo Ann Lutz, Greg McLeod, Marilyn Schiermeier, Dan Teague, and Karen Whitehead. We also thank former faculty member Steve Davis, who was especially helpful to us. We thank the school director, John Frederick, and the principal, Steve Warshaw, for sharing documents, granting permission to interview teachers and visit classes, and agreeing to be interviewed themselves.

We are grateful to the following mathematics teachers at the other sites: Sandy Adamek, Jeff Floyd, Mary Ann Lecesne, Becky Myers, Paul Myers, and Mike Wylder at Woodward; Karen Falkenberg, Beverly Johnson, and Grier Novinger at Webb; Jackie Capell, Jerry Carnes, Landy Godbold, Chris Harrow, Charlotte McGreaham, Jean Milnor, and Hulan Webb at Westminster; and Verne Bakker, Sandy Christie, Collette Heffner, and Rick Jennings at Eisenhower. For

granting permission to interview these teachers and observe classes, we thank
Paul Myers, mathematics and science department chairman at Woodward;
William Pfeifer, president at Webb; William Clarkson, president at Westminster,
and David Betzing, principal at Eisenhower.

We thank the following mathematics teachers for providing additional data
about their experiences with NCSSM and the precalculus course: Tom
Seidenberg and Joe Wolfson of the Phillips Exeter Academy and Barbara Nicoll
of Triton High School in Erwin, North Carolina. Special thanks go to Henry
Pollak, of Teachers College, Columbia University, for his recollections of the
early days of NCSSM as well as his valuable observations about related reform
activities. For information about grants to NCSSM, we gratefully acknowledge
the assistance of Mary Kiely of Stanford University (formerly of the Carnegie
Foundation), and Florence Fasanelli of the Mathematical Association of America
(formerly of the National Science Foundation). Barbara Janson and Eric
Karnowski of Janson Publications generously shared information and observations about the textbook, as did Richard Askey of the University of Wisconsin-
Madison and Bill Roughead of the Georgia Department of Education. Our
thanks, too, to Sally Berenson of North Carolina State University for sharing
data from her evaluation study of the NCSSM calculus reform and to Lisa Byrd
Adajian and Walter Secada of the University of Wisconsin-Madison for sharing
data from their study of school-based reform.

We benefited enormously from feedback at numerous meetings over the
course of the U.S. case studies project from our colleagues working on other
cases. We thank, in particular, Bob Stake, Doug McLeod, Norm Webb, Dan
Heck, Mike Atkin, Michael Huberman, and Senta Raizen for supplying insight-
ful comments and references to valuable sources. Our advisors, Tom Kieren of
the University of Alberta and Donna Long of the Indiana Department of
Education, helped with both guidance and criticism. Not least, we are indebted
to our colleagues at the University of Georgia for sharing their reactions with us
and to our students at Georgia who gave us useful feedback on the textbook and
on mathematical problems taken from the course.

It is customary to preserve the anonymity of case study participants. We
have not done that in this case. The only pseudonyms we have used are for the
names of students we talked to and observed. It was impossible to cloak
NCSSM in anonymity, and the identities of the other schools—and thereby most
of the teachers—would have been difficult to disguise. We saw no point in setting puzzles for our reader. Moreover, these teachers are our colleagues in mathematics teaching. They deserve credit for sharing their forthright opinions and
professional accomplishments. We recognize that some teachers might have different perceptions about their work than we do. We have taken as much care as
we could to see that they were comfortable with what we said without changing
the story we wanted to tell.

As a final note, we have struggled at length over how to make subsequent
references to the teachers we observed and interviewed once we had given their

full names. It is customary, especially in the South, to attach honorifics to last names when referring to teachers: Mr., Ms., Dr., Coach. Although that practice is well-established as formal address and has the advantage of acknowledging the professional status of the person referred to, it also places her or him at a distance from the writer. We did not like that. These were our friends and colleagues. For that reason, we liked even less the academic practice of using last names only. That may work for reference to people outside the school classroom, but did not seem right to use with these teachers. Ultimately, with some trepidation, we settled on using first names. Even though that choice implies a familiarity that may jar some readers, we see it instead as signaling our acceptance of the common form of address used in the communities these teachers inhabit. In those communities, we were given not just access but a kind of temporary honorary membership. For that, we shall always be grateful.

Examining Innovation

The history of change in the secondary school mathematics curriculum in the United States is a story of small victories amid strong inertial forces and despite an increasingly inflated rhetoric of reform. The forces of inertia in U.S. secondary education include a system of college entrance requirements bearing little relation to curriculum content; large-scale programs of college entrance testing that cannot be readily revised, owing to the cost of changing them and to the tests' increased use as indicators of educational quality; and the prevalence of commercially published textbooks whose content is driven more by market forces than by professional judgment. The absence of a national curriculum and of examinations signaling the completion of secondary school education means that curriculum change cannot be mandated nationwide and must instead be stimulated by other means. In the absence of large-scale programs to assist teacher development and to provide creative curriculum materials, calls for reform from professional organizations may set the tone and direction for innovation but cannot bring it about.

The United States is the home of the curriculum development project that first appeared after World War II—an analogy to large-scale projects to cure disease, build weapons, or venture into space. During the so-called new math era of 1955 to 1975, some mathematics projects were able to create text material and conduct teacher education programs that eventually modified both actual curriculum practice and the curriculum embodied in commercial textbooks. Despite the common perception of the new math as having failed, curriculum development projects of various types have continued to provide fresh ideas for mathematics teachers to consider.

Most attempts to change the U.S. school mathematics curriculum, whether they consist of instructional materials, teacher education programs, or publications setting a direction for reform, come from some organization. University-directed projects and commercial publishers produce innovative textbooks, software, videotapes, and teacher manuals. Universities and school districts provide workshops and institutes for teachers. Professional and governmental task forces produce curriculum proposals and frameworks. The voice of the classroom teacher may be heard in these activities, but it is ordinarily a faint voice. Teachers are seldom the instigators of reform efforts. This case study is the story of teachers who, by attempting to change a course they were teaching, ended up changing their professional practice as well.

A Grassroots Curriculum Innovation

At the annual meeting of the National Council of Teachers of Mathematics (NCTM) in New Orleans in April 1991, a small, recently established publishing house—Janson Publications—made available for review a textbook titled *Contemporary Precalculus Through Applications: Functions, Data Analysis and Matrices* (Barrett et al. 1992; see appendix A for the table of contents). The book did not resemble the typical U.S. secondary mathematics textbook. It had 330 pages rather than 700 or 800. It had a straightforward two-column format, with narrow margins and long paragraphs of prose. Although it had plenty of figures and photographs, everything was in black and white. There were no cartoon characters or boxes to highlight rules to be remembered, no slick full-color photographs of fractals or spaceships, no lists of terms to be memorized, no vignettes of mathematicians, no review exercises, no sample chapter tests. Clearly, this was a text for serious study, more like a college textbook than a schoolbook.

The book adopted an innovative approach to what had come to be called precalculus mathematics. The precalculus course had emerged as a sequel to a three-year sequence of courses in algebra and geometry in secondary school and as a transition to and preparation for a first year of calculus. The "traditional" precalculus course, ordinarily offered in the 11th or 12th grade, had grown out of the elementary functions course for 12th grade mathematics that was developed during the era of the new math to replace assorted senior-year courses in trigonometry, solid geometry, and advanced algebra.[1] The usual precalculus course concentrated on reviewing and extending students' understanding of rational, trigonometric, and exponential functions. It emphasized algebraic manipulations and the properties of functions. Few authentic applications were

---

[1]The School Mathematics Study Group (SMSG) produced the first elementary functions textbook in 1960. It was designed for a one-semester course whose second semester would treat either introductory probability and statistics or introductory matrix algebra. A member of one of the SMSG writing teams, Mary Dolciani, later wrote, with several colleagues, a best-selling series of high school mathematics textbooks published by the Houghton Mifflin Company that reflected SMSG's approach to the new math. The textbook in that series that was most closely related to the SMSG 12th grade course was *Modern Introductory Analysis* (or *Introductory Analysis*). Several teachers in the case study had used one edition or another of this book and usually contrasted it sharply with what they were currently doing in precalculus.

treated in the usual course, and there was almost no attention to discrete phenomena. Mathematics was put in context through somewhat contrived word problems intended to illustrate the use of continuous functions.

The Contemporary Precalculus course took a different approach—one we dubbed "cross-country mathematics," a phrase coined years earlier by applied mathematician Henry Pollak.[2] The course stressed mathematical modeling of real phenomena, an activity with challenges analogous to those of a cross-country hike. More specifically, the course contained a number of applications and made extensive use of data analysis, with the data taken from actual observations or measurements. Matrices, which had not always been considered as part of precalculus, were introduced through applications rather than through their algebraic structure. The emphasis on numerical methods throughout meant that the use of graphing calculators or computers was not an option but an essential part of the course.

Perhaps the most unusual feature of the textbook, however, was neither its appearance nor its content but its provenance. Although many authors of commercially published textbooks are teachers, they are usually invited by colleagues or a publisher to join an author team. This textbook, and the course from which it emerged, was developed by a group of teachers on their own. The 10 authors of the textbook were all members of the mathematics and computer science faculty of the North Carolina School of Science and Mathematics. This case study reports how the course was developed, what it became, and how other teachers and students responded to it.

Plan for the Study

None of the eight innovations in science and mathematics education in the United States chosen for the original case studies (OECD 1993) had been produced by a group of teachers. Because teachers in other countries are often the sole source of curriculum innovation in mathematics, science, and technology, the advisory board for the U.S. case studies decided that the second phase of the case studies should include a "grassroots" effort. The NCSSM Contemporary Precalculus Through Applications course was chosen to replace a statewide curriculum reform effort in mathematics.

The original plan for the NCSSM case study was to look first at how the course had developed and was being implemented at the school itself. We on the research team would choose a set of three or four additional sites where the course was being taught by teachers having varying degrees of involvement with NCSSM. The schools would range from those in which the curriculum materials had been tried out in draft form to those that had purchased and were using the textbook without having any acquaintance with NCSSM or its activities.

---

[2]Pollack was a member of the first NCSSM Board of Trustees and was influential in shaping the Contemporary Precalculus course.

To identify the schools for further study, we mailed a questionnaire in October 1993 to131 schools on a list of purchasers of *Contemporary Precalculus Through Applications* supplied by Janson Publications. The questionnaire asked about the number of precalculus classes in which the textbook was used at the school, how it was used, and how long it had been used. Information was also requested on characteristics of the school, the number of teachers using the book, the previous involvement the teachers had with workshops at NCSSM, and why the book had been chosen for use at the school. Respondents were asked to indicate their willingness to be contacted further. (See appendix B for details of the questionnaire and the responses.)

As far as we could determine from available information about the schools, the 84 questionnaires returned were reasonably representative of the total group using the textbook. An unusually large number of the schools were private (usually independent, but in some cases affiliated with a religious denomination). Although only one American high school student in every nine is enrolled in a private school, roughly half of the schools surveyed, and of those responding, were private. The schools in which the *Contemporary Precalculus Through Applications* textbook was used tended to be located in states along the East Coast, with concentrations in New England and the South. Most were high schools containing grades 9 to 12, but many of the private schools included elementary and middle grades. In public schools, the textbook was used as a secondary or supplementary textbook as often as it was the primary textbook for the course; in private schools, it was more likely to be the primary textbook. Just over a third of the public school teachers using the book had attended an NCSSM workshop; two-thirds of the private school teachers had done so.

The survey data suggested, and subsequent interviews confirmed, that the textbook was being disproportionately used in private schools for a variety of reasons. Public schools are often restricted to using textbooks appearing on an official list of books approved by the state department of education. Such lists are revised only every five years or so, and it took time for *Contemporary Precalculus Through Applications* to appear on the lists. Moreover, private schools devote relatively more of their resources to college preparatory courses. The private schools responding to the survey were much more likely than the public schools to offer a large assortment of high-level mathematics courses, including multiple tiers of precalculus. Private school teachers appeared to be more likely than public school teachers to receive encouragement and support for travel to professional meetings and to engage in other professional development activities: It may also be that private school teachers were more likely to be oriented toward seeking out such opportunities.

The selection of sites for the case study did not follow the neat plan we had proposed at the outset. We were not able to identify a school that had no connection with NCSSM before using the textbook in its precalculus classes, and we had difficulty finding public schools easily accessible to us that were teaching the course. Furthermore, with limited resources available to us, it seemed important to study schools that had been using the materials long enough to have institutionalized them in various ways. In the end, four schools, in addition to NCSSM, were selected and agreed to participate.

The Sites

Woodward Academy, in College Park, Georgia, was selected because it was nearby, had sent no teachers to an NCSSM workshop, and appeared to have had no connection with NCSSM. After the school had agreed to participate, however, we found that the teacher who had been instrumental in getting *Contemporary Precalculus Through Applications* used in that school had been acquainted through other channels with NCSSM faculty members and the ideas they were putting into their course. At the time of the study, *Contemporary Precalculus Through Applications* was the primary textbook in three sections of a sophomore honors course entitled Data, Models, and Predictions. It was also used as a supplementary textbook in three sections of a precalculus course for juniors and seniors. The school is an independent coeducational college-preparatory day school. At the time of the study, it had about 2,400 students from prekindergarten to grade 12, with 925 on the grade 9 to 12 campus.

The Webb School of Knoxville is in Knoxville, Tennessee. It was chosen because five teachers from the school had attended NCSSM workshops, but the school had not been among the pilot schools acknowledged in the preface to *Contemporary Precalculus Through Applications* as having tried the materials in unpublished form. The teachers had, in fact, had access to the unpublished materials and had been teaching the course for some years. The school offered regular and honors precalculus courses for junior and seniors in which *Contemporary Precalculus Through Applications* was the primary textbook. The Webb School is an independent coeducational college-preparatory day school; at the time of the study, it had about 750 students in grades 5 to 12.

We chose the Westminster Schools, in Atlanta, Georgia, because it was nearby and had been identified by the NCSSM faculty in the preface to *Contemporary Precalculus Through Applications* as being involved in trying out the new curriculum "in its earliest forms." The school was no longer using the book as a primary textbook in precalculus courses, but it was used as a supplementary text there and in other courses. It was not used as a primary text because the ideas from it were seen by the faculty to have permeated the entire mathematics curriculum at that school. The school is a Christian coeducational college-preparatory day school; it had about 1,700 students in kindergarten to grade 12 at the time of the study.

Eisenhower High School in Yakima, Washington, was chosen as a public school with the reputation of having incorporated the innovative materials rather thoroughly into its mathematics curriculum. In fact, the school was seen by the NCSSM faculty as a "second generation" site in which the ideas from the course had influenced not just existing courses but new courses as well. Teachers at this school, too, had been acknowledged in the preface to *Contemporary Precalculus Through Applications* and were identified by the NCSSM faculty as influential contributors to the course. Precalculus at Eisenhower was termed Honors Precalculus, and *Contemporary Precalculus Through Applications* was the primary textbook for the course. It was also one of two texts for a course entitled Introduction to College Mathematics, a senior course at a less advanced level than Honors Precalculus. At the time of the study, the school was one of three public coeducational high schools in the Yakima School District. It enrolled 1,500 students in grades 9 to 12.

As the initiator of the Contemporary Precalculus innovation, the North Carolina School of Science and Mathematics in Durham, North Carolina, was the central focus of the study. It offered two precalculus courses: Contemporary Precalculus Through Applications, for most students; and Contemporary Precalculus Through Applications With Topics. "Topics" courses were for students who were better prepared and more interested in mathematics. At the time of our observations, each course met four times a week, with one of the meetings a 90-minute laboratory session. An experimental version of the course was offered by television to schools in the Durham area. *Contemporary Precalculus Through Applications*, although used in these courses, was supplemented with a variety of other units and activities developed at the school.

NCSSM, the first of its kind to be established in the United States, is a public, residential, coeducational high school for juniors and seniors who show high intellectual ability, a commitment to scholarship, and a special interest in and potential for science and mathematics. The students are drawn from across the state and are fully supported for their tuition, room, and board. Established by the North Carolina General Assembly in 1978, the school opened in September 1980 with its first class of 150 juniors; as of fall 1994, it had 550 11th and 12th graders in residence. Governor James Hunt, who was instrumental in getting the school established, saw its mission as twofold: (1) to develop a new pool of North Carolina students who would become leaders in mathematics and science, and (2) to have an outreach program of assistance in mathematics and science to all schools in the state.[3]

---

[3]The idea for NCSSM was conceived by novelist John Ehle, who, as consultant to Governor Terry Sanford in 1962, had been responsible for initiating the North Carolina School for the Arts along with several other special schools in the state. Ehle proposed NCSSM in 1972 to a subsequent candidate for governor but had to wait until Hunt was elected in 1976 to receive a sympathetic hearing. Hunt "determinedly supported the school in spite of early opposition from the Department of Public Instruction and educators from the state's major colleges and universities" (Banner 1987, p. 385). By 1995, the school's outreach program included summer workshops, conferences, public service activities conducted jointly with the North Carolina Department of Public Instruction, and the operation of 1 of 10 centers of the Mathematics and Science Network of the University of North Carolina (NCSSM 1994).

At the time of the study, NCSSM offered a unique educational environment that was far from that of a typical U.S. high school. Its students were selected on the basis of aptitude and the desire to achieve. Although not all were highly talented in science or mathematics, they did have special interests in these subjects and came to NCSSM looking for challenges beyond those available in their own schools. Precalculus was required of all students. Some were able to take the course at an advanced level on entrance; others came with much less preparation in mathematics and needed to take another course first.

The five sites selected for the case study turned out to be special in many respects. Our sample was heavily skewed toward independent private schools and toward schools in which there was extensive contact with NCSSM and its activities. What gradually became apparent as we visited these schools, talked with the mathematics faculty, observed classes, and attended conferences and workshops at NCSSM was that the precalculus teachers at all of these schools were members of a loosely defined community of teachers around the country who were seeking to improve their instruction and their curriculum. What had begun as a study of curriculum innovation grew into a study of teacher change as well.

In the remainder of this report, we attempt to tell both how the innovative NCSSM Contemporary Precalculus curriculum was implemented and evolved in each school and how a community of professional interest grew up around this and related innovations. In the next section, we give a historical account of how the course came to be developed and the textbook written. The third section characterizes the changes in mathematical content and approach that the innovation reflects. The conditions that appear most salient for promoting the changes that occurred in curriculum and in teaching at NCSSM are analyzed next. Then we describe the various ways the curriculum was adapted at the different sites, discussing influences on and consequences of those adaptations. We also discuss the communities of teachers that contributed to and emerged from the innovative curriculum. The final section offers our view of how change was seen by the NCSSM faculty and by many other teachers in the communities we studied.

Building a Faculty and a Course

Precalculus at NCSSM

On a Monday morning in January 1994, the grounds of the former Watts Hospital in Durham that now serves as the home of NCSSM are bleak and cold. Inside remodeled Watts Hall, however, the atmosphere is cheery, warm, and inviting. On the first floor of Watts are the administration offices for the school; upstairs are the classrooms and laboratories of the Department of Mathematics and Computer Science. In one classroom, Helen Compton's third period precalculus class meets during a 90-minute lab session.

The students are seated at desks arranged in groups of four. They are beginning their study of exponential and logarithmic functions. The class starts with students talking among themselves about the day's homework assignment. Helen leads the class in discussing a problem on compound interest that seems to be causing particular difficulty. Several students explain how they have approached the problem, although no one ever mentions or asks for a final numerical answer. After a few minutes, all seem satisfied that they can at least continue with their work. Helen addresses the class:

> I'd like to turn our sights in a little bit different direction. Problem 5 is a good one to think about—what we're doing over and over again here is taking a constant value to different powers. So what we are talking about here is something called the exponential function. What I'd like for us to do in our lab time today is to give you the opportunity to explore what a function of that type is like. We're adding it to our tool kit.

Helen distributes a sheet with six questions about exponential functions. The first question reads:

> 1. The Exponential Function is defined as a function of the form: $y = b^x$. Explore and then explain what non-negative values of b can be used in the function. How does your choice of different b-values affect the resulting graph? Be sure to explore all possibilities and catalog the resulting graphs.

She continues:

> Because we have been playing with money, our bases have been things like 1.08. We had a squirrel population problem in which the base was 0.8, because we had a decreasing population, right? So, we have a fairly narrow view right now of what the bases could be. One of the questions I'm after today is what kinds of numbers make sense for b. You and your calculator are welcome to play with this. Certainly that is an important piece. But do remember that the screen your calculator presents you sometimes does not tell the whole story. So, don't leave your brains at home on this. The next question I am interested in is why is this function important to us?

For the remainder of the class period, the students work in their groups while Helen goes around the room listening to discussions and talking with students about what they are doing.

Helen would not have approached a precalculus lesson in quite this fashion 10 years before, or even 5. The way students worked in groups, the framing of the questions about exponential functions, the use of "tool kit" functions,[4] the exploration with the graphing calculator—these were teaching techniques she had

---

[4]Tool kit functions are described in the next section of this report.

developed as she and her colleagues put together a new course in precalculus as well as an approach to teaching the course that worked for her. Speaking to a group of teachers at the yearly winter conference on Teaching Contemporary Mathematics, hosted by the department in February 1994, Helen described some changes the department had recently made in the course:

> We felt like things needed to have a big picture—very similar to calculus. That kids had to wade through some tough problems and put together pieces, sort of a synthesis of ideas. And that we wanted to get away from the skill stuff, where, you know, you give 10 questions that involve logarithmic statements and the kids are supposed to change them into exponential form. Because that doesn't really give a very good view of mathematics in the sense of math helping you do things. I mean, it gives you that little segment.

The previous May she had reflected in an interview on her experience, not only in developing the course, changing her teaching, and writing a textbook, but also in working with teachers in other schools who wanted to grow and change: "It has made me realize what education can really be like, what life can really be like for teachers, if the right kind of support—I mean, there are all these little things that have to be in place. Over the years, having watched the hundreds of teachers we've worked with, how much they grow. It's incredible."

Helen's colleague, Dot Doyle, in a separate interview, spoke about why she had stayed in the department teaching mathematics. Dot cited the opportunities the mathematics faculty had to challenge one another and for her to challenge herself, the excitement of talking about mathematics with one's colleagues, the trust they had in each other that allowed them to discuss mathematical and pedagogical issues openly. "We are learning new mathematics, too. We will put ourselves into situations in which we may not know the answers. I am constantly learning new things. It's intellectually challenging."

The satisfaction Helen, Dot, and their colleagues in the department felt at changing their curriculum and changing themselves from isolated purveyors of mathematical ideas to collaborators and instructional leaders had a long history. It was a history of a new school, a new department, and a new way for these high school mathematics teachers to work together and with others.

Assembling a Team

"You had a very good group of people who were told that they weren't dreaming and took that as a challenge."—Steve Davis

The first faculty member to be hired at NCSSM when it opened in 1980 was Steve Davis, who was also the first head of the school's Department of Mathematics and Computer Science. Steve was to provide the catalytic leadership that eventually sparked the teachers in the department to try new ideas in instruction and curriculum. He had done his doctoral research in complex variables under Walter Rudin at the University of Wisconsin-Madison where, as he put it, he "fell in love with very good mathematics, mathematics that is elegant." He also decided, however, that he did not have the creativity or talent needed to do sustained work in pure mathematics at the high levels he had seen at Wisconsin. Steve had been a member of the mathematics faculty at the University of Hawaii for two years before coming to Durham with his wife, who had accepted a medical research position at Duke University. He was a teacher and department head at Durham Academy, a selective private school, for five years before being hired by NCSSM.[5] Not only did he bring to NCSSM an unusually strong mathematics background and experience with computers, but—unlike most secondary mathematics teachers—he was familiar with the practice of funding projects with grant money. Through his experiences in Hawaii and from conversations with the people with whom his wife worked, Steve knew that "grants were how you got things done."

Steve wanted to get things done. He had not finished his education with a conscious desire to teach, but he was driven "to use my Ph.D. to make a difference. I did want to make a difference." He found a way to make that difference by teaching high school and by working with other high school teachers. He saw the establishment of the new school in Durham—which had, in effect, "backed a battleship up to my back porch"—as providing a unique opportunity to shape a program of mathematics. He was also drawn by the school's outreach mission, articulated by Governor Hunt and promoted by the school's first director, Chuck Eilber, to influence mathematics education across the state: "When I came here, I felt it was important the school do more than teach the students in the school. And that was kind of the agreement under which I came." He considered his new position to be "the best job in high school education" because of the high degree of autonomy granted him as head of the mathematics department: "It was whatever I wanted it to be. If I wanted this to be a regular high school, that's what it would be. If I wanted it to be a curriculum development thing, I could make that happen too—if I could get funding."

As department head, Steve was given the initial responsibility for hiring the mathematics faculty. (This later became a group effort involving the entire department.) For the first year of the school, 1980-81, Steve hired Dot Doyle and Kevin Bartkovich. The three of them rolled up their sleeves and launched the exemplary high school mathematics program for the state of North Carolina in a dusty old building with no equipment or special funding. The next year, with the school expanding, Steve hired Jo Ann Lutz and Helen Compton. Early in that year, Steve went to Eilber and made a strong case that five classes a day was too heavy a load for these teachers. He got the load reduced to four classes each and hired an additional faculty member, Dan Teague, in the middle of the year to take the extra classes. The following year, John Goebel was hired to help cope with the expanding enrollment. After grant money produced more released

---

[5]Steve remained with NCSSM until 1987, when he became a software consultant at Digital Equipment Corporation. He is now back teaching at Durham Academy.

time, Gloria Barrett and Julie Graves were added to the faculty. Lawrence Gould was a subsequent addition as the school's resources and enrollment continued to increase. When the department moved into a remodeled building, Steve was instrumental in designing the classrooms, laboratories, and offices for the larger faculty.

Steve described his biggest contribution to the school as the quality of the faculty members he hired. He looked beyond resumes and lists of accomplishments to find teachers who, like him, wanted to make a difference. He also tried to find teachers with different talents who could work together. He likened the building of the department to the creation of an orchestra, with each member playing a different instrument well. The teachers later chuckled at the notion of Steve orchestrating the department but acknowledged that they did bring different strengths to the enterprise. Helen agreed with Steve's metaphor for the department: "I mean, Steve really did have an orchestra. He knew what everybody's role would be and how they would interface with each other and what he would tell them their jobs were."

Dot Doyle brought extensive public school experience and a knowledge of how to operate in the environment of public education. Kevin Bartkovich had worked with precocious students and could program computers. Jo Ann Lutz had a strong mathematics background and a pragmatic approach to problems. John Goebel had organized statewide mathematics competitions and had been the 1986 North Carolina Presidential Awardee in Mathematics. Helen Compton had administrative experience and had worked with a group developing learning activity packages. In her former career with the Navy, Gloria Barrett had been a statistician; she had spent much of her time working in teams. Julie Graves had previously worked as a high school textbook editor. Dan Teague brought a unique perspective on teaching because his background included high school coaching as well as mathematics teaching. Lawrence Gould, who died before the book's publication, was an older, experienced mathematics teacher hired, in part, to serve as a role model for African-American students at NCSSM, which did not have many faculty members to serve in that capacity. According to Steve, "When I hired people, they were very different people. That's what makes it work, because they learn to get together."

Steve said that, though the teachers were very different, the one thing they had in common was that they were all very good teachers. He knew that they individually had talents that exceeded his own. He saw his role as one of getting the teachers started on the track to doing something new, then leaving them alone.

Steve was a powerful force behind the development of the innovative course. He had a capacity for long hours and hard work which he attributed to his background of teaching at a university and in a private school:

I did not come from the public sector. I came from an environment where teachers did whatever needed to be done. Whatever needed to be done. Grateful that you were not at a boarding school where that applied for 24 hours. So, if what you did ended at 7 or 8 at night, you thought it was great, life can't be any better than this.

From the beginning, Steve realized that he saw the role of the department differently than did the other teachers. He was willing to provoke and to push the others toward his vision of how the department should operate.

I would tell them there [were] certain things that I demanded. If that meant shortchanging the students who were here, well then, so be it. Well, but they weren't the personalities that would shortchange the students that were here. I knew that. So, it was tough on them. I think what made it fair is that I worked harder than anyone else. That was fair, but I was demanding.

The most valuable trait Steve brought to the faculty was his talent for team building. The precalculus course was developed not because he drove the other teachers to carry out his ideas, nor because he persuaded them to his way of thinking. It came about because he was able to bring out the best in the other teachers, showing them what worked and then handing them the ball. His work in the department was an extension of how he approached teaching: "[I run] a very question-oriented class. I use the Polya approach when I teach.[6] I'm considered a dominating person in the classroom. [But] if anyone ever taped it, they would be astounded: the students did all the talking. I was always proud of that." Steve characterized his contribution at NCSSM as follows: "I spot talented people, put them on the project; I go away."

Getting Grants and Getting Known

Steve Davis had first seen an Apple computer in 1977 and realized that it could be used to change instruction. Almost as soon as he was hired at NCSSM, he got a small grant from the Apple Foundation to purchase some computer equipment for the school. In the second year of NCSSM, he got another grant from a joint National Science Foundation (NSF)-National Institute of Education program that built on the Apple funding and allowed him to work with teachers in the science department to write computer software tools for graphing functions. These tools eventually led to the precalculus course. Steve described them as follows: "'Supports of data analysis' is too fancy a term. It was essentially least squares and printing graphs. Getting graphs on a page and being able to graph functions so kids would believe you. And they could do it themselves."

When the new software was used in mathematics classes, he noted the effect on students: "We discovered that they learned to ask far better questions once

---

[6]The mathematician and educator George Polya was known for his approach to teaching mathematics by demonstrating and helping students acquire heuristics for solving challenging problems. (See Polya 1945 and 1981.)

they learned to use the software. The questions in class got much better." Later, a large grant from IBM to develop and distribute software and install a network in the school, supplemented by grants from other computer companies, allowed the purchase of enough personal computers to equip a computer lab.

Steve took seriously the outreach function of NCSSM; it was one of the strongest attractions the school had for him. The summer after the school opened, he obtained funding from a private source (Lowe's Foundation) to conduct two-week summer workshops for teachers from other schools on how to use computers in teaching. The first workshops were devoted to helping the teachers learn to use computers. They studied BASIC and Pascal programming and learned how to use the Apple tools for teaching mathematics. The response to the workshops was, in Steve's words, "overwhelming, both in applicants (there were too many) and in the evaluations after the workshop." He saw in this first effort an opportunity to offer residential workshops to North Carolina teachers. It was not long before out-of-state applicants were accepted, too. Under Steve's leadership, the mathematics faculty became the first, and for some time the only, department to conduct an outreach program. The department's activities ultimately came to include a newsletter, winter conferences, summer workshops, and distance-learning courses on the North Carolina information highway.

Eager to have the department become a site for innovation, Steve began to share his ideas about mathematics and teaching with other members of the department. He invited other teachers to sit in on his classes and shared his experiences during weekly department meetings. Yet the others were reluctant to join in because many of Steve's ideas seemed too far out. One year he used interest-rate problems to motivate almost every unit in the precalculus course. His colleagues viewed this with more than a little skepticism. He did a unit on continued fractions, and some of the other teachers were not even sure what they were. Kevin recalled, "We would just roll our eyes. What's he talking about now? For the first year of the school, he was out there where no one else was." The other mathematics teachers listened to him politely but ignored his invitations to try some of his ideas in their own classrooms.

Steve began to question the mathematics curriculum he was teaching by asking himself, "Why are we teaching this stuff?" He realized that computers had made much of the content of the traditional precalculus course obsolete. To follow up his work on software tools, he decided to outline a new precalculus course that would take advantage of the tools and of the school's new computer lab. He wanted a data-driven curriculum in which students would use computers as tools for mathematical explorations.

Realizing that a new curriculum would take time and money to develop, Steve submitted a grant proposal to the Carnegie Foundation in November 1983. The proposal was for funds to write a syllabus for a precalculus course that would incorporate his ideas about the use of technology to analyze data.[7] When Carnegie turned the proposal down, Steve realized that he had not managed to say clearly what he wanted to do. He sat down with several years of back issues of journals in mathematics and mathematics education trying to find support for his arguments, as well as the language he needed to use. From articles published by the Mathematical Association of America, National Council of Teachers of Mathematics, and other groups, he selected what he considered to be valid statements.

To help teachers at NCSSM locate and apply for outside funding, the school had established a development office. Steve turned to Mark Lichtenberger, the assistant director of that office, for help in writing a revised proposal, sharing the statements he had found: "I sat down with Mark and said, 'Okay, this is what people are talking about. Here's the outline of what we want to do and need to do.'" Lichtenberger was able to use his writing talent to help Steve put together a proposal in words the people at Carnegie might better understand. Steve wanted to design a course syllabus, use it at the school, and teach other teachers about it. He was careful not to say that he was seeking funds to produce a textbook.

The rest of the mathematics faculty had the opportunity to review the new proposal, but they did not give it much attention before signing off on it. After all, Steve often applied to outside agencies for funding. This time, however, Carnegie responded to Steve's proposal with a list of questions. Steve had only a week to answer, and he knew he had to get the rest of the department involved. In reading the questions, the other teachers realized that Steve had given the people at Carnegie the impression that his experiments in teaching were taking place throughout the department. Some tense moments followed. Several of the faculty remember an emotional confrontation in the hallway between Steve and the rest of the department members. Helen described it: "I think everybody wondered, 'How in the world are we going to do this?' Steve was like, 'Y'all aren't really going to let me get here and then not support me?' We were like, 'What are these words? We don't even know what this means.'" But Steve reminded his colleagues that the grant had each of their signatures on it and that the only way it could succeed was through a team effort.

The other teachers decided that they had no recourse but to support Steve on the Carnegie grant, and they offered suggestions for responding to the questions and further revising the proposal. The finished proposal went to Carnegie in February 1985. Almost six months passed before the Carnegie grant came through, and by then there was much excitement in the department about the

---

[7]The proposed course was tentatively titled Analysis and Finite Mathematics. By April 1985, it had become Introduction to Modern College Mathematics; this was changed to Introduction to College Mathematics the following year. The first *Teaching Contemporary Secondary Mathematics* newsletter in February 1990 announced the change of title to Contemporary Precalculus Through Applications.

project. To flesh out the proposed syllabus, the teachers began writing lesson modules. Steve, however, was not happy with their initial efforts:

> They didn't seem to want to think forward. They wanted to talk in terms of what they were already doing in the classroom . . . They were starting to do things, and when I looked at what they were trying to do, I felt it was very much the same stuff. So I ruled that we weren't doing anything that would stay the way it was, and that it would last for one year only. Everything was a tentative first draft and no more . . . And I'd be disappointed if the drafts were what we actually ended up doing. There was open rebellion.

Although "rebellion" might be an overstatement, the teachers had several reasons for being disturbed by what Steve was asking them to do. Many of them had little experience with computers and were unfamiliar with numerical algorithms and methods of data analysis. All of them had been successful in the classroom and considered themselves very good teachers. Steve's insistence that they work a different way was unsettling. Steve recognized the problem: "They were uniformly good at their teaching. So it was going to be hard for them to grow." Also, for the first time, the teachers realized that working on the precalculus project would take them out of their classrooms. The grant money had provided some released time from their regular instruction so that they could write materials. "They just felt they would shortchange the students who were here because there was going to be time away," said Steve. "You understand, these are very good teachers. I mean really superb teachers. They had never had this like half-time release and support before." John recalled: "Steve told us that what we were doing, the curriculum, was more important than those kids in the classroom. And I don't think lots of us wanted to hear that."

As the teachers tried out Steve's ideas in their classrooms, they began to better understand what he was challenging them to do and responded with a new enthusiasm for the project. They began to meet nearly every day to discuss their latest efforts and to plan new ones. Steve gave them a condition for experimenting: "If you can find at least one other person willing to try it out, then go ahead and see how it works." The faculty responded to Steve's challenge and started to exploit the freedom they had to experiment. Helen, for example, taught an entire year of trigonometry without using the unit circle in order to settle a debate about fundamental concepts of trigonometry. Dan took up Steve's use of least squares lines to analyze data and developed the treatment of data analysis techniques that was to become a major theme of the course. The teachers began to hold sessions to get together and learn mathematics. The entire school year of 1985-86 was devoted to trying out ideas in the classroom. Steve saw that year as crucial:

> [It was] the year we really didn't do anything, except we learned and people went to meetings and people tried things. Instead of being encumbered by having to execute what you've tried, you'd go try something else. That was the real benefit. It was just a full year of just trying. You try things and you file it away. The next year, we'd go and do that stuff right. That was great.

Every topic in the course was given a trial run in the classroom. If the idea was not successful, it was thrown out. According to Steve, "If it didn't succeed here, it wasn't going to succeed anywhere else." The teachers found that a surprising number of ideas did not work when tested in the reality of the classroom. By this time, the whole mathematics faculty was enthusiastically participating in the project. Changes were taking place in every teacher's classroom.

The debates within the department changed from whether to rewrite the precalculus curriculum within the existing framework to what the curriculum should become. A major controversy surrounded the issue of mathematics content for the course. According to John, a lot of time was spent "sitting around the table in the big boardroom just arguing amongst ourselves what should be done in the course, what would have to go." Steve strongly believed that applications of mathematics made the subject more interesting to students because they could see its usefulness. The teachers used what came to be known as the Steve Test: "If we can't introduce a concept with an application, then we won't teach it."

The Steve Test became another source of tension, because many of the teachers believed in the importance of pure mathematics and thought that worthwhile topics were being discarded. John explained: "I had come from a very traditional, Dolciani math background where there were hardly any applications.[8] At the time, I could not see applications for most of the stuff I taught. So I just sat there and shook my head for most of the first year."

Each of the teachers, however, eventually came to agree with the importance of good applications. Kevin talked about the effect of the Steve Test:

> When we'd argue about some topic being in the curriculum, [Steve] would pull out the Steve Test, which was "Show me an application. If you can show me an application, I'll agree that we should keep it in." I remember we were arguing about conic sections. Then he came up with the test. We had to give [the topic] up.

Gloria remembered how the teachers had felt as they began to include more applications in the course:

> We were always asking ourselves what we were giving up to make room for this stuff that was so neat. What we had to realize was the new stuff was more important than the old stuff. It took teaching it to come to that realization. It was hard to buy into that until you had actually experienced it and seen how much value data analysis had, for example, in helping kids understand functions.

Many of the ideas the teachers proposed for the course syllabus were shared with other teachers in summer workshops at the school and in sessions conducted around the state. From these experiences, the teachers realized that what they were doing was so different they would have to write more detailed text material. An outline syllabus was not sufficient for others to understand the meaning of

---

[8]Footnote 1 describes the Dolciani textbooks.

their ideas. Also, they found that teachers, working by themselves on a precalculus course, were unable to make much use of a bare syllabus. At the end of the year of trying out their ideas in the classroom, they decided to spend the entire summer (1986) writing precalculus materials. For the first time since the school began, they did not conduct summer workshops for teachers—there simply was not time. Enough materials had to be completed to allow them to open school in the fall. Steve equipped a classroom with computers, and pairs of teachers worked on writing the early chapters in the course. Helen remembered it as "the most glorious professional experience I think I have ever had."

During the 1986-87 school year, the mathematics teachers taught the precalculus course with their new materials. The materials consisted mostly of questions. Often, handouts to be used in that day's classes were taken straight off the copy machine and distributed to students. The grant had not bought enough released time for writing during the year, so the teachers' workload was very heavy. They came in early in the mornings and met after school almost every day. They all recalled the toll the work had taken on their time and energy. Helen said, "One of the big issues of the project is that everybody taught all day, then wrote all night. I can still remember getting up at 4:30 in the morning to write a couple of hours before school 'cause it was really hard to do anything at work. I think for two years everybody here worked two jobs."

In the early years of the project, the teachers were unsure how widespread interest would be in the course they were developing. They still believed it to be "a totally wild course." Few of the North Carolina teachers to whom they presented workshop sessions showed much interest in the course. The workshop participants were mostly drawn to the function-graphing and other software that had been developed at the school under one of the IBM grants and was being used in the course. Such software was not widely available at the time. NCTM was looking for materials that incorporated computer use, and the NCTM Instructional Materials Committee asked the NCSSM teachers to write modules (with software) on geometric probability, data analysis, matrices, and algorithms. Because of time constraints, the algorithms module was never completed. The other three were published separately by NCTM (NCSSM Department of Mathematics and Computer Science 1988a, 1988b, and 1988c). Although the teachers found NCTM's attention to be encouraging, they realized that NCTM was not interested in publishing their entire course.

Writing a Textbook

In the spring of 1987, Helen Compton "wangled an invitation" to a meeting Zalman Usiskin held in Illinois on a fourth-year course for his University of

Chicago School Mathematics Project (UCSMP).[9] As she listened to the "big wigs" talking, Helen realized that she and her colleagues had designed a course that would fit at the end of Usiskin's materials and that was at least as innovative as what UCSMP was planning to develop. Helen found that realization to be pivotal in her thinking about the precalculus project. She returned to NCSSM convinced that the department's work was something they should be more aggressive in telling others about. The teachers began to discuss the possibility of putting their precalculus materials together in a single textbook.

That spring was a time of many changes in the department. In May, Steve resigned from the school and took a position with Digital Equipment Corporation. At about the same time, a grant to support teacher workshops that Steve had applied for came through from NSF. Helen became principal investigator on the Carnegie grant, Dot became principal investigator on the NSF grant, and John became department head. The summer workshop was sponsored by NSF and was attended by Urban Mathematics Collaborative (UMC) teachers from Durham.[10]

NCSSM teachers continued to think about disseminating their materials. During the summer, Dot, Kevin, and Helen attended a conference on mathematics and technology at Phillips Exeter Academy in Exeter, New Hampshire. As it turned out, more sections were needed of a seminar course on teaching with computers. The Exeter people offered NCSSM teachers a chance to teach them. The teachers decided to use their own precalculus materials and found them to be exceptionally well-received. Also that summer, John met someone at a Mu Alpha Theta convention who put him in touch with an editor at a large publishing firm. Throughout the following school year, interest grew within the department about publishing a textbook. Preliminary discussions took place with Stewart Brewster of the innovative textbook division of Addison-Wesley.

In the fall of 1988, Jo Ann became the mathematics department chair. By chance, Jo Ann spoke with someone at a meeting of the Mathematical Association of America who introduced her to Barbara Janson. Janson was interested in the precalculus materials, even though her relatively small publishing house had not handled a textbook before. The teachers were excited about working with Janson, because they felt she would be more willing to allow them to do things their own way than might a larger publisher. Moreover, Addison-

---

[9]UCSMP began in 1983 with a grant from the Amoco Foundation and continued for over a decade with funding from that and other sources. Its director, Zalman Usiskin of the University of Chicago, also codirected (with Sharon Senk) its secondary school component, which resulted in the publication by Scott-Foresman of textbooks for a six-year mathematics curriculum. Two books in the series—*Functions, Statistics, and Trigonometry and Precalculus and Discrete Mathematics*—were used by some case study schools in various precalculus courses.

[10]See the case study on Urban Mathematics Collaboratives in chapter 3 of this volume for details on that program.

Wesley appeared not to be interested in pursuing innovative materials such as theirs.

A contract was signed with Janson Publications in spring 1990. The NCSSM teachers were very pleased that the only compromise Janson asked them to make concerned the length of the book. Janson was concerned that the textbook not be too thick, and so asked that the materials be cut down. Other than that, the contents of the textbook were left entirely to the authors.[11]

Contemporary Precalculus Through Applications was published the following year. A year or so later, John completed a manual for using the Texas Instruments TI-81 and Casio 7700G graphing calculators with the textbook (NCSSM Department of Mathematics and Computer Science 1993b). Helen headed up work on an assessment resource (NCSSM Department of Mathematics and Computer Science 1993a), and Karen Whitehead (a new faculty member who left during the 1993-94 school year) wrote an instructor's guide and a supplementary resource (NCSSM Department of Mathematics and Computer Science 1993c and 1993d). Such materials were not just helpful to teachers; they were essential when a textbook was published commercially. Many school districts required these extra materials. If a textbook was to be adopted, it had to be part of a complete package.

Changing Precalculus

The Use of "Big" Problems

The Midge Problem
In 1981, two new varieties of a tiny biting insect called a midge were discovered by biologists W.L. Grogan and W.W. Wirth in the jungles of Brazil. They dubbed one kind of midge an **Apf** midge and the other an **Af** midge. The biologists found out that the **Apf** midge is a carrier of a debilitating disease that causes swelling of the brain when a human is bitten by an infected midge. Although the disease is rarely fatal, the disability caused by the swelling may be permanent. This is no insect to mess with! The other form of the midge, the **Af**, is quite harmless and a valuable pollinator. In an effort to distinguish the two varieties, the biologists took measurements on the midges they caught. The two measurements taken were of wing length and antenna length, both measured in cm.

---

[11]Such freedom is rare in the context of school textbook publishing in the United States, where anonymous editors are notorious for reworking, replacing, and supplementing the material attributed to the names on the spine.

Af Midges

Wing Length (cm)	1.72	1.64	1.74	1.70	1.82	1.82	1.90	1.82	2.08
Antenna Length (cm)	1.24	1.38	1.36	1.40	1.38	1.48	1.38	1.54	1.56

Apf Midges

Wing Length (cm)	1.78	1.86	1.96	2.00	2.00	1.96
Antenna Length (cm)	1.14	1.20	1.30	1.26	1.28	1.18

Is it possible to distinguish an **Af** midge from an **Apf** midge on the basis of wing and antenna length? In your work, determine a method for distinguishing the two varieties of midges. Write a report that describes to a naturalist in the field how to classify a midge he or she has just captured.

(From Teaching Contemporary Secondary Mathematics, vol. 5, no. 1 [October 1993]: p. 5.)

The midge problem was developed by Dan Teague for the Applications Reform in Secondary Education (ARISE) project.[12] It is based on a question from the 1989 Mathematical Contest in Modeling.[13] Students at NCSSM used several approaches to solve it. One was to graph the data and try to construct a boundary in the region separating the two species. Some students used a least squares or median-median line as the boundary. Others, worried about the misclassifications those lines produced, tried using "outermost" midges in each group to define a line and then used the mid-line between the two as the boundary. Another group approached the problem by finding the ratio of antenna length to wing length and choosing a middle value in the interval between the nonoverlapping sets of ratios as the boundary. All groups had to worry about the relative importance of safety over accuracy of classification.

In a discussion of assessment with a group of teachers, Helen Compton described the midge problem as a "big problem" that allowed her to introduce several mathematical concepts:

---

[12]ARISE is an applications-based curriculum project funded by NSF and developed in conjunction with the Consortium on Mathematics and Its Applications in Lexington, Massachusetts.

[13]The Mathematical Contest in Modeling was a yearly contest open to teams of high school and college students and sponsored by the Consortium on Mathematics and Its Applications. In 1992, one of the years in which NCSSM had a winning team, there were over 290 teams from 189 schools and colleges in six countries ("Eighth Annual MCM Winners Announced" 1992). Dan, who coached the NCSSM team, remarked that their toughest competition was coming from NCSSM graduates who were on teams at several universities.

One issue in assessment is that a lot of problems that we ask are ones that say more about what I know than what my students know. Meaning, when I ask the question, I sort of build in guides to what they are going to produce because I have this sense of what the right answer is. So, if I ask a question, a lot of what I perceive as the right answer is the voice with which the question is asked. The midge problem is one that doesn't have that voice. I gave my students this problem in their groups during the third week of school and then walked away. In their portfolios for the first semester, every one of my students chose to include this problem. They said things like, "It was so incredible. I did this problem totally different from somebody else and I think mine is right."

The midge problem illustrates the type of problem that the developers of the Contemporary Precalculus course sought: a deep, yet manageable question arising from a realistic situation and offering ample opportunities for exploration. That the problem is not taken from the textbook illustrates the teachers' continuing search for ways to extend the modeling and data analysis themes and to use technology creatively. That the problem came to be used in assessment illustrates the teachers' openness to reforming all aspects of their instruction. This section discusses how the course evolved and the characteristics it came to have.

The Changing Course at NCSSM

The first year or so after their textbook was published, the NCSSM teachers were so pleased to have it in hand that the course did not change very much. As their excitement wore off, however, they began to restructure the course as they saw new ways to explore topics, as well as new topics to explore. They experimented with new versions of modules, including modules that had been omitted from the textbook. They began to pose problems in which statistical analysis software could be used. They introduced activities requiring the use of spreadsheets and other capabilities of the graphing calculator. For example, they used spreadsheets to introduce ideas of iteration that could lead into exponential functions.

From the early days of the course, students had been encouraged to create, represent, and use functions in a variety of ways. The students might begin by exploring the behavior of a function through data analysis and graphing, and then be asked to guess its closed form when they encountered it algebraically. They were expected to understand how changes in a function's algebraic representation would be reflected in its graph and vice versa. Virtually no class time was spent practicing manipulative skills. Instead, skill development was handled almost exclusively in the context of problem solving. The advent of the graphing calculator and the availability of statistical and other software simply accentuated these features of the course.

The teachers stopped teaching data analysis in two separate units, as they had when they produced the textbook. Instead, under Dan's leadership, they tried to incorporate data analysis and modeling activities throughout the course. Trigonometry received less emphasis. The course also changed, in that students did many more extended investigations of problems from real life, which kept the teachers continually on the alert for fresh data and new sources of ideas for problems that could be explored over several class periods. Students were expected to do a considerable amount of writing—writing up their investigations of problems; writing about their reactions to an activity, what they needed to do, and how they felt about their learning. Assessment gradually became more diversified, with a new emphasis put on portfolios of collected work.

Asked to look back at the textbook they wrote, the NCSSM teachers stressed the advances that the availability of new technology had brought to their thinking, the better problem material they had located and developed since writing the book, and the connections they had come to see between various topics in the course and how those topics fit together. They had a better grasp of the big picture of what they were trying to accomplish. They were more conscious of the "big ideas" of the course—interpreting data, developing mathematical representations, transforming and using functions—and how those ideas needed to be handled if students were to make sense of them. A comment from Dan captured the approach to mathematics these teachers were trying to convey to their students:

> It used to be it was unfair of me to ask a student a question I hadn't taught them to solve. And I don't believe that anymore. Because that was always looking back. I mean, I'd learn something and I'd say, "Can you remember?" It's sort of mathematics from memory—the way you do mathematics is you remember things. I think now my focus is much more [that] the way you do mathematics is you think about things. That the whole point of this is to answer questions that you don't know how to answer. What do you do when you don't know what to do?

Shift in Orientation

The Contemporary Precalculus course reflected a shift in curriculum emphasis from pure mathematics to applied mathematics. At NCSSM, the shift began with Steve Davis's simple question, "Why are we teaching this stuff?" The teachers began to use the Steve Test criterion for determining what to include in their precalculus curriculum ("If we can't introduce a concept with an application, then we won't teach it"). This approach contrasts with the common pattern found in courses in abstract mathematics. Ordinarily, a new concept is introduced, more or less out of the blue, with a definition or an example. Theorems involving the concept are presented, and their proofs are either shown or requested of the student. Additional content topics are then treated for the sake of completeness of exposition. In the Contemporary Precalculus course—at least as it was being taught at NCSSM—the content was constantly being modified as new applications were found that might introduce concepts more effectively. Grounding the course in data analysis and modeling, together with the Steve Test, shifted the instructional sequence in precalculus from definition-theorem-proof-example-practice to situation-problem-data-model-solution.

This shift was in line with a general movement that had been occurring in the school mathematics curriculum since the 1970s, as the impact of computers on society increased and more applications of mathematics to the social sciences became accessible to schoolchildren. During the early 1990s, mathematics was increasingly justified as a school subject because it prepared students for occupations and increased the country's economic competitiveness, and not because of any presumed aesthetic or intellectual value. Documents such as the *Curriculum and Evaluation Standards for School Mathematics* published by NCTM (1989) mirrored this transformation.[14]

One of the teachers at the Webb School, Karen Falkenberg, summarized the response of many mathematics teachers to the movement away from theory:

> The thing that I will say is that I don't think any child needs to leave here and be a theoretical mathematician. I've taught geometry out of Dolciani, and I've taught algebra 1 the same way. And sometimes my eyeballs would just cross looking at what they had put in those books. I thought, "What does a student need to know this for?" And if [students] do need to know it, they don't think they need to know it. Do we really need to burden them with some of this stuff? I was real pleased with the fact that some of that stuff was not in there, the theoretical math.

The teachers who had developed the Contemporary Precalculus course became dissatisfied with the claim that students should be taught abstract mathematical structures in order to learn later the specific mathematics they might need as adults. Curriculum sequences should be determined by what students can use now rather than by what follows logically. The argument seemed phony that the purpose of teaching mathematics should be to prepare students for the next course they might take, in the hope that someday they might be able to apply their knowledge to complex problems and real situations. The teachers became convinced that teaching students to apply mathematics to real-world problems was at least as important as teaching them mathematical concepts and procedures. They wanted students to solve interesting problems from daily life and various professional endeavors using the mathematics the students had already learned or were learning.

Some of the textbooks in school mathematics of the early 1990s, following the NCTM *Standards* and other calls for greater attention to relevance in school mathematics, included what they termed "applications." The common approach, however, was to begin with the mathematical concepts and procedures to be taught and then to demonstrate how they might be applied. Application came after one had learned the relevant mathematics. Instead of introducing a mathematical topic and then casting about for some way to apply what had been taught, the approach in the Contemporary Precalculus course was to begin with

---

[14]See the case study on the NCTM *Standards* in chapter 1 of this volume for details on this document and its content.

an applied problem, which then would become an opportunity to learn the mathematics that could be used to solve it. Asked what he liked most about the course in Contemporary Precalculus, Verne Bakker, a teacher at Eisenhower High School, said he liked the approach. Asked to characterize that approach, he said, "If you have a need to learn how to do something, then you learn how to do it when you have the need."

In many cases, the first step in the Contemporary Precalculus course when a new topic was introduced was to reformulate or make more precise a problem arising from an ill-defined situation. For example, a section of the chapter on modeling in *Contemporary Precalculus Through Applications* began with the following situation:

> A stretch of Interstate 40 is being widened just outside the Research Triangle Park in North Carolina to accommodate increasing traffic from Raleigh. Unfortunately, to widen the highway, only one lane is left open to traffic for several hours each day. The Department of Transportation would like to have the traffic move as quickly as possible along the one available lane, but does not want to have a lot of rear-end collisions (Barrett et al. 1992, p. 173).

Students were asked to consider what speed limit should be set to maximize the flow of traffic and still ensure safe travel. They were also asked to consider whether such a speed limit even exists. To obtain a solvable problem, students needed to consider what was meant by the flow of traffic. They also needed to make a variety of simplifying assumptions. In other situations in the Contemporary Precalculus course, data might need to be gathered or interpreted.

To begin with assumptions and data rather than with definitions required a major change in pedagogy for most mathematics teachers. Fortunately, the availability of technology—and particularly the graphing calculator and associated sensing instruments—made the handling of data much easier. Technology could be used to display and analyze the data obtained under various assumptions. In some cases, a problem could be put aside as students addressed or reviewed the mathematics they needed to tackle it. In every case, students were expected to learn skills in problem analysis and in the application of mathematical models to real data.

A course in applied mathematics requires that students work with real data. The applications they see should not have been "cooked" to yield precise and thereby unrealistic answers. Real data do not ordinarily yield exact solutions. When such data are used in instruction, they require the fitting of simple models that can never account for all sources of variation. Graphs of functions do not fit the data points exactly, and the data themselves are not easily manipulated mentally or with pencil and paper.

In applications of mathematics, the data are seldom well-behaved. A host of unmeasured and unmeasurable factors keep any set of data from satisfying a mathematical model exactly. Lack of precision and errors of measurement yield numbers that fit awkwardly into neat formulas. One way to characterize the world of applied mathematics is to say that it is numerically "bumpy." That term was used by Dan Teague/in describing the reactions of some teachers to using calculators in mathematics instruction: "They were talking about the calculator as too bumpy, and so they don't use the calculator. My response is, 'Well, in fact, the world is bumpy. And the calculator may, in fact, be giving you a better representation of the world than your abstract mathematics, which is telling you the world is smooth.'"

In addition to the Contemporary Precalculus course, NCSSM offered a semester-long Mathematical Modeling course that followed precalculus and that had been directly influenced by it. The teachers at NCSSM saw some of the outcomes of the modeling course as validating their approach to precalculus. Dan offered the following explanation:

> For me, a lot of [the change in my thinking about mathematics and mathematics learning] comes from the modeling contest. There is this collegiate contest in mathematical modeling that we've had students from our Mathematical Modeling course enter. They've done incredibly well. We've had three winning solutions in seven years. We've had all but 3 of our 14 entries in the top 15. So, I ask myself, why is that? I mean, how is it possible that students in high school—many of whom have only had one semester of calculus, precalculus, and this modeling course—can do better on real-world problems than mathematics majors at Harvard? It's not because they know more. It's because they can use what they know.

Elementary Functions

After the elementary functions course was created as part of the new math reform, problems requiring the use of polynomial, trigonometric, exponential, and other elementary functions became essential content for any course immediately preceding the first course in calculus. The Contemporary Precalculus course at NCSSM continued to treat such functions as central. Rather than beginning with the properties of these functions and posing problems that used those properties, however, the course encouraged students to investigate situations using graphs and spreadsheets to explore these functions and their variations. Instead of memorizing definitions of the functions, students were expected to be familiar with their graphs and their properties.

In the Contemporary Precalculus course, functions were valued because they could be used in representing real phenomena symbolically. Elementary functions were presented as a means of making sense of data and of arriving at reasonable solutions to complex problems (when simplified appropriately). The functions were to be seen not merely as sets of ordered pairs, but as entities that could be manipulated to fit realistic data. The problems involving these data were intended to represent some of the complexity of real life and not the artificial complexity induced by difficult symbol manipulations. This approach to teaching students about functions underscored the idea that functions result from human efforts to describe and understand the world.

As students proceeded through the Contemporary Precalculus course, they built up a collection of elementary functions whose unique qualities allowed them to represent a variety of phenomena. This collection of functions was called a *toolkit*. The toolkit functions, such as *sin x*, *log x*, and x^2, were used repeatedly throughout the course, not only to model problem situations but also to introduce increasingly complex functions. One teacher at Westminster, Charlotte McGreaham, described the advantage of this approach: "We immediately started using a lot of the function [material], the whole idea of geometric transformations . . . The idea is that if kids know these [tool kit] functions and know transformations, they can graph virtually anything."

Students learned to transform the toolkit functions to fit specific data sets by flipping, rotating, shifting, stretching, and compressing them. They also learned how to control a transformation by manipulating the symbolic form of the function and to picture the graph of the result. As a result, functions began to take on almost a physical quality in the minds of Contemporary Precalculus teachers and students. Peggy Craft, of the NCSSM mathematics faculty, des-cribed a concrete image she liked to use: "I really think of it as, when you take a known function and transform it, it's sort of like taking this long, skinny Silly Putty and stretching it, or squooshing it, or shifting it up or down or left or right."

Landy Godbold, a teacher at the Westminster Schools, described how the NCSSM approach to functions came to permeate the entire secondary mathematics curriculum at his school:

> In the main, run-of-the-mill march from algebra 1 to calculus, there was no mention of data. So the students dealt with functions. They had no reason to understand where a function really came from. If someone said, "Okay, suppose that this population is growing exponentially," the only way you knew that was the book. Every book says, "Suppose this is growing exponentially." Populations grow exponentially. God said it. That's the only way they knew it. They had no way to take data and decide whether it was really growing exponentially or not. So the whole idea of taking one step backwards—rather than being given the function, to discover the function—that to me is not just a content change, it's a philosophical change. That was one of the main strands that got sort of sucked out of the book and put into other places [in our curriculum].

Mathematical Models

In a classic discussion of applications of mathematics, Henry Pollak argued that "we are, in fact, giving a dishonest picture of mathematics if we do not allow the student to participate in finding the right problem or theorem" (Pollak 1970, p. 328). He contrasted the usual course in mathematics with one that allows a student to engage in mathematical modeling:

> A carefully organized course in mathematics is sometimes too much like a hiking trip in the mountains that never leaves the well-worn trails. The tour manages to visit a steady sequence of the "high spots" of the natural scenery. It

carefully avoids all false starts, dead-ends, and impossible barriers, and arrives by five o'clock every afternoon at a well-stocked cabin. The order of difficulty is carefully controlled, and it is obviously a most pleasant way to proceed. However, the hiker misses the excitement of risking an enforced camping out, of helping locate a trail, and of making his [or her] way cross-country with only intuition and a compass as a guide. "Cross-country" mathematics is a necessary ingredient of a good education (Pollak 1970, p. 329).[15]

The Contemporary Precalculus course, as noted earlier, attempted to engage students in "cross-country mathematics" by providing activities in which they posed problems, collected data, found a model for the data, tested their model, made predictions based on it, and analyzed its limitations. Although the students did not go through each of these steps with every problem, they nearly always were expected to build a model for the data on their own rather than having the model suggested to them by the text.

Including applications in a school mathematics course was not so unusual in the early 1990s, but giving the students a taste of the modeling process was. The Contemporary Precalculus course linked applications to modeling and functions. John Goebel described its attraction for him:

> The approach from an applications standpoint gives a context to problems I didn't have before. The emphasis on gathering data and having the mathematics come from real data motivates learning the mathematics involved. Instead of telling students about the characteristics of various functions, we start out with data and talk about modeling the phenomena found in the data with functions. Coming up with the functions is an approach that just really intrigued me.

Although teachers might have been familiar with much of the mathematics behind the applications used in the course, the modeling approach was likely to be unfamiliar. They had ordinarily taken few mathematics courses in which modeling had been done. Introducing students to modeling, without giving them heavy guidance, required that teachers venture into unknown territory. Grier Novinger, a teacher at the Webb School, found the first week-long summer workshop she took at NCSSM to be "fabulous." She was struck not by the mathematics she was learning—indeed, she claimed not to be learning any—but by the approach to teaching it: "I can't say that the math was new. It was just the idea of modeling and using data analysis. I'd never done anything like that, so in that sense, the topics were new, but the math wasn't."

The authors of *Contemporary Precalculus Through Applications* attempted to incorporate into the text the approach taken in classes at NCSSM. The give-and-take of a classroom discussion and the idiosyncrasies of students' solutions could not be duplicated. The authors did succeed, however, in moving away from the approach taken in comparable textbooks that had incorporated

---

[15]See Lampert (1990) for another reference to Pollak's image of cross-country mathematics as showing how the teacher of mathematics can model an approach to problem solving, thereby representing "what it means to know mathematics."

applications and modeling. When Paul Myers, a teacher at Woodward Academy, was asked what he liked about the *Contemporary Precalculus Through Applications* textbook, he replied:

> It doesn't overdo mechanical kinds of things. [There are] really good models, really good applications in there. You know, ones that are not typical in most— even most pretty good—textbooks. You know, Foerster's textbooks have really interesting kinds of problems, but in a sense they're somewhat contrived.[16] They're cute and they're good, but they're kind of set up for you. I think a lot of the ones in *Contemporary Precalculus Through Applications* are just much more open-ended. I also like it that they have a section on modeling.

Asked to compare *Contemporary Precalculus Through Applications* with the Foerster precalculus textbook, Rick Jennings, a teacher at Eisenhower High School, said that although he liked the Foerster book, he did not think it had "the richness of topics that the North Carolina book does." It contained similar sorts of problems, as well as somewhat more drill-and-practice exercises. But he did not find in it the problems he wanted to use for extended investigations involving modeling:

> As far as taking a problem and working with it for two or three days, a lot of Foerster's material doesn't do that. There are some projects that you could do that with. And in fact, we did some of the modeling out of Foerster when we first started using the North Carolina materials. But my sense is that the kids would probably feel more comfortable with Foerster, because it's more like a traditional mathematics book.

Students could see that courses based on the Contemporary Precalculus approach were different and that modeling was one of the differences. Asked to describe to an incoming freshman the sophomore honors course Data, Models, and Predictions, in which *Contemporary Precalculus Through Applications* is used as the primary textbook, a Woodward Academy student said: "It's kind of what it says it's about. It's like making models and that type of thing. The way he teaches it is kind of neat because he teaches you—He'll give you a situation, and he makes you figure it out rather than him telling you."

Not all students found these courses "neat," however, and quite a few students we talked with complained about the textbook's failure to provide sufficient guidance. Moreover, teachers themselves, perhaps especially those in independent schools, often had to contend with parents and administrators who did not see the value of a modeling approach to precalculus mathematics. Jerry

---

[16]Paul Foerster, a high school teacher from San Antonio, Texas, wrote a series of textbooks for the Innovative Division of the Addison-Wesley Company which were among the first to incorporate applications and modeling problems while simultaneously keeping much of the traditional course structure. His book, *Precalculus with Trigonometry: Functions and Applications*, was used in several case study schools as a primary or supplementary textbook for precalculus.

Carnes, mathematics department chairman at the Westminster Schools, recalled some of the problems he faced in reconstructing the curriculum:

> In that whole change we went through, we have been roundly criticized—Landy [Godbold] and I, in particular . . . Many people thought mathematical modeling and the ideas of precalculus [were] a waste of time. We had parents tell us this . . . What they meant was, "It's not calculus. Calculus helps you get into college."

Those teachers who had attempted to include and preserve modeling as part of precalculus mathematics seemed largely satisfied with whatever accommodation they had made, whether it was to make modeling the centerpiece of the course or to give students only a taste of what modeling might be. In either case, they tended to be somewhat uncertain as to what notion of modeling their students had formed. Rick Jennings seemed to speak for many of the teachers in responding to the question, "Do you think the students come away knowing what mathematical modeling is?":

> In a formal sense, I'm not sure. In an informal sense, I think yes. I think they're much more receptive to trying different things, looking at a process, seeing if it fits a particular natural phenomenon, trying to use that process to describe that phenomenon, and, if it doesn't work, going back, recycling through the process. That's done all year. But as far as the formalization of that, I'm not sure that they have a concept of it.

As of 1995, the Contemporary Precalculus course, no matter how it had become institutionalized in the curriculum, was still a work in progress at the schools we visited. The six "spiraling themes" cited in the preface of the textbook (Barrett et al. 1992, pp. ix-xi)—mathematical modeling, computers and calculators as tools, applications of functions, data analysis, discrete phenomena, and numerical algorithms—could be seen in the instruction and distinguished the course from more traditional versions. But teachers were still struggling with the justification for a modeling approach and with finding an appropriate balance between concepts and skills. They were seeking new and better ways for students to explore realistic problem situations and to assess the quality of their work. They were using technology not only as a tool for doing old tasks more easily but also as a springboard to a changed curriculum.

Conditions for Change at NCSSM

"It's the academic environment here. Everything is so electric where you have students who are all here because they want to learn, and they want to learn more than what their older school can offer them. In turn, it allows the teachers to be more excited about their field because their students want to learn about more and more math or science or French or whatever."—Greg McLeod

The North Carolina School of Science and Mathematics was, unquestionably, a very special place. Almost every one of its graduates had entered a college or university. Teams from the school had won state and national academic competitions. Individual students had designed space shuttle experiments, won science talent searches, earned perfect scores on national mathematics examinations, represented the United States in the International Physics Olympiad, and become Rhodes Scholars. Greg McLeod, one of the newer members of the Department of Mathematics and Computer Science, rightly characterized the atmosphere at the school as electric.[17]

Because the students and the school were so special, any visitor might have been inclined to see the teachers as special people, too. Reflecting on the group who devised the Contemporary Precalculus course, Jo Ann Lutz had a different perception:

> I think everyone was good, and they were strong math teachers, but in some ways they weren't very special. [Steve Davis] didn't go out and hire people from across the country who were big "names." I mean, no one had a name. Most were teaching in Durham and Raleigh. People came in from wherever they were. They were experienced teachers, but in very standard kinds of things.

The teachers came to the school with a wide variety of backgrounds and experience. And, as Steve said, "They are very different to this day, even after all these years of working together." Reflecting on the team he had assembled, Steve realized that he had benefited from his association with Fred Brooks at the University of North Carolina.[18] Brooks had taught Steve several things:

> One of these was the importance of personnel and that one should hire people of character with diverse talents and backgrounds. My goal was to hire people of character (first). Second, my goal was to hire very different people who would enjoy contributing in different ways. So, the bottom line is that I did not focus on people who were team players but on people whose talents would form a good team. I felt that the character aspect would make for a good team where the talents could be used.

Steve was not familiar with any of the scholarly literature on "team building," but he did read the *Wall Street Journal*. He enjoyed reading articles that told how executives formed teams of talented individuals. He wanted to set goals for his team's work much as business executives do:

---

[17]See Davis and Frothingham (1985) for a further characterization of the school, as well as details of the school's mathematics curriculum when the precalculus course, then titled Analysis and Finite Mathematics, was still in its early stages of development.

[18]Brooks, known along with Gene Amdahl as the "father of the IBM 360 computer," managed hardware and software projects at IBM. He subsequently founded the computer science department at the University of North Carolina in Chapel Hill, where he served as Distinguished Kenan Professor of Computer Science.

I tried to translate the idea of "goal sheets" to a school setting (no small task). When I left, I was beginning to understand better how this could be done. I had focused on each teacher letting me know what their goal was for the year and how I could help. I believe I was headed toward a more formal approach, when I decided to leave.

He had learned about goal sheets from a conversation with a manager at Data General, a computer company that had been one of the contributors to the school, who had told him how the company evaluated its employees. Later, when Steve joined Digital Equipment Corporation, he saw that goal sheets were a standard operating procedure in business and that they could be used very effectively.

So how did such a diverse group of, perhaps, "not very special" teachers come to work together to produce an innovative mathematics curriculum? The answer seems to lie in a combination of three factors: the innovators themselves, the departmental setting at NCSSM, and the position of the precalculus course within the secondary school mathematics curriculum.

Characteristics of the Innovators

For all their differences, the members of the department did have at least two things in common. One was their commitment to being at the forefront of mathematics teaching in the state. As Steve described it, "In hiring faculty, I looked for people who wanted to make a difference. Now, usually I would look for— . . . and they would talk in terms of—making a difference for students. But I tried to sense whether they would be willing to make a difference on a larger scale. That was difficult."

Also, the teachers came to the school with strong mathematics backgrounds—all had at least one advanced degree in mathematics or mathematics education.[19] Though the mathematics they had studied and taught was usually quite different from the applied mathematics in the Contemporary Precalculus course, they were all confident enough in their own mathematical abilities to be open to learning mathematics from each other.

When asked about the importance of their advanced degrees to the success of their curriculum project, the teachers pointed out that they did not always tell others about them. They had found that their extensive teaching experience was their most important credential for establishing credibility with other classroom teachers. They had also found out, however, that their ideas carried much more weight with university people, and sometimes with funding agencies, once their doctorates were mentioned. Steve had emphasized from the beginning of their project that they must all work hard at polishing their instructional material so

---

[19]As a condition of employment at NCSSM, all teachers were required to have at least a master's degree in their field.

that it did not look as though it "came from high school teachers." To assist that process, the teachers spent tedious hours learning a page-formatting computer program that gave their material a more professional appearance.

Certainly Steve Davis's leadership was crucial to the project. His experiences with teaching and technology gave him the vision for a new approach to precalculus that motivated the project's initial funding. However, from the point at which the other members of the department accepted Steve's challenge to work on the syllabus funded by the Carnegie grant, the project was very much a team effort. Jo Ann described the transition of project responsibilities from Steve to the other teachers:

> For a while, [Steve] wrote all the proposals and he did all the contacting and he did all the stuff, but as time went along he was no longer the best person at each of these things. And I think he knew that. When I read some of these old grant proposals, I mean, they were horribly written. Almost anyone in the department would write better than that now. As that happened, he knew he could give that stuff up. He was always willing to give things up and give people full credit for what they did. It was wonderful. You see that not happening in life. And people don't grow and change because they don't have that offered to them. I think that's part of it. He knew everyone could help, and it could get done.

As the work on the project progressed, Steve took a smaller and smaller role. When he eventually left the school, other teachers took on leadership roles in the department and on the grants that funded the project work. The script Steve followed was one he would later use as a software consultant: he found talented people, put them to work, and went away.

Characteristics of the Department

The mathematics teachers at NCSSM also benefited from what Helen Compton called "the right kind of support" in their department and school. This section describes characteristics of the school setting that the teachers identified as most helpful in sustaining their professional activities.

Outreach Mission. A unique feature of NCSSM was its responsibility to reach beyond its campus to serve education across the state and beyond. The NCSSM Board of Trustees approved the following statement in March 1990:

> The mission of the North Carolina School of Science and Mathematics is to help meet North Carolina's need for responsible leadership in the development and application of science, mathematics, and technology. The NCSSM community should offer a comprehensive, challenging, and innovative academic program and act as a catalyst for educational improvement in the state and nation (NCSSM 1990, p. 1).

The outreach mission had first attracted Steve Davis to the school, but it appeared that many of the other teachers in the department had initially discounted outreach in favor of the mission of educating students. When a team from the National Institute of Education's Study of Exemplary Mathematics

Programs visited the school in December 1983, the departmental faculty had been reluctant to think about "exporting" what they were doing. Their caution was due in part to the young age of the school (four years) but also to "the fear that an ivory tower may be closing around them, that they are too isolated from regular public and private schools" (Driscoll 1987, p. 92).

That fear seemed to have dissipated a decade later. As the teachers were drawn into curriculum revision and teacher development projects, they began to find considerable satisfaction in turning outward and reaching other teachers. John Goebel observed: "The math department takes very seriously the idea that we're not here just to teach the kids. I think all of us in the math department have bought into the idea that we are here to set the bounds, to look at what's new, to disseminate new ideas."

The outreach mission had not only forced the teachers to think about mathematics education beyond the boundaries of their classrooms, but also provided them with a common goal toward which to work. In addition to conducting distance-learning courses for North Carolina teachers, they had received grants from NSF to set up a support network of teachers who were using the Contemporary Precalculus materials and train lead teachers who would then return to their high schools and train other teachers to use the materials. This effort was dubbed the Lead Teacher Project.

Collegiality. In most high schools, precalculus courses are few in number and often small in enrollment. There are likely to be, at most, one or two teachers who teach a precalculus course at any one time. At NCSSM, however, there were numerous sections of Contemporary Precalculus every semester and numerous faculty who taught these. The precalculus teachers met once a week at lunch to discuss how they were doing, sharing ideas for teaching and articles related to the topics on which they were working. Much of the conversation dealt with mathematical questions raised by the problems the students were studying or by ways in which computers or calculators had performed. The teachers also held a debriefing session at the end of the school year.

Every teacher in the NCSSM Department of Mathematics and Computer Science either had taught, was teaching, or would soon teach the Contemporary Precalculus course. All the teachers said they liked to teach the course and requested it as part of their teaching assignment—which had surprised some newcomers to the faculty who expected to be "stuck with teaching precalculus." The teachers felt that their common experience of teaching the precalculus course had helped develop a spirit of cooperation among them. According to Steve Warshaw, NCSSM principal: "There is not the territoriality about courses you find in some schools. There is a lot of sharing of ideas and information, mutual support." John Frederick, NCSSM director, referred to the "family orientation in the department."

Working toward common goals such as developing a distance-learning course or collecting materials for precalculus course revision contributed to the strong sense of collegiality among the department's teachers. When asked what they enjoyed most about working at NCSSM, the teachers' responses were similar to Dan's: "The first thing that comes to mind are the colleagues. The willingness to help and teach you and work with you and give you ideas and share so that your class becomes better, I think, is extraordinary."

Teacher Autonomy. Because Steve Davis was the first staff member hired at NCSSM, he was relied upon to help establish the school. Then-director Chuck Eilber recognized that Steve's experience on the faculty of a private school made him a valuable resource. Steve reveled in the opportunity to have a hand in planning and organizing the school and its program. He did his best to achieve autonomy for himself and his faculty. For one thing, he managed to get and keep firm control over the hiring process, at least at the beginning. According to Steve, the state system required that the final decision on hiring a new faculty member could not be made by anyone who had served on the committee to interview applicants. Consequently, he kept himself off interviewing committees. He never sent forward more than one name for a position after the committee of his colleagues had made their recommendation. Strong teacher input in hiring decisions continued after Steve left. The hiring process at NCSSM continued to include an interview with the members of the department.

The teachers also had opportunities to contribute to other types of decisions about the school. For example, they helped the school architects design offices, classrooms, and laboratories when the new mathematics wing of the school was built. Dot returned from a year's visit at another school with the idea of seating students and the teacher around a conference table to help create a classroom atmosphere of cooperative learning. Two classrooms in the mathematics wing were subsequently furnished so that they contained one large oval table surrounded by chairs instead of student desks. When laboratory work with calculators and computer equipment became a more conspicuous part of the course, the mathematics teachers were instrumental in making changes to their class schedule so that each class would have one 90-minute lab session a week.

Grant Money. One of the most noticeable differences between the work of the NCSSM mathematics teachers and that of teachers at other schools was the amount of time and effort they put into obtaining funding from outside sources. Steve Davis was able to open up this avenue of resources to the department because of his familiarity with what grant money enabled people to do. To help teachers at NCSSM locate sources of funding and apply for funding, the school had established a development office. Steve used the facilities of that office while writing his original proposals to Carnegie and NSF for funding of the Contemporary Precalculus project. Over the years, outside funding provided the department with resources that were crucial to the success of their project: Grants from major computer companies bought computer hardware for use in classrooms and workshops; the Lowe's Foundation sponsored summer workshops and a program in which six teachers from state schools spent a year teaching on the NCSSM faculty; various small grants supported summer workshops and other activities; and the multi-year grants from Carnegie and NSF supported the development of materials and accompanying workshops for teachers.

Time. Early in the school's history, Steve Davis realized that the teachers in his department needed more time outside of their teaching duties to develop new ideas for their classes and to meet their outreach obligations. He was able to convince the school's director to reduce the regular teaching load to four classes. Later, grant money was used to buy teachers the time they needed for their work on the Contemporary Precalculus project. The teachers believed that the extra time away from teaching afforded them opportunities for professional growth. John Goebel explained: "Our normal teaching load is a lot less [than most teachers have], so we do have time during the day for subgroups of the department to get together . . . And the grants give us time to really think through some things . . . One year, I spent half a summer working on one lab."

One of the activities the teachers spent time on was meeting with each other. Frequent meetings in the department were a tradition that goes back to the early days of the school. Steve said, "We had regular meetings, even when they didn't want a meeting . . . [I] never canceled the department meeting. And then it was always amazing sometimes what would turn up." When Greg McLeod was asked about the differences he noticed in the NCSSM faculty, he responded,

> The math department has a very open communication line between them. Everybody is very accessible. They're very supportive of one another. Every week, we have departmental meetings and subject meetings. The precalculus teachers get together every Tuesday during lunch. We discuss what we are doing and we share with each other articles that are related to that particular subject area. The calculus teachers get together every Wednesday. It allows for more learning to take place among the teachers.

In fact, mathematics was often the topic of the meetings, especially when the teachers were trying to learn about new approaches to mathematics through data analysis. Julie Graves recalled her first year at the school:

> I remember feeling pretty amazed by [the number of meetings] . . . But the sole purpose of those gatherings was not to discuss ways to teach something or talking about an exam; they were "here's some new information, here's a new way of thinking about something" . . . My first year here it was set up so that every three weeks we had a class taught by one of our own, teaching the other teachers in the department.

Helen Compton pointed out that teaching each other mathematics was a regular feature of the lunch-time meetings of the precalculus and calculus teachers.

In addition to frequent meetings throughout the school day, the teachers also took time for occasional faculty retreats before and after school terms. Helen described the end-of-year meeting:

> The precalc teachers meet for several hours at the end of the year and talk about how the year went and what changes could be made for the next year. We sit around a big table somewhere away from campus and spend the day talking

about how the year has gone. We discuss mainly what has happened between us and our students. It is a very student-focused day. It obviously is a very important part of this [precalculus course]. We talk with each other about what we thought happened for our students in the school year and where do we think they are. The most important part of the conversation is thinking about next year.

The meetings helped to foster the strong sense of collegiality that developed within the department during the Contemporary Precalculus project. The teachers also found the frequent meetings important in helping the department nurture its new members.

Visits From Henry Pollak. Steve Davis had been encouraged in his efforts to get grant support for a new precalculus course by Henry Pollak, one of the original members of the NCSSM Board of Trustees. Pollak, a world-renowned applied mathematician also noted for his contributions to mathematics education, was director of the Mathematics and Statistics Center at Bell Telephone Laboratories in Murray Hill, New Jersey. Ever since his participation in the School Mathematics Study Group in the 1950s and 1960s, Pollak had been concerned with efforts to improve secondary mathematics instruction. When he was in Durham for a board meeting, he would stop by the mathematics department office after the meeting ended and talk about department activities. Describing the early relationship between Pollak and the department, Helen said,

> Pollak was on the board. You knew when the board met, and you knew Pollak would hang around and ask us what we were doing, how things were going. He did lots more question-asking than telling. "Have you thought about?" "What is fundamental and what is not ?" He talked mostly to Steve, then to Dan after Steve left.

The teachers who were on the department faculty when Steve Davis was chair attributed many of Steve's ideas about mathematical applications to Pollak's influence. Kevin and Dot recalled hiding in their offices rather than joining the conversations because they could not understand what Pollak was talking about. Kevin explained:

> Henry had these crazy things to tell us. Rather than have to sit there and nod your head and pretend you agreed, we would just as soon avoid it . . . Our heads would start spinning because we didn't understand any of this . . . Henry underestimates the impact of those hallway or office conversations he had a couple times a year. I think that had a profound influence on Steve. With us, a couple of years later we would come back to that and say, "Oh, that's what Henry was trying to tell us all along."

Pollak can also be credited with helping to establish the connection between the NCSSM mathematics teachers and the Woodrow Wilson Summer Institutes for Teachers.[20] He sent a notice about the 1984 summer institute to Steve, who persuaded Dan to attend:

> At the end of a department meeting I handed this to Dan, and I said, "Dan, if you can squeeze this in, I think it would be great, because no one else has the time to go. We're working on the project." Well, Dan went there and came back knowing far more than any of us. And going to go learn more. He went back to help with computer stuff and ended up as an actual instructor.

Steve termed that summer experience "the greening of Dan Teague." Dan's work with data analysis that summer was to have a profound effect on the Contemporary Precalculus course. He not only continued to serve as a Woodrow Wilson instructor but also, along with Dot, Helen, and other NCSSM faculty, played a key instructional role in the summer conferences that began in 1985 at the Phillips Exeter Academy.

The timing of Pollak's casual reference to the first Woodrow Wilson summer institute was propitious. The NCSSM mathematics teachers were not following the schedule for course development they had submitted to the Carnegie Foundation. Instead, they were taking time off to learn more about data analysis, modeling, and applications. When Dan went to the subsequent Woodrow Wilson summer institute (on precalculus mathematics) to help with the computing, Jo Ann attended, too, as a participant. Their experience in those institutes, according to Gloria Barrett, shaped not only the summer institutes being conducted at NCSSM but also an informal network developing among high school mathematics teachers interested in reforming their curriculum and their practice.

Students. Although the students at NCSSM were not all gifted or even much interested in mathematics, they did, as a group, provide a challenging and stimulating milieu for mathematics instruction. Because the school contained only grades 11 and 12, the student body was more cohesive than that of the usual high school. Because the school was residential, there was an esprit and camaraderie visible in every class. Steve Warshaw, a former biology teacher from Texas who was the NCSSM principal, thought teaching at NCSSM was "like you died and went to heaven":

> The caliber of students that you work with . . . and I'm not talking about how brilliant they are, because that's not really the situation. Most of them are good, bright kids, but not geniuses. Some of them have had very impoverished back-

---

[20]The National Leadership Program for Teachers of the Woodrow Wilson National Fellowship Foundation was begun in 1982. The first institute was in chemistry. The first mathematics institute, in 1984, had data analysis as its theme. Pollak was a member of the team planning the 1984 institute and advised the institutes held in 1984, 1985, and 1986. In 1987, upon his retirement from Bell Laboratories, Pollak became academic director for the institutes, and served as director or codirector until 1993. He characterized the 1984 institute as "pivotal in getting data analysis started in high schools." See Wick, Westegaard, and Wilson (1994) for program details.

grounds academically. They come from very small communities with limited resources. Others come from wealthy backgrounds, have traveled a lot with their families. Their parents are associated with universities. Some of them have taken a number of university math courses before they come here. So you've got a wide range of students. Relatively few of them are truly outstanding in terms of how academically gifted they are. But most of them are very oriented toward bettering themselves educationally. A lot of them are teacher pleasers but learn here to work more for themselves. They like learning.

The special qualities of the student body made it, in Steve Davis's words, "a very intensive place to teach." He saw the students as very demanding. They asked many questions in class, came in to teachers' offices after hours, worked late in laboratories, and sought other opportunities to learn. They also demanded a level of instruction that required the teachers to prepare extensively and thoroughly. Beyond the special courses that were offered in such topics as data structures, geometry, discrete mathematics, fractals, and chaos, the school had a Special Projects Week each year when students were expected to explore a self-initiated and self-directed project. Some students ended up doing additional modeling problems or studying topics in statistics for which there had been no time in class. In such an atmosphere, teachers were easily drawn toward the development of innovative curriculum materials.

Students from the school contributed to the development of the Contemporary Precalculus course not only by providing feedback when materials were being tried out but also by showing teachers the potential of various forms of technology. Students even helped by writing programs. In one of the grants that the school obtained from IBM, money was included to hire students to do programming. One student, over several summers, wrote the software—some of it later distributed by NCTM—that would do function graphing, matrix operations, and data analysis.

In what may seem an irony for a school devoted to science and mathematics, very few of its graduates actually went on to major in mathematics in college. Students were not necessarily drawn to the further study of mathematics. Steve Warshaw thought that phenomenon did not especially reflect well on their experience in NCSSM mathematics courses and attributed some of the problem to the demands on the teachers due to outreach activities:

> I remember my third year here there was a poll done for the local [newspaper's] teen page which asked students to name their favorite subjects. Not many of them chose math. Not many students at the time who I talked to had positive things to say about the precalculus course. I think there are several reasons for that. One of them is that everybody here has to take the precalculus course, so it's just a mundane requirement for the students. Also, because other departments weren't as involved in outreach, they could focus more on the kids. There is a real fear that we undermine our residential program when teachers are pulled out of the classroom to work on other things. Probably the students didn't feel as royally treated by the math department. That's probably not as true as it was at one time.

Gloria saw the phenomenon differently:

> Most of [our students] turn into math users. Very small numbers of our kids are math majors. I don't think it's because they get turned off, but I think they, to some extent, see so much while they're here, and they love the math . . . It's amazing the number of kids who come back and thank us for the data analysis skills they learned. They get off in their chemistry or biology class at the college that they go to, and all of a sudden they're fitting curves to data. They know how to do it, and the kids around them have never heard of it before.

Given that the number of undergraduates majoring in mathematics at most U.S. colleges and universities was rather low in the 1980s and early 1990s, it is not clear how much the NCSSM Department of Mathematics and Computer Science might have been influencing their students toward or away from further mathematical studies. NCSSM students, however, did help establish a climate in which their teachers could innovate. Gloria noted, "We could afford to take some chances with these kids because they were smart enough to be able to pick up the pieces." John made a similar observation: "One of the really nice things about this environment is that I don't think much that we were going to do with or to the kids was going to hurt them."

Characteristics of the Precalculus Course

The 11th grade precalculus course at NCSSM turned out to have been a most felicitous choice as a site for innovation. The faculty might have begun with the intermediate and advanced algebra courses that they offered to entering students who either had not studied the subject or needed to review it. In that case, however, the course would not have been offered in so many sections and would not have involved so many of the faculty. Moreover, the advanced algebra course at NCSSM was seen by the faculty as somewhat remedial in nature. Both courses seemed to lack the challenge of precalculus and failed to attract the best students.

Calculus might appear to have been a better option than precalculus. It was a high-prestige course taken by the best-prepared students. Virtually everyone saw it as a college-level course that was central to the study of advanced mathematics. Many mathematics educators felt that calculus was ripe for reform in the 1980s, and, in fact, a reform movement began in American colleges at that time (Douglas 1986 and Steen 1987).

Calculus posed some problems, however, as a site for reform. For one thing, it had a well-entrenched syllabus packed with topics seen as essential. For another, its preeminence as the capstone of the school curriculum continued to be threatened by the rise of discrete mathematics. Steve Davis was not convinced that calculus was a suitable target: "I wanted to stay away from the calculus because I think there's still going to be controversy as to whether calculus is even the right class. I think it will always be named that. But I felt that was always the domain where the colleges would have the advantage." He wanted

NCSSM to be able to make a unique contribution, and he felt that calculus was not a suitable venue for that. He would have preferred to move on to Algebra 2 (intermediate algebra) after reworking the Contemporary Precalculus course.[21]

When the NCSSM teachers subsequently turned to calculus reform (NCSSM in press), they encountered some problems in getting schools to agree to try out their materials. The schools they approached tended to offer Advanced Placement (AP) Calculus AB and BC courses.[22] Both courses had syllabuses that did not offer much leeway for experimentation. Teachers who wanted to incorporate data analysis and modeling activities along the lines of the NCSSM calculus course found themselves torn between using the new materials and preparing their students for the AP examination. Many were able to do both, but others decided that the AP examination and the accompanying credit students would receive were too important to set aside. School administrators and parents, especially those parents paying high tuition fees, did not look kindly on experimentation with courses they saw as high-status ones offering students advancement into college work.

The precalculus course was burdened with neither a high-stakes examination nor a well-defined syllabus. Precalculus was lodged in the U.S. college-preparatory mathematics curriculum between two year-long courses—intermediate algebra and calculus—whose content was well-known and presumably fixed; whereas it had a weak identity. It was defined only, and even then rather vaguely, in terms of the course for which it was "preparation." Textbooks for precalculus courses sported titles as varied as *Advanced Mathematics, Introductory Analysis*, and *Precalculus with Trigonometry*; content, obviously, was rather fluid. It might or might not have included trigonometry, complex numbers, theory of equations, polar coordinates, or derivatives of elementary functions.

In retrospect, the precalculus course seems to have been just about optimal for reform: high in status, but not college-dominated; open to syllabus change; and unencumbered by examinations or by heavy parental and community expectations (except that somehow it prepared young people to study calculus). Moreover, in most high schools it was a course that was generally taught by experienced, knowledgeable teachers and populated by students with enough mathematical ability to have surmounted the hurdles of algebra and geometry. It presented a good opening to innovation in the mathematics curriculum.

---

[21]In January 1988, Kevin Bartovich and Dan Teague submitted a proposal to NSF to fund a three-year project to develop a new Algebra 2 course; it was not funded.

[22]The AP program conducted by The College Board offers courses in many subjects that enable high school students to earn college credit by taking a national examination set and marked by college and high school teachers. Calculus AB corresponds to a year-long course in calculus with elementary functions in which over two-thirds is devoted to differential and integral calculus. Calculus BC is also a year-long course, but it covers the calculus content of Calculus AB plus advanced topics such as sequences and series and elementary differential equations.

Adapting an Innovation

Precalculus at Woodward Academy

As Paul Myers's third period class enters the room, they find a cup of water, a thermometer, and a stopwatch on each table. As class begins, Paul explains that the students are going to conduct an experiment and record the results. The experiment involves recording the temperature of the water every 10 seconds after ice is put in it.

The students are asked to sketch their prediction of what the graph would look like. They begin to collect data. Paul instructs them to take an initial temperature reading of their water. He then places a handful of crushed ice in the cup on each table. The students work in groups of three; one running the stopwatch and announcing when it is time to read the temperature, another reading the temperature, and the third recording it.

After all groups finish recording their data, Paul instructs them to draw a scatterplot on their TI-82 calculators. He then asks the class what their scatterplots look like. Jennifer notes that her graph "goes back up," and Paul asks why. She responds that eventually the ice will melt, and the water will return to room temperature. Lester says, "My graph looks like it has an asymptote."

Paul sketches a graph on the board and asks if the students' graphs looks something like his. They nod. He asks the class if it is reasonable for the graph to have an asymptote. After some discussion, he asks what kind of functions have asymptotes. Several students respond that rational functions have asymptotes and are of the form $f(x) = k/x$. Paul asks if there are any other types of functions that have asymptotes, and a student replies that square root functions and exponential functions have asymptotes. Others argue that this is not the case. The students finally agree that square root functions do not have an asymptote and that exponential functions do. Paul asks what kind of exponential functions look like the one he has drawn on the board. Madeleine says, "Negative, like a^{-x}." Paul writes $g(x) = a^{-x}$ on the board and says, "If you think this might be exponential, how could you check?" Madeleine replies, "Re-express it."

Paul suggests that students re-express their data and look at the residuals. He reviews where actual values and predicted values go in the calculator and what to do with the re-expressed values. He wanders around the room, picking up students' calculators, examining the displays, and helping the students with problems and questions. Thomas is having difficulty (his graph runs above all of the data points) and asks for assistance. Paul picks up the calculator and says, "Tell me about your lists" (meaning tell me how and where you put your data and what you did with it). Paul questions Thomas until he figures out his own mistake and is able to correct it on his own.

> The period ends with a class discussion in which the group concludes that the students with the best-fitting curves had actual temperature readings that were close to 0° C at the end of data collection.

At Woodward Academy, Paul Myers was chairman of the 14-member mathematics department, which included a computer science teacher. The mathematics classes were separated into three tracks: academic, college preparatory, and honors. The academic track was for the lower level students (either in terms of achievement or work ethic), and the honors track was for the "gifted" students. Approximately 25 percent of the school's students were enrolled in each of these two tracks. The remaining 50 percent were in the college-preparatory track. All students were required to take four years of mathematics in grades 9 to 12.

The only course at Woodward in which *Contemporary Precalculus Through Applications* was the primary textbook was the sophomore honors course entitled Data, Models, and Predictions. The book was used even though the students enrolled in the course had not studied many of the traditional intermediate algebra and trigonometry topics that the book assumed. As juniors, these students took a course called Mathematical Sciences that focused on rates of change, and then they took AP Calculus (either AB or BC level) as seniors. In the academic and college-preparatory tracks, which differed only in the pace and depth of study within each course, the courses followed the textbook series from the University of Chicago School Mathematics Project, which was used from grades 7 to 12. *Contemporary Precalculus Through Applications* was used as a supplementary textbook in the precalculus courses in those tracks in which the primary textbook was the UCSMP *Precalculus and Discrete Mathematics.*

Paul was the first teacher at Woodward to use the *Contemporary Precalculus Through Applications* book. He chose to use it because his "main focus was functions motivated through data analysis." He also wanted to emphasize the use of graphing calculators. In 1987, he attended a Woodrow Wilson institute on mathematical modeling, where Dan Teague was an instructor and Gloria Barrett was a participant. Through that association, Paul received draft copies of the *Contemporary Precalculus Through Applications* materials. He used bits and pieces of those materials in several of his classes, shared them with colleagues, but did not find them suitable for any one course. Therefore, Woodward was not involved in the pilot testing.

In 1992-93, Paul began to use *Contemporary Precalculus Through Applications* as the primary textbook in the sophomore honors course. He was the only teacher to teach all three sections of the course for the 1992-93 and 1993-94 school years. In 1994-95, he decided to involve other faculty in teaching the course. Paul selected Jeff Floyd and Mike Wylder to teach one section each of the Contemporary Precalculus course. Jeff had served as the assistant department chairman for several years and was familiar with the book. Mike was in his third year of teaching and had no prior experience with the book. Paul gave careful consideration to the selection of teachers for the course. He chose Jeff and Mike because they both had teaching styles that were congruent with an investigative approach to mathematical modeling.

The focus of the Contemporary Precalculus course at Woodward was "functions through statistical understanding." The goal of the course was for students to understand that functions represent real-world situations and that there are many ways to approach a problem (e.g., numerically, analytically, geometrically). The teachers at Woodward were drawn to the book because they thought it tied in with an emphasis they placed on transformations. Paul particularly liked the use of real-world data for modeling.

Jeff noted that the *Contemporary Precalculus Through Applications* book reflected the NCTM standards well and had an appropriate treatment of mechanical skills. He said:

> That's the raging debate—How important are mechanical skills? Your traditional kind of precalculus curriculum used to be extremely mechanical—rational expressions, adding rational expressions where you have to factor the denominators, and all that stuff. I'm not sure that anyone argues that any of that stuff is terribly important any more. But I think I draw the line a lot further to the left than a lot of other people do in terms of what mechanics are important.

Whenever their students needed to use a mechanical skill in which they lacked proficiency, Paul and Jeff would take time to address the skill at that point. Rather than emphasizing mechanics throughout the course, they addressed them on an as-needed basis.

In the mathematics classes at Woodward, the development of mathematical communication and reasoning appeared to be priorities for the teachers. The students in classes using the *Contemporary Precalculus Through Applications* book sat in groups of two, three, or four at round tables and conversed freely with each other and with the teacher about the mathematical topic under discussion. The students directed as many questions and comments to their peers as they did to the teacher.

All three teachers used an investigative pedagogy when teaching the Contemporary Precalculus course. Paul had adopted a more investigative style prior to teaching the course:

> [The *Contemporary Precalculus Through Applications* text] just fits into kind of the style that I've gotten to. I really think that seven or eight years ago things just started to happen, in terms of the way I thought about teaching. I think the Woodrow Wilson program really kicked it into high gear in '87, and after that, the way I teach, I think, changed drastically. I think I've always . . . taught in a relaxed style. The ways of approaching things, the multiple perspectives, came into play a lot. The North Carolina book just fit into my scheme of things really nicely.

Jeff's teaching style had also evolved over his career. He described how it had changed: "[My teaching style is] much more inductive now than it ever used to be. It used to always be very deductive. Even in an algebra 2 or a precalculus course, I was proving things constantly. Now I value the inductive approach almost constantly. It's so much better for a lot of reasons—mainly for the sake of the students."

Mike had less teaching experience than Paul or Jeff. He had earned a bachelor's degree in engineering and, after working a few years as an engineer, decided to pursue a career in teaching. He began by substituting in his home district and eventually got a job at Woodward. He noted a difference between his teaching style in the Contemporary Precalculus course and how he taught his other courses. The selectivity of students in the precalculus course made teaching easier for him: "You don't have to prepare a lesson, so to speak, because [the students] will carry it—if you have a few ideas, know where you want to go, and lead them that way . . . The most interesting part about teaching the honors [students] is that they generate so much. So many ideas pop up."

At Woodward in the early 1990s, the appropriate and efficient use of graphing calculators was a major component of mathematics instruction. The students brought Texas Instruments graphing calculators to class and used them freely, asking for technical help when necessary. After asking for help, the students were routinely asked to explain what they had done that had led them to the point of the question or problem. Rarely, if ever, did a teacher take a calculator from a student and push buttons to locate or correct the problem. All students at Woodward were required to purchase a TI-82 graphing calculator at the beginning of their freshman year. Each teacher had an overhead projection device for use with a graphing calculator in the classroom, and in each classroom both the overhead projector and projection device were set up at the front of the room as though they were used frequently. A computer lab equipped with eight DOS-based computers (equivalent to IBM 486 machines) was opened during the 1993-94 school year.

Most of the students at Woodward had generally positive perceptions of both the Contemporary Precalculus course and the *Contemporary Precalculus Through Applications* book. The students noted that the book did not provide them step-by-step instructions for solving problems. Some students found the lack of guidance frustrating and said they had to rely on the teacher for explanations that the book did not provide. Others found that the examples in the book provided adequate information to allow them to solve problems on their own.

Paul and Jeff both noted that because the students in the course were sophomores who had not had intermediate algebra or trigonometry, they were sometimes frustrated since the book assumed that they knew materials they had not studied. Jeff elaborated that point:

> [The students] tend to say [the book] is hard to read. I don't agree with that myself. I read it, and I think it's very readable. But it may be a matter of where we've chosen to use this book in the 10th grade year. Here's a good example. In one paragraph in this book, it talked about the factor theorem and the remainder

theorem. It said by long division you accomplish these things. That was all stuff that these students had not been exposed to at all. See, this is really a precalculus book. It's meant to be used after some kind of an algebra 2 course. And [these students] haven't done any of that. So of course [I] take the time to do that stuff. But that first night when they're reading that section, they're going "I don't understand any of this," and rightly so because they haven't done any kind of division with polynomials, whether it's long division or synthetic or whatever. I think those are the issues that have come up: where [the students] interpret it as "The book is hard to read," and the real reason is that it's just talking about things that it expects them to have already done when they haven't—which is not that big a deal. You just take some time and do those things, and then they make sense.

The continued status of *Contemporary Precalculus Through Applications* at Woodward Academy was uncertain as of 1995. An Advanced Placement Statistics course was to be introduced in a few years, and it was probable that Woodward would revise its curriculum to allow students to take both AP Statistics and AP Calculus as year-long courses. Because the Contemporary Precalculus course focused on data analysis, it would likely be subsumed into the AP Statistics course, and the textbook would no longer be used as a primary text. However, Paul saw the book continuing to be used as a supplement.

Precalculus at the Webb School

"Okay, each group gets two art papers," directs Karen Falkenberg as she walks around the room, handing the five groups of students pieces of blank standard-sized paper. "On one sheet you work the problem out. Just write the equation to be graphed on the other, so that you can exchange it with another group." The groups start work.

"Let's do sin or cos," suggests Ashley.

"How about $x + \cos^{-1}(fx)$? It would look like this." Edward traces a curve sloping downward to his right in the air with his finger.

Without looking up from his calculator, James mutters, "$2 + \cos^{-1}$ . . . No, $3 + \tan^{-1}2x$."

"Hmm . . . They'd never get a tangent one." Edward writes down James's second suggestion.

Ashley begins writing too. "I know what the tool kit x and y's are." She begins a list on her paper,

$$(x, y) \longrightarrow (\tfrac{1}{2} x, 3 + y)$$

$$(-1, -\tfrac{\pi}{4}) \longrightarrow (-\tfrac{1}{2}, 3 - \tfrac{\pi}{4})$$

Edward begins to sketch a graph using Ashley's points. James hasn't looked up from his calculator yet. He is having trouble locating the graph because of the setting for the range of x and y he had used on a previous problem.

Edward looks at the TI-85, then back at his own paper. "That *is* pretty wild. I've got a point out of place somewhere here." He indicates his sketch of the graph. "Double check me on the calculator."

"What's the equation again?" says James.

"You suggested it." Edward turns his paper so that James can see the equation.

"I think you've got a point out of place here, Edward." Ashley is comparing her sketch of the graph to Edward's graph and points at a portion of his graph.

"She's right." James hands his calculator to Edward. He begins writing: "I did it graphically because after that I can do it analytically."

"Pass your paper to another group now." Karen walks by. "Are you ready?"

The groups exchange papers, and the lesson continues with students sketching the graphs of the equations they've been given on dry-erase boards without using calculators. Near the end of the period, when one student claims her equation has no graph, the others take their calculators and try graphing it. Eventually, Karen asks Harley how he graphed it and displays his graph on the overhead projection calculator.

"Let's try *y* equals . . ." Karen is cut short by the bell. "Try number 4 and 5 for homework."

In 1994-95, there were five mathematics department faculty members at Webb. Two levels of precalculus were offered: Honors Precalculus and College Preparatory Precalculus. Karen Falkenberg discussed how the two courses differed:

> The difference is subtle in some ways and significant in others. One of the obvious differences you'd see if you sat in on a lecture-type class is that for my Honors kids, I give them maybe one day of lecture and then give them an opportunity to work together . . . In my CP [College Preparatory] classes, there's a lot more hand holding, and there's a lot more eyeball rolling on their part. You know, when you have to do something fairly technical, they just kind of sigh and then they grouse a bit, that it's a little bit harder . . . This book is real slim on multiple examples, and so for my CP classes, I've really got to say, "Okay, let's do all of these. And let's talk about why each one of these is in here."

In the early 1990s, there were usually about four precalculus sections (two of them honors) each year, and all four sections used *Contemporary Precalculus Through Applications* as the primary textbook. Grier Novinger, the department head then and the first teacher at Webb to attend NCSSM workshops and use the book, explained that Webb chose *Contemporary Precalculus Through Applications* because "we have committed to teaching math as applications rather than skills. This was the first text, other than Foerster's [*Precalculus With Trigonometry*], that had this orientation."

Grier had heard about NCSSM conferences from colleagues who went to the Exeter summer program. She had attended an NSF geometry workshop in Arizona and was interested in innovating within her classroom. The changing roles of teachers and students in the NCSSM approach appealed most to her. She commented on the first NCSSM summer workshop she attended: "For me, it was somebody giving me permission—being with other teachers and learning from other teachers. What I meant by that is giving me permission to think that I could do things differently. I just hadn't had that opportunity. I had felt stagnant."

After her first summer workshop at NCSSM in 1989, when she had been introduced to the *Contemporary Precalculus Through Applications* materials in draft form, she could not integrate them into her courses because the existing curriculum was too crowded. A year later, she was invited to participate in summer workshops as part of a grant NCSSM was awarded to develop their materials further. At that time, Webb committed to the purchase of a computer and adoption of that textbook for her precalculus class so that she could accept the invitation.

During the early 1990s, Karen and Grier taught the precalculus courses at Webb. Grier had started teaching at the school in the late 1970s, and Karen began in 1992. Grier was the only teacher using *Contemporary Precalculus Through Applications* until Karen came. Karen had previously been a chemical engineer, had taught mathematics and science at a community college, and had taught science at a private school. That same year, another new faculty member arrived, Beverly Johnson, who had taught computer courses at the South Carolina School for Science and Mathematics. The combination of Grier's experience with the *Contemporary Precalculus Through Applications* approach, Karen's knowledge of science, and Beverly's computer expertise set the stage for an exciting period of growth in the department. Subsequently, the other teachers in the mathematics department attended winter conferences at NCSSM. Three Webb teachers attended NCSSM summer workshops.

The school had a computer lab that was used regularly by the precalculus classes. Classes were small, and the students typically worked cooperatively in groups of two or three. The students made extensive use of a commercial spreadsheet program and TI-81 and TI-82 calculators. In Honors Precalculus, portfolios were part of the teachers' year-end assessment. The students selected eight entries based on any work they did for the course and representing high points in their progress through it. These entries, accompanied by an explanation of the rationale for selecting the entry, were submitted to the teacher as a bound journal. The portfolio assessment, first tried in a limited way, became an important part of assessment in the course.

After two years of teaching Contemporary Precalculus together, Grier and Karen made some adaptations so that the course would fit their students' needs better. They decided to rearrange the units on data analysis and trigonometry; take matrix applications out of precalculus; and add work on identities, sequences, and series. They broke up the data analysis section and integrated it throughout the year because they felt that students tired of doing so many data analysis problems at one time. Work on proving identities was supplemented with materials from other textbooks. Karen gave two reasons for the emphasis on proving identities: practice in doing proofs and enabling students to learn identities better. Matrix applications were moved to the algebra 2 course section on matrices to make room in the precalculus course for sequences and series. Trigonometry was moved to the beginning of the year. Grier explained their reasoning:

> Students are fresher [in the fall]. Trig is probably the hardest part of precalculus, and we're trying to get them when they're fresh. The seniors sometimes call it a day in February, and they're not going to put in the effort it takes to learn trig. By [February], we've just started trigonometry, so it feels like pulling teeth to get them to do anything. That was the reason behind [the change], and the other reason is the honors kids. So many of them are so bright that you start in on analyzing functions and they've already done that . . . They think, "Oh, ho hum. We've already done this." That was the second reason—to get their attention from the beginning of the course.

Asked about their precalculus course, most Webb students commented on the textbook instead, perhaps because of their difficulty in separating the course from the textbook or the teacher. Many students mentioned problems in reading the textbook and the shortage of worked examples. One student commented on her difficulties in using the textbook: "If we don't understand something in class, the book's not equipped—You can't go home and look at the book and try to get it. If you don't understand something in class, then you're not going to be able to do it that night for homework just by looking at the back of the book for [answers]."

Another Webb student noted: "The thing about algebra is that they give you, like, an example for every type of so many types of problems. Whereas this book just kinda, if you don't get that basic method, then [you're stuck] . . ." Grier explained her students' complaints about the lack of examples:

> That's what really bothers them. It's "where are the example problems?" and there are plenty of example problems. It's just that there isn't an example modeling every type of problem. [The examples] show them how to think through these problems, and then the students are left to figure out how to think through the other ones.

One student commented on what she felt was different in precalculus compared to other mathematics courses: "In other courses you just, like, summarize this formula, work out the problems, and substitute in numbers and stuff like that. But this wasn't like that. You sort of reason it out. It was more explore and more peer work rather than individual work."

The mathematics teachers at the Webb School indicated that they would strongly consider other precalculus texts when they next adopted textbooks. Their reasons included students' problems in reading the *Contemporary Precalculus Through Applications* text and the desire to adopt a precalculus textbook that was part of a series of texts. Webb had been one of the schools that field tested the initial drafts of NCSSM's calculus curriculum too. After that year, the Webb teachers decided to adopt a more traditional calculus text, in large part because a different teacher was to teach calculus the next year.[23]

Precalculus at the Westminster Schools

The students in Hulan Webb's honors precalculus class have their graphing calculators out on their desks and are attempting to graph two functions: $y1 = \ln(1 + x)$ and $y2 = x$. Hulan is at the overhead projector, using the calculator display. "We're talking about rates, so your window should be what? For x?"

"Zero to one," answers Simon.

When the students and Hulan have graphed the two functions, they observe that the functions are approximately equal in the range $0 < x < 1$. The students have been using logarithmic functions to estimate the time required for money deposited in a savings account to double under various compounding frequencies and interest rates.

Hulan points out how the students can use the linear function to estimate the logarithmic one. "This problem turned out to be interesting because it explains the old rule: Divide the interest rate into 72 to get the number of years you'll need to double your principal."

He explains that next they will be considering mortgages and annuities. The previous night, he says, he had been thinking about the Georgia lottery. The class discusses the prizes and the chances of winning.

Hulan says, "If we win $6 million, we'll get 20 equal payments of $300,000 each. How much will it cost them if they pay you one payment each year?"

A lively discussion ensues. Sarah observes that the lottery commission would need to invest some money so that they could have the $5,700,000 needed to make 19 additional payments. "We need an interest rate," she says. "What about 6 percent?"

"Yeah, it'll be like a de-annuity." Matt enters some numbers into his calculator.

"Or an anti-annuity," says Hulan. "At the 19th payment, I'd like to have $0 left. Let's pick an amount. I don't want a formula."

---

[23]They selected the Finney, Thomas, Demana, and Waits book, *Calculus: Graphical, Numerical, Algebraic*, published by Addison-Wesley in 1994.

At Esther's suggestion, that they start with $5 million, Hulan asks how much they would have after the first year. "5,000,000(1.06) - 300,000," replies Sarah.

"Isn't your calculator suited to do repeated—?"

"Let's plant the seed," says Kevin.

The students find that the guess of $5 million is too high. They try $4 million—also too high. They have almost $2 million left after 19 payments. Hulan gives them a few minutes to try other values for the seed money. "If any of you get a letter published in the Atlanta Journal and Constitution, I'll give you an extra 100 quiz grade. Who's got the seed?"

The students' estimates are still too high. Hulan reminds them of the recursive formula they used in algebra 2 and asks what would happen if they worked backwards from zero. Using the overhead projector, he finds $3,347,434.95 after 19 iterations. He asks the students to try working it forward to see if they get zero. He gets -1.1×10^{-17}, a number very close to zero, on the overhead projection calculator.

"There's a formula in the book," says Steve.

"Yes, but 10 years from now, I can remember this," says Hulan. "Ten years from now, if I just have that formula in the book memorized, I haven't a clue."

The class ends with a discussion of house payments, amortization, and the payments they would make when buying a new stereo for $3,000.

The curriculum and the teaching in the mathematics department at Westminster in the early 1990s reflected various aspects of the NCSSM approach to precalculus. Technology, especially in the form of TI-82 graphing calculators, was an integral part of many lessons. In every classroom, a calculator projection panel for the overhead projector was set up and ready each period. In addition, the students were comfortable operating their calculators, interpreting graphs, and testing conjectures. They expressed interest when the teacher would explain new ways to use calculators or offer tips on using them efficiently. Hulan Webb noted the change the calculator had made in his teaching. In the days before graphing calculators, "you simply could not have worked [the lottery problem] without a formula. I obviously didn't use any formulas. You come up with some idea, and now that you've got the calculator, you can implement that idea."

In the mathematics classrooms at Westminster, traditional student desks had been replaced by small round tables. Teachers proposed problems in precalculus and nonprecalculus classes alike that were not well-defined or traditional. Occasionally, projects were assigned; and, in some instances, the projects were open-ended and required that students gather data. There was an emphasis in the lessons on mathematical reasoning and problem solving, with more than one method of solving a problem often discussed. A teacher would sometimes say he or she wanted a method rather than a formula because "in the real world" or "10 years from now" being able to figure out how to solve a problem was more important than being able to use a formula. The mathematics teachers at Westminster had changed their program, and they gave much of the credit to the NCSSM course, materials, and approach.

In the late 1970s, when the Westminster Schools had gone from classes segregated by sex to coeducational instruction, the boys' mathematics program ("a high-powered math major program") and the girls' mathematics program ("the finishing school approach to math") were combined. The teachers got together and designed a common mathematics curriculum in which all teachers gave the same assignments and the same tests. At that time, the program was quite traditional and had a rather strong orientation toward the development of manipulative skill.

Landy Godbold, who had joined the Westminster faculty in 1975, attended the first Woodrow Wilson institute on mathematics and met Dan Teague. Landy described the mutual influence these colleagues had on one another:

> Dan Teague and I were both at Woodrow Wilson in the summer of '84 together in statistics. One of the things that the Woodrow Wilson program does is [that it] encourages people to become more active in outreach; that is, doing workshops and things for other people. I did a workshop in Milwaukee the following fall, and Dan was also there doing the same thing . . . He sat down in my session, and one of the things that I did was a lot of stuff with data analysis and things like that. He got interested at that point in trying to actually construct a curriculum . . . I saw him again at the national [NCTM] meeting in Washington, which I guess was maybe in '86, somewhere around there, and the [NCSSM faculty] had begun writing some things.

Landy liked the NCSSM materials so much that he asked Jerry Carnes, the department chairman, to let him pilot test them during the 1986-87 school year in a new course for seniors who had completed the regular precalculus course but had not been recommended for, or were not otherwise planning to take, AP Calculus. Jerry later said that Landy "turned on kids that [first] year who had never been turned on, in their lives, to mathematics." The course was taught for two years directly from NCSSM materials.

Jerry, who had joined the faculty in 1965 and had become the chairman of the mathematics department in 1970, attended the 1987 Woodrow Wilson institute. The topic that summer was mathematical modeling. "In 1987, none of us knew what it was. What do we mean by mathematical modeling?" We thought it meant applications, not real modeling when you start from scratch and build in a process." The institute, as well as Landy's experience with NCSSM materials, convinced Jerry that Westminster's curriculum, which trained students in the skills needed for calculus, "had entirely the wrong focus." He decided that he wanted the mathematics department to change its curriculum "from a list of formulas, a list of procedures, a list of skills that need to be accomplished, to a way of thinking" so that students would be able to solve problems that were unfamiliar to them. The working draft of the NCTM *Standards* appeared in October

1987, just as the Westminster faculty began work on reorienting their mathematics program.

NCSSM materials were used at Westminster in 1987-88 for a year-long series of workshops for the mathematics teachers. Once or twice a month, the school administration released the mathematics faculty from duty at scheduled assemblies to attend in-service workshops. The workshops served, in Jerry's words, "to retrain the faculty to think about and approach mathematics in a different way." The teachers went through most of the material in the course that was new to them. Landy described the in-service work, which he directed:

> Developing models, starting with the context and working toward the mathematics rather than being plunked down in the middle of a well-formed problem. Lots of work with data and computers. "What is this telling me? What should I see? What questions should I ask?" We had support [from the administration], not just permission, and I think there is a difference there that really mattered. I would not have been willing to stand in front of my colleagues and do all those workshops had I not known that somebody else believed that it was worthwhile.

Initially, some of the teachers were apprehensive about dealing with mathematical situations in which there were no right answers. One of them, in fact, told Jerry that the reason she had gone into mathematics was that everything had an answer. It was disturbing for her to find out otherwise. The most difficult task was getting the teachers to change their instructional style. Jerry remembered working almost the entire year on the problem of finding the volume of a box formed by removing squares from the corners of a rectangular sheet and then folding the resulting figure. Together, the teachers considered how the idea might be developed with students in a two-day lesson. Then the teachers went back into their classes and did it in 15 minutes. Rather than having the students generate the problem and a solution, the teachers had essentially shown them how to do it. Eventually, however, several of the teachers began to move further in the direction of letting students do more of the mathematical investigation themselves. In the May 1991 *Teaching Contemporary Secondary Mathematics* NCSSM newsletter, Landy wrote, "To a large extent, what we teach, how we teach, and maybe even why we teach have changed. This is largely attributable to the adoption of the philosophy underlying the CPTA [*Contemporary Precalculus Through Applications*] query program."

In the early 1990s, the school adopted the UCSMP textbook series for all of its secondary mathematics courses because it provided a "complete program" for grades 6 to 12. It included more topics as well as some of the ideas of the Contemporary Precalculus course, albeit in a more "cautious" way. The UCSMP books offered, in Hulan's words, "a more comprehensive way of teaching." *Contemporary Precalculus Through Applications* went from being a primary text to being a resource. During the 1994-95 school year, there were 14 sections of college-preparatory mathematics at Westminster using *Contemporary Precalculus Through Applications* as a supplementary text.

As of 1995, all 12 of the mathematics faculty members at Westminster had attended workshops at NCSSM as well as conferences at the University of Chicago. The school continued to support further education for its faculty, allowing each teacher up to $600 a year to attend conferences and workshops. The teachers were encouraged to make short presentations at departmental faculty meetings to share information from these activities with their colleagues.

In a sense, the Contemporary Precalculus course at Westminster had become a victim of its own success. As the teachers became more familiar with the ideas and approach advocated in the NCSSM course, they began to use the textbook in other ways. Jerry said:

> We started taking issues from [the Contemporary Precalculus] course and putting those in other areas of the curriculum. For instance, data analysis, which we knew nothing about in those days, we now teach as part of what we do all the way through. We now teach mathematical modeling as part of units in what we do in all of our curriculum.

Another example was matrices. Teachers at Westminster found that they could take the module with a computer disk that NCTM had published (NCSSM Department of Mathematics and Computer Science 1988c) and use it in their algebra and geometry courses. Charlotte McGreaham, a teacher at Westminster, observed:

> Instead of saying, "Here's how you multiply matrices," [the materials] start off with a problem. "How would you organize this information?" It leads very naturally into using a matrix . . . [The module] is used in Honors Algebra 2, and when we did it this year, the kids said, "We've already seen that." Apparently, they [had done] it in Honors Geometry with transformations—represented a transformation with a matrix—and the teachers went ahead and talked about those problems.

The *Contemporary Precalculus Through Applications* book gave the Westminster teachers a language with which to discuss functions, as Landy noted:

> One of the ideas that they made formal—I think we had been doing it for awhile, but without the language to go with it—was the idea of the tool kit of functions. We had been teaching transformation for years and years and years, and drawing graphs based on fundamental principles and properties of a few functions and how to transform them rather than having to do the new analysis every time you see a problem. But they formalized that, I think, in a pretty nice way. The terminology that they used certainly changed the way we talk about functions.

Charlotte observed that the idea of the tool kit, along with the transformations, was immediately incorporated into courses ranging from algebra 2 to calculus as soon as NCSSM modules were published by NCTM.

Jerry felt that the Contemporary Precalculus course did not offer sufficient preparation for calculus. Landy, however, saw the problem as being more that the *Contemporary Precalculus Through Applications* textbook was very short. If a teacher could expand on what was hinted at in the book, Landy thought that the course would be adequate. But Landy, too, saw the content of Contemporary Precalculus moving into earlier courses at Westminster:

> [We noticed] that these ideas in the book were not necessarily of a depth or sophistication that it had to be only used for seniors. Good ideas were being pulled out and stuck into algebra 1 or algebra 2 and various places like that. After a couple of years, it got to the point where you couldn't use it as a senior course because a lot of the good stuff had already been done in other courses, which is, sort of, the best way a book can die, I think. It's been a source for other things.

Characterizing how the Westminster students had responded to the changed approach, Landy said:

> Their first reaction [when they do applications] is to say, "Wait a minute, this isn't math." Having to write sentences and explanations is very foreign to them. The precalculus course very definitely was the catalyst in making the change. Had there not been text materials that pointed in that direction, we never would have done it, because a teacher doesn't go in with nothing in her hand and invent all this stuff, except in isolated cases . . . You are not going to get a whole school curriculum to change without some written materials somewhere . . . Even though we're not using the book as a text, we steal problems from it left and right, whole sections, projects—almost all of that book is being taught."

The Contemporary Precalculus course as a second year of precalculus eventually gave way to one-year precalculus courses at three levels: Precalculus and Discrete Mathematics, Precalculus (honors), and Precalculus (high honors). After the initial positive reaction to the elective course, parents had begun to complain because their children were taking a second precalculus course instead of going on to calculus. They thought that calculus would be of more help to their children in getting into college. The students' reaction, however, according to Jerry, had "always been positive. They see the material as being valuable, that they are learning something that is both new and worthwhile."

In 1991, Jerry was called before the school's board of trustees to defend the new mathematics program. They told him that after three years they would look at the students' scores on the Scholastic Assessment Test and on AP Calculus examinations to "see if this program is working." When the board looked at the scores, they were satisfied. Jerry said, "Our scores are better than they have ever been because the kids are thinking."

The faculty members were pleased with the preparation for college they believed their precalculus curriculum gave students. Recalling the reaction of her students to college courses in mathematics, Charlotte said that they were often "amazed at how old-fashioned the presentation is by the professors." She recalled one student telling her how his professor was using the entire chalkboard to solve a problem in inequalities when, the student realized, it could have been handled easily with a technique learned in precalculus class. She said, "We're talking dark ages there. Some of the kids said they were literally showing their professors how to use the graphing calculator and saying, "Hey, we need to be doing this." Landy, too, believed that students considered themselves well-prepared for college mathematics courses because of the approach used in precalculus courses at Westminster:

> Those who have bothered to come back and talk to me after the fact have all had positive things to say about it. It may mean that the ones that didn't have anything positive to say stayed away. But generally the sense was that it was helpful more than it should have been, I guess. I can think in particular of one student who left here and went to Princeton . . . He had not taken AP Calculus and was therefore put into a calculus class his first year. He came back and told of having to tutor all these other people in his class because he understood functions so well that, you know, he knew what was going on. They knew all the formulas but had no clue what was happening. I thought, "That's the best thing in the world a student can tell you."

Landy also saw increased enrollment in calculus courses as another outcome of the changed approach:

> One of the consequences, I guess, here is that the change in the focus has changed the number of people who are taking calculus here. There are people taking calculus now that would not have been in the AP program a few years ago . . . That's a measure of the effectiveness of looking at functions from a different way from early on.

In 1993, Westminster began using materials from the Harvard Calculus Consortium in its calculus courses.[24] The teachers had looked at NCSSM calculus materials and decided that they were not appropriate. They did not like the laboratory-based approach. They argued that because NCSSM was a residential school, the students had access to the computer lab many hours a day, whereas that would be difficult to arrange at Westminster. The Westminster teachers, however, did borrow some ideas from NCSSM calculus materials.

In the 1993-94 school year, Landy taught a course on concepts of calculus that used *Contemporary Precalculus Through Applications* as the main textbook:

> What I did was to take that book as the main thing and use it as a branching-off point for other ideas. [I] actually taught what amounted to a calculus course without the recipes for "here's a formula for the derivative" . . . But it was using data analysis and labs to discover the derivative rules and things like that.

---

[24]This NSF-sponsored project, based at Harvard but involving the faculty of nine institutions, attempted to revitalize calculus by focusing on a limited number of core concepts; showing what calculus can do rather than its pathologies; and emphasizing its graphical, numerical, and analytical aspects throughout. After several years of testing, the project published a commercial version of its materials (Hughes-Hallett et al. 1994).

It was really a lot of fun . . . We included about a month or month and a half of very formal statistics in there, which is not in the North Carolina book, and then pretty much followed the book as far as topic selection goes. We used that as an introduction and then followed it with more lab activities. I felt at the time it was one of the best courses I think I've ever done in terms of what I thought was being accomplished.

Some Westminster teachers appeared to have been relatively uninfluenced by the NCSSM precalculus course and materials. Others, however, felt the influence to have been profound. Certainly, Landy and Jerry could see its effect. Jerry said, "I don't know of any project, any program in this country, that's had the sort of impact that did." Another member of the faculty, Chris Harrow, could see the influence on his teaching: "[*Contemporary Precalculus Through Applications*] has become our way of looking at mathematics. The philosophy that was taken there has become our way . . . Even though we are using the Chicago series in our regular courses, my approach involves a lot of what North Carolina is trying to do."

Precalculus at Eisenhower High School

Rick Jennings's second period Introduction to College Mathematics class has only nine students present today; the other seven are in Seattle for Math Day at the University of Washington. Sergei, who recently arrived from St. Petersburg, Russia, comes in late, goes off to a table at the edge of the room, and works alone all period on some mathematics problems Rick has given him.

"Okay, today I'd like you to work on problems 4, 5, and 6 as a group." Rick starts the class working on the problems, which deal with graphs of exponential functions. The students get into two groups, each at a separate table. They open their *Contemporary Precalculus Through Applications* textbook, with their other book, *For All Practical Purposes*, nearby for possible reference.

The students work largely by themselves, occasionally conferring with a neighbor, while Rick attends to other matters. Janie and Marie are attempting to plot the function $f(t) = 10(1 - e^{-t/4})$ on Marie's TI-82 calculator. The function supposedly shows the number of cakes a bakery worker can decorate per hour after t days on the job. They cannot decide on an appropriate domain for t. Rick comes over. He asks what they have tried and what they think is happening, but does not suggest what they might do. After some discussion, Janie and Marie take their initial domain to be $0 < t < 30$, with t an integer.

Rick sits down to work with Brian, who is working alone trying to determine the peak concentration of aspirin in the bloodstream using a graph he has on his calculator. Brian cannot get the calculator's cursor to do what he wants it to do. Rick asks Brian to explain what he has been doing and what the problem seems to be. He offers a few suggestions and waits while Brian tries various techniques and is satisfied that one works. Rick then asks Brian some questions about other parts of the problem. He spends perhaps 15 minutes with Brian to make sure that Brian sees the point of the problems and how he might attack it. Rick moves on to help Janie, who is having trouble with the last problem. Meanwhile, the other students continue working, asking each other questions and helping each other with their calculators.

With 10 minutes remaining in the period, Rick asks for one best paper as a summary of what the "group," meaning the whole class, has been doing. By common consent, Stephanie is recognized as likely to have the best paper. She offers her paper to Rick, who takes it and continues talking to Sam, who, like Janie, is struggling with the last problem. Most of the other students have finished the three problems and either put their books and papers away or begun working on assignments from other classes. As the period ends, Rick is still working with Sam.

In 1993-94, there were two Honors Precalculus sections and one section of Introduction to College Mathematics (ICM) at Eisenhower, each taught by a different teacher in his or her own classroom. The classroom in which Rick Jennings taught the ICM course contained round tables at which the students could discuss their work. The other two had rectangular tables seating two students each. In all three classrooms, most students worked in pairs or threes most of the time. Students tended to work most of the period without much interference from the teacher. This was in contrast to classes offered by other teachers in the department, in which the teacher usually did most of the talking at the chalkboard or overhead projector, working problems with or without the aid of the students. In the precalculus classrooms, the teacher might interrupt the seatwork to explain a problem or introduce a new topic, but more often than not, the teacher simply roamed the room helping groups of students and monitoring their progress.

Most of the precalculus students used graphing calculators or scientific calculators from time to time in their classwork. One or two in each class might own their own; the others borrowed from the class set on the teacher's desk, returning them at the end of the period. The two class sets of TI-82 calculators that were obtained in the middle of the 1993-94 year, under a grant for work on an applied mathematics course, were seen by the teachers as having made a great difference in the precalculus courses. The calculators were available at the end of each day for any student to check out overnight. A computer laboratory was available for use by the mathematics department on the floor below the precalculus classrooms.

Rick's ICM course was designed for seniors who had completed an advanced algebra course. It was an alternative to Functions, Statistics, Trigonometry, a course that used the UCSMP textbook with the same title; and was less advanced than the Honors Precalculus course, which used *Contemporary Precalculus Through Applications*. Topics such as probability and statistics were given more attention in the ICM course than, say, rational functions. Problems in social choice and decisionmaking were especially prominent. In creating his ICM course, Rick had wanted to replace the traditional mathematical analysis with a senior mathematics course that would better serve those students who were not prepared for or interested in taking calculus. He had learned about the *For All Practical Purposes* project of the Consortium on Mathematics and Its Applications (COMAP) while attending a Woodrow Wilson institute on decisionmaking in the summer of 1988.[25]

The ICM course typically alternated from one textbook to the other. For example, in 1993-94, the course began with data analysis from *Contemporary Precalculus Through Applications*, continued with statistics and probability in *For All Practical Purposes*, returned to functions and their graphs from *Contemporary Precalculus*, moved back to the social choice material in *For All Practical Purposes*, returned to *Contemporary Precalculus* for exponential and logarithmic functions, and continued in a similar fashion through the rest of the year. In contrast, the Honors Precalculus course stayed with the *Contemporary Precalculus* text throughout the year, with supplementary problems drawn from various sources, including NCSSM conferences and newsletters. At the end of each trimester of the ICM course, the students did an independent project. The first trimester, they all worked on the same topic; later, they got some suggested topics to work on in small groups. Each student produced a written paper and presented an oral report. They were evaluated on these as well as on their work in the group. As in all of Rick's classes, students could, for extra credit, keep a daily journal of their reactions to the mathematics they were learning.

The first two courses in the honors mathematics sequence at Eisenhower were ordinarily taken by 9th and 10th graders and dealt with geometry and advanced algebra, respectively. In 1993-94, the Yakima district had three high schools. The *Contemporary Precalculus Through Applications* book was used only sparingly or not at all at the other two schools, and the ICM course did not exist there. The district had adopted the UCSMP series of textbooks, which began with *Transition Mathematics* in grade 7. Although Eisenhower offered several courses for ninth graders, including a Transition Mathematics course, entering freshmen who received either an A or B grade in eighth grade algebra could take the Precalculus (Honors) geometry course. Those who received a B or C could take the Integrated Algebra course, the first course in a so-called integrated sequence for students not able to keep up with the honors sequence.

---

[25]COMAP developed the textbook *For All Practical Purposes* (published by W.H. Freeman in 1988), together with a series of 26 half-hour videotapes, to demonstrate to nonspecialist high school and college students how contemporary mathematics and its applications are used in modern society.

UCSMP books were used as texts in both sequences. Students in the honors sequence could take the third Precalculus (Honors) course in grade 11.[26]

Functions, Statistics, Trigonometry was offered for the first time in 1993-94 as an alternative for students who had completed the Precalculus (Honors) advanced algebra course or the Integrated Advanced Algebra course. The new course was seen as having attracted some 12th graders away from ICM, although the faculty also thought that the senior class that year was perhaps not as talented as previous classes. The sixth book in the UCSMP series, *Precalculus and Discrete Mathematics*, was unlikely to be used at Eisenhower to replace *Contemporary Precalculus Through Applications*, or even as an alternative to it; but some teachers in the department, and many students, obviously preferred the UCSMP approach.

Compared with UCSMP textbooks, *Contemporary Precalculus Through Applications* was seen by teachers and students alike as presenting more of a challenge. Some students in both the ICM course and Honors Precalculus expressed considerable dissatisfaction with the difficulty of *Contemporary Precalculus* and contrasted it unfavorably with the UCSMP books. They said that *Contemporary Precalculus* was "hard to read" and "all math—words and numbers." They did not like to read, do five or six problems, and then read again. Most said they preferred the UCSMP books with their collections of problem sets. They claimed that they had made "better progress" in the UCSMP courses and that the books were "easier to understand." They liked the way the UCSMP books explained how to do problems.

The mathematics department at Eisenhower had 11 faculty members in 1993-94. They elected a chair each year; that year Rick was co-chair of the department along with Collette Heffner. They and Verne Bakker taught Honors Precalculus. Like Rick, Collette had taught the course when the materials were in draft form; Verne was only in his second year of teaching it. Collette and Verne were participants in the Lead Teacher Project at NCSSM, which took them to the school for two-week workshops during the summers of 1993, 1994, and 1995. Both had attended the winter conferences at NCSSM as part of the project. A fourth teacher, Sandy Christie, had been to NCSSM in 1991 for a two-week summer workshop on precalculus, but did not teach the course until the 1994-95 school year.

The 1993-94 school year was Collette's fourth consecutive year of teaching the Honors Precalculus course. Before teaching it for the first time, she had been to North Carolina for one of their two-week precalculus seminars. Her experience there helped her change her way of approaching the material. She especially liked the way NCSSM teachers taught the other teachers "lessons in the man-

---

[26]Only two years of mathematics were required for high school graduation in Yakima in 1993-94, but colleges routinely expected three or four years. Forty percent of the students at Eisenhower planned to enter a four-year college; 90 percent planned some postsecondary education.

ner that they would teach their classes." In her own precalculus classes, she said she liked to give her students more freedom than in her other courses: "I do a lot less talking. I'm kind of there as, I think, the 'clean-up agent.'" She used something of the same approach in her other courses, but said:

> It kind of depends on what classes I'm teaching. I find that some of my classes need more structure at the beginning. My freshman classes—they need more structure. But all of my classes—I have tables in my classroom where there's just two students at a table, although they can opt to work with more than two. So they all know that they can work together and that they should talk to each other.

Collette found that many students had trouble adapting to the different instructional style in the precalculus course:

> When [the students] first come in, in the fall, we start right off with data analysis, and they no longer have their 30 problems a night . . . It bothers them that we don't tell them everything. It does, and their anxiety level is definitely increased because these are students that have been very successful in traditional-type programs. And now they're in an untraditional program . . . It takes them a little while to start getting used to things. And we see some that really start to just go ahead in this as they work more together.

Like Rick and Verne, Collette found considerable support for her changed teaching methods in the responses of students who had graduated and gone off to college. They may not have appreciated the precalculus course at the time, but looking back they could see its value. Collette cited an example:

> I was talking to the mother of one of my students that graduated last year, at Christmas time, and she's going to the U [University of Washington]. And she said, "Oh, you'll be glad to know that Carrie said she's just doing great in her math courses. And she was really pleased because the program just had her totally prepared."

Although Verne was relatively new at teaching the course, he had become acquainted with it early, while the draft units were being tried out by Tom Seidenberg (the former department chair) and Rick: "When they were doing that, I was kind of going through the materials on the side because I had an interest in it, in what was going on there. I hadn't been teaching math that long at that time. So it was really kind of exciting material."

Verne's students, too, thought that *Contemporary Precalculus Through Applications* was difficult to read. He attributed that to what they had become accustomed to doing in earlier mathematics courses and was optimistic about the effects of using the UCSMP textbook series in the courses preceding precalculus: "They're used to not having to read in a mathematics course. And we've hopefully changed some of that by using the University of Chicago materials." Verne had came to Eisenhower after teaching junior high school algebra. He had been to several NCSSM conferences before being selected for the Lead Teacher Project.

In the early 1990s, two calculus courses were being offered at Eisen-hower—usually one section of each. One was a rather traditional AP Calculus course following the AB syllabus. The other, called Honors Calculus With Technology, covered the same syllabus but with the assumption that technology would be used to do much of the algebraic manipulation.[27] In 1993-94, enroll-ment was low in most senior mathematics courses. Only one section of calculus was offered, and textbooks and materials from both courses were used. Some students in that section claimed that the precalculus course had not prepared them adequately for calculus. Teachers and students agreed there was something of a disharmony between precalculus and calculus at Eisenhower, which seemed to stem from differences not only in syllabuses and instructional materials but also in teaching styles.

A new calculus course in which the NCSSM calculus materials were used was tried out by Collette in 1994-95, when Rick was the chair, but it was not taught the following year even though it had seemed to go well and was support-ed by Collette, Rick, and Verne. Sandy, who was the chair for 1995-96, and oth-ers on the faculty did not approve offering NCSSM calculus. As of 1995, it was an open question how many different calculus courses the school could or would continue to be offered. The majority of the department favored abandoning NCSSM calculus for the traditional version so that other members of the depart-ment could teach it. That sentiment raised the prospect that the precalculus course might ultimately regress to a more traditional version.

The decision by the Eisenhower faculty to stay with the traditional calculus was especially disheartening to Rick, who had worked with Tom Seidenberg in the early years to develop new styles of instruction, new materials, and new courses. Tom, who had been department chair at Eisenhower in the 1989-90 school year, left in 1990 for Phillips Exeter Academy. He had attended the 1985 Woodrow Wilson institute where Jo Ann Lutz had been a participant and Dan Teague had been the computer coordinator. Tom, Jo Ann, and two other partici-pants were asked to spend the summer of 1986 as part of the Woodrow Wilson "Traveling Road Show," to take what they had learned the previous summer around to teachers in various parts of the country.[28] While on the road, Jo Ann asked Tom when he was going to teach the Contemporary Precalculus materials, and he agreed to teach them in pilot test form during the 1987-88 school year. He got Rick to help him:

> We decided to do it with the two pre-AP Calculus classes. These kids had a background in Algebra 1 and Geometry, but no Algebra 2. We asked if we could simply teach [Contemporary Precalculus] in lieu of Algebra 2 because our program was Algebra 1, Geometry, Algebra 2, Calculus. There was really

---

[27]The text for the regular AP course was *Elements of Calculus and Analytic Geometry* by Thomas and Finney. The Honors Calculus With Technology course used *Calculus; Volume I* by Dick and Patton, with HP48S calculators and MATHCAD software.

[28]This effort was also known as TORCH, for Teacher Outreach (Wick et al. 1994, p. 231).

no true precalculus. We asked to have the courses offered the same period so we could team teach. So, we put all the kids in one room, and then he and I kind of held each other's hand, because this was new to us, too—the data analysis and modeling. That's basically how it got started. And we had to xerox the pages and hand them out to the kids.

Tom had been interested in changing his teaching practices. He had heard about cooperative learning at a talk Neil Davidson had given the previous year at the Northwest Mathematics Conference:[29]

> I thought: This is interesting; I want to try this . . . So I came back, pushed the desks together in groups of four, and said, "This is what we're going to do." And it was a disaster. The kids—It was just awful." Finally, one day I said, "All right, fine. We're going to move the desks back in rows." But I did decide that really was the way teaching ought to be, that the kids should work at a table like this, and work together on a problem. I could walk around and help. So I asked the principal to actually get some tables. And I felt if I had tables, I couldn't split them up and put them back in rows.

When Tom and Rick began their precalculus course at Eisenhower, everyone was concerned about the transition to calculus. What would happen to these students when they went on to take the AP Calculus course? Tom saw many gaps. The students would not have studied such topics as complex numbers. If the students failed the AP examination, the precalculus course would have been seen as a failure. To keep that from happening, Tom himself taught the AP Calculus course the next year. Because he wanted the precalculus course to be a success, he worked hard at adapting the AP Calculus appropriately:

> I wanted it to succeed, so I approached it that way rather than wanting it to fail, as I think some other people might. It turned out that they were the best results that they'd ever had in AP Calculus. When we talked to the kids about what they felt about the precalculus that had helped them, they just said their understanding of functions was so solid that much of the AP Calculus work early on—finding derivatives—was easy . . . They just couldn't do the algebra involved with the functions.

At Exeter, *Contemporary Precalculus Through Applications* was used only in a course for students who had already graduated from high school, having taken three to four years of high school mathematics, and had come to the school to improve their preparation for college. Instead of repeating precalculus, they could encounter material they had not seen before.

---

[29]Davidson's talk was based on cooperative learning ideas documented in his handbook (Davidson 1990).

In 1995, Tom and his colleagues at Exeter were developing course materials that were influenced by his experience with the NCSSM curriculum.[30] He characterized those materials in a way that suggested not only how his ideas had been influenced but also how they might have subsequently evolved:

> The stuff we're writing, there is no text. It's problems . . . [In the materials for the sophomore-level course] there are 95 or 96 pages of problems. There are 10 to 15 problems on a page, and they're all complete sentences. There's no "Solve the following" and A, B, C, D, or something. It's very frustrating for the kids at first. And the definitions are made within problems. So I tell the kids if there's an italicized word, it's probably a word that we're defining. And then there'll be a problem with it. "Now that you know what this means, do this problem."

> I'm convinced that that's the way it ought to be. I'm convinced that the kids ought to create their own textbook, write their own notes, do their own color highlighting, and draw their own pictures. [In our materials] there are no or very few pictures. Every now and then there'll be a diagram, but it's very bleak looking. But I like it. I think it's great because we do a lot of stuff by assigning the kids a problem, and then they come back and we talk about the problem. Instead of saying, "Now, here's how you do this," and then you assign these problems . . . we're just completely opposite that.

Tom's conception of instruction was far removed from the view manifested in the Eisenhower faculty's vote not to go forward with the innovative calculus course. The argument that students should create their own mathematics textbooks might strike many teachers as unworkable and foolish. For the majority of the mathematics teachers at Eisenhower, it would have contradicted their recent choice of a noteworthy textbook series for their college-preparatory sequence. Abandoning the textbook as the basis for instruction is, in some ways, a logical culmination of the approach developed at NCSSM, but it is not a step many U.S. mathematics teachers—including some at NCSSM—have been prepared to take.

Communities of Teachers

Tracey Harting's eighth period precalculus class at NCSSM is meeting this Tuesday at 9:40 in the morning. One of the questions on yesterday's homework assignment asked what various interest rates would yield on $1.00 after one year when compounded annually, quarterly, monthly, daily, hourly, and by the minute. The students have used their graphing calculators to answer the question. Mary, James, Sue, Waldo, and Shareef volunteer to put

---

[30]The Exeter materials were also strongly influenced by various COMAP materials and by the textbook *Advanced Mathematics: Precalculus with Discrete Mathematics and Data Analysis* published by Houghton Mifflin and written by Richard Brown, a member of the Exeter faculty.

their calculations on the board. Mary has also calculated the yield when the given interest rate is compounded by the second.

Tracey begins to lead the class in a discussion of what happens to yield as the compounding period decreases. Mary raises her hand and asks, "Why, when I did it to the second, did the amount go down? See the very last two numbers?"

"Well, I don't think it should," Tracey replies.

Several students have noticed the same phenomenon: "I did it a couple of times." "Mine did too."

Tracey picks up her calculator, "Okay, let me see." She manages to convince the class that they must have made a calculation error, but the problem surfaces again when they check the calculations for a different interest rate and discover that compounding by the minute yields a larger amount than continuous compounding.

Tracey says, "I think our calculators are wrong. The problem is that we have run out of accuracy on our calculators. They are making mistakes because of the high power here."

The students are silent. "I'm not being very convincing in my argument, am I? Let me show you something I did on my spreadsheet."

She turns on the overhead display of her computer spreadsheet calculations and shows students that yield continued to rise when the computations were done on her computer. "The problem we are having is a technology problem," she concludes. "I'll investigate and find out more about exactly what is going on with your calculators."

The class continues and Tracey uses the interest calculations to introduce the concept of limit to the students.

As soon as class is over, Tracey steps across the hallway to talk with Helen Compton about the interest calculations the students had done on their calculators. As it turns out, the same question had come up in Helen's class the day before. Helen had spoken to John Goebel about the errors and found that he had been doing some recent investigations into machine error on his own. Later that morning, Helen reports that she is "hot on the trail" of finding out what was going on with the calculators.

At the regular Tuesday lunch meeting of the precalculus teachers, the discussion turns to the topic of machine error. Helen observes,

> There has been a fun kind of conversation going on—I don't know whether you've hit it in your classes or not—about what the calculator is doing when you invest a dollar at 100 percent, and you go and look at it monthly and daily and hourly. Then you do the second, and all of a sudden it starts crashing? In fact, when you go to look at compounding every second, it gives you a smaller number. I had a student who went away and played with a spreadsheet on this. Then, last night, I sat and tried making my compounding periods move up from a minute by 10s instead of going by 60s. And you can watch it crash. It's just amazing.

NCSSM mathematics teachers regularly engaged in a "fun kind of conversation" that made questions arising in class and in other aspects of their work the theme of collective discourse and inquiry. A matrix problem that one teacher had seen at a conference would be posed in class by the other teachers in varied ways and the results discussed when the course team met over lunch. A new piece of statistical software would circulate around the department and be tried out and evaluated by faculty and students alike. A draft chapter of a text would be hashed over in a faculty meeting with the teachers referring to their marked-up copies, exchanging frank critiques, proposing alternatives, and arriving at consensus. A new faculty member would hear the others talking about showing students "how to straighten out an exponential curve" later in the week and, having no idea what that might mean, would stop by someone's office and ask for an explanation.

The approach to precalculus taken in *Contemporary Precalculus Through Applications* required, not only that teachers and students adopt new views of mathematics and technology, but also that teachers change their views of teaching and learning. It led to new roles for teachers and students. Among the new roles that teachers had to play were learner of mathematics, teacher of mathematics to peers, self-reflective teacher, and pedagogical experimenter. These roles required the teachers to take considerable personal and professional risks. For example, some teachers were reluctant to admit that the content of a high school mathematics course was unfamiliar to them. Taking one's own teaching practices as an object of private or public scrutiny was also a risky endeavor.

Three Levels of Communities

To cope with these new roles and risks, teachers at each site sought support from their peers. The various groups providing support were true *communities of teachers* in the sense that the teachers within them were linked by a common desire to change their teaching and a need for support in doing so. Each community was marked by a spirit of cooperation and fellowship. Communities of teachers came in a variety of sizes and forms, but three distinct levels could be seen. First, at each school in the case study, there was a *local* community of mathematics teachers who were being influenced in one way or another by the NCSSM course, materials, and activities. A second, *NCSSM-based*, community of teachers was larger. It was centered at NCSSM, but included teachers outside the NCSSM Mathematics and Computer Science Department. It might be defined as those teachers who had attended a workshop or conference at NCSSM and who received the departmental newsletter. The community at the third level, the largest, comprised several ill-defined *national* networks of high school teachers who had met and kept in contact through various professional development opportunities beyond those at NCSSM. This third-level community was quite amorphous; it would be impossible to make a list of who belonged and who did not. Figure 1 suggests how communities at these levels can be seen as both separate and interrelated.

Figure 1.   Levels of teacher communities

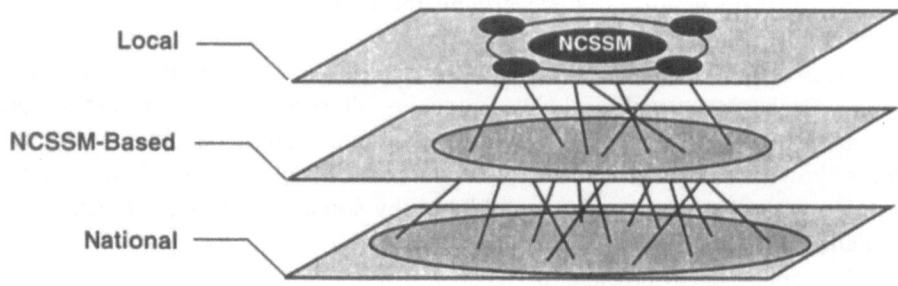

It is not possible to delineate exactly the origins of these communities because there was a great deal of interplay among them. After the local community comprising the NCSSM faculty had appeared and had begun to engage in outreach activities, it became the nucleus for the community of teachers attending NCSSM workshops and conferences. The local communities at the other sites developed as the NCSSM-based community was forming. Several of the networks comprising the national community existed prior to the emergence of the site-based community at NCSSM; others arose later. These networks served to bring NCSSM teachers into contact with other teachers around the country who later became part of the NCSSM-based community.

Evolution of the NCSSM-Based Community

The local community of teachers at NCSSM evolved by design. When Steve Davis built the department, he did so with the intention of creating a cohesive group of people who would support and challenge each other. Steve instituted weekly department meetings, which were never canceled, even when the other teachers could see no need to meet. As Steve's plans for revamping the precalculus course progressed, the meetings became a necessary and vital part of the reform process. The teachers used these meetings to discuss new ideas, challenge old ideas, and modify evolving ideas about precalculus. A significant feature of the meetings was the mathematics teaching and learning that occurred. During the first year of the revision, the teachers took turns teaching precalculus content to each other. These teaching sessions also served to familiarize the teachers with the technology available to them. When NCSSM teachers turned to the development of a reformed curriculum for calculus, they continued the practice of peer teaching during faculty meetings.

As the teachers at NCSSM became involved with redesigning the precalculus curriculum, they began to share their work informally with colleagues at professional meetings and at activities of the national community (e.g., Woodrow Wilson institutes). They became more aware of other reform efforts and grew increasingly convinced that their work might be of value to a larger audience. Other precalculus teachers showed an interest in their work, and they began to distribute draft modules for colleagues to pilot test. The community of teachers that was based at NCSSM evolved from such contacts.

To acquire more technology resources as well as to gain more time to work on the precalculus curriculum, NCSSM teachers continued to explore possibilities for getting grant money. Beginning in the summer of 1981, following the first year of the school's existence, the department had held two-week summer workshops for teachers. The workshops for the first two summers focused on the application of computers to science and mathematics. In the summer of 1983, NCSSM teachers conducted workshops at sites across North Carolina, using former participants as instructors. Workshops during the summer of 1984 at NCSSM included a two-week workshop in Logo for teachers of grades 7 to 12, as well as two workshops on Logo and manipulative materials for elementary school teachers.[31] A workshop on programming with a precalculus component was held in the summer of 1985, along with several workshops for elementary school teachers. All of these workshops were popular with teachers and drew many applicants. Steve Davis described them: "We did workshops differently than universities by a long shot. [The participants] realized we were teachers . . . We were different. They said they'd never been worked that hard. Our schedule is pretty impressive, and I believe in [the participants] moving around."

The original proposal to the Carnegie Foundation had earmarked funds to conduct workshops for teachers on campus and around North Carolina, but those funds were eliminated from the revised proposal that received funding. When a proposal to NSF's Division of Materials Development and Research in fall 1985 to field test the precalculus course was turned down, Steve visited NSF and found a more receptive audience in the Division of Teacher Enhancement. During the summer of 1986, no workshops were held. The NCSSM faculty was busily writing materials for the precalculus course. Steve sent off proposals to various funding sources, including the Department of Education and NSF, to conduct leadership workshops. He wanted to hold two-week summer workshops on the precalculus materials each year for two years for 44 teachers from the 11 districts participating in the Urban Mathematics Collaborative program. The workshops in the second year would be conducted by alumni of the first-year workshops at NCSSM and would be held in the UMC districts.

---

[31]Logo, a computer language developed by Seymour Papert at the Massachusetts Institute of Technology, enabled young children to program computers and explore ideas in mathematics and science.

Steve recalled how the experience in conducting workshops had led to the Contemporary Precalculus project and to the workshops that were connected with it:

> Instead of proposing [to the Carnegie Foundation] a curriculum project where we would do "original" work, I had proposed a project where we would take the work of others, use our experience as teachers to come up with supplementary material to be included in a course, and train teachers in how to use the material . . . I sensed that funding sources did not believe that high school teachers in North Carolina could do original work. I believed that a practical, implementation-oriented approach stood the best chance of success (and it was our strength). The NSF funding was in response to our experience with workshops, the proposal to Carnegie, and an announcement from NSF listing workshops as being something they might fund. Hence, we wrote a proposal [to NSF] to do workshops.

The NSF proposal was successful. Although the school was told in December 1986 that it could expect funding, the official announcement of the award did not come until late in the school year. NCSSM teachers had to scramble to get participants for the workshop from around the country. Fortunately, NCSSM had been working with the UMC network, which was able to locate applicants quickly. One of the UMCs was the Durham Mathematics Council (DMC); Jo Ann Lutz was its executive director.[32] NCSSM was able to conduct not only NSF-funded workshops in the summers of 1987 and 1988 but also a simultaneous workshop for North Carolina teachers in 1987. DMC, through its six subject matter networks (Algebra II/Precalculus, Geometry, Middle School Mathematics, Algebra I, Calculus, and Basics), brought some additional teachers into the burgeoning NCSSM community. A two-day conference at NCSSM on Mathematics Reform and Teacher Professionalism in June 1988, sponsored by DMC with support from NCSSM and the Education Development Center of Newton, Massachusetts, attracted more than 100 mathematics educators from 19 states. The conference not only informed teachers from DMC about current national proposals for reform in school mathematics but also raised the visibility of NCSSM among mathematics teachers around the country.

In July 1989, Helen Compton and Dot Doyle received funding from NSF for what had originally been termed the Introduction to College Mathematics SHARETEXT Network Project, but then became known as Preparing for the New College Mathematics: An Innovative High School Course. The purpose of the project was, in addition to preparing the precalculus materials for publication, to strengthen and expand what was termed "the existing loose network" of

---

[32]Jo Ann was the executive director from 1985 to 1988, when Helen Compton took over for a year, after which Jo Ann returned as project director. Karen Whitehead, another NCSSM teacher, was executive director from 1991 to 1994. During all that time, DMC was headquartered at NCSSM. Thereafter, the headquarters moved to the North Carolina Museum of Life and Science in Durham.

users of the precalculus materials through conferences, workshops, a newsletter, and an electronic bulletin board.

Project funds helped support two-week summer workshops in 1990 and 1991 at NCSSM on precalculus mathematics. These workshops, along with a similar summer workshop in 1992, were also supported by corporate funds raised by NCSSM and by participant fees. The workshops, which were attended by 40 to 50 participants each, consisted primarily of NCSSM teachers getting the participants to work on mathematics problems in much the same way that they would involve their students in a precalculus class. Time was also allotted for discussions of pedagogical issues associated with teaching precalculus in this innovative manner.

SHARETEXT was to have been the name of a network organized and conducted by NCSSM and operating through an electronic board on Mathlink, an electronic network for mathematics teachers developed by NCTM. The 52 teachers attending the summer 1989 workshop on precalculus at NCSSM were to have been the nucleus of the network. Ultimately, however, the network was set up as an electronic conference, Mathtalk, on the Iris bulletin board (operated by the Minnesota Educational Computing Consortium), with 14 teachers provided with accounts and modems.[33] Problems with hardware and software, as well as in obtaining access time, helped make this aspect of the project the most disappointing in Helen's and Dot's eyes. Their experience with this electronic network, however, was valuable in making their subsequent effort to link teachers electronically in the NCSSM Lead Teacher Project rather more successful, although still not entirely satisfactory from many teachers' perspective.

In 1992, additional funding was received from NSF for a project to disseminate not only the Contemporary Precalculus but also the reformed calculus course that was under development at NCSSM.[34] In the Lead Teacher Project, 25 pairs of experienced teachers from schools around the country were selected to attend summer workshops at NCSSM each year from 1993 to 1995. Each pair was to consist of one teacher who had considerable experience teaching Contemporary Precalculus and who had attended a previous NCSSM workshop, and one teacher from the same school or district who was new to teaching the Contemporary Precalculus course. The workshops were designed to prepare these teams of teachers not only to teach precalculus and ultimately calculus but also to return to their schools and support each other and teach their peers. To carry out this mission, each participating team was asked to design an

---

[33]These teachers included Landy Godbold at the Westminster Schools, Rick Jennings at Eisenhower High School, and Grier Novinger at the Webb School.

[34]Work on a revised calculus course had begun in 1990 in conjunction with Duke University under NSF support. The NCSSM faculty, however, quickly realized that high school students seeing calculus for the first time would need a rather different approach than the Duke project was taking, so NCSSM sought separate funding from NSF to develop its own calculus materials.

implementation and dissemination plan that they would use to share what they had learned with their colleagues. In this way, the summer workshops helped to ensure the development of local communities at other sites.

Another activity that supported the NCSSM-based workshop was the winter conference held annually at NCSSM, usually in February. The winter conference, begun as a users' conference when the precalculus materials were being developed, was a two-day meeting designed to give teachers who were using *Contemporary Precalculus Through Applications* a mid-year boost in their teaching. The sessions at the conference were presented almost exclusively by teachers who were using the *Contemporary Precalculus Through Applications* materials. Typical sessions involved sharing problems and data sets that teachers found to be particularly fruitful and exploring the capabilities of new technologies such as graphing calculators and calculator-based laboratory instruments. Teachers from several schools commented that the winter conference was the most professionally rewarding meeting they attended because there was never a bad session. Almost without exception, the participants found the conference to be intellectually stimulating and directly relevant to their teaching.

In addition to providing summer workshops and a winter conference, the NCSSM department also published a quarterly newsletter, *Teaching Contemporary Secondary Mathematics*. Begun in 1990, the newsletter contained information about upcoming events at NCSSM, new mathematics problems, and discussions of pedagogical issues. The newsletter was produced by NCSSM, but teachers outside of NCSSM were frequent contributors.

Teachers at the other four sites in the case study had various levels of connection with the NCSSM-based community. Many teachers initially made contact with NCSSM teachers at events associated with the national community of teachers. Through connections made at these events, some of the teachers gained access to the *Contemporary Precalculus Through Applications* materials when they were in draft form. The more NCSSM teachers interacted with colleagues in the national community and shared their work on precalculus, the more they could see that the ideas they were developing were comparable to ideas being put forth by others. This realization gave them confidence in the quality of their ideas and materials and a conviction to push ahead with their project and share it. NCSSM teachers began to provide the colleagues they had met with copies of the materials and asked them to help pilot test and provide feedback. Teachers at Eisenhower, Webb, and Westminster piloted draft materials. Teachers at Woodward obtained copies of the draft materials but were unable to pilot them because they did not coincide with the existing course structure. After pilot testing materials, teachers from Eisenhower and Westminster were invited to make presentations at summer workshops held at NCSSM, and other teachers from Eisenhower and Webb attended summer workshops.

Evolution of Local Communities

As the teachers from each of the four sites other than NCSSM became more familiar with the NCSSM precalculus materials, a local community of teachers began to evolve at each site. There was considerable variation in the nature of those communities. No other site had the level of interaction or structure that existed in the community at NCSSM. At Westminster, the in-service activities with Contemporary Precalculus mathematics helped the merged faculties of the boys' and girls' schools develop a greater cohesiveness around and satisfaction with their mathematics program. Subsequent experience by the faculty in using units and individual problems from *Contemporary Precalculus Through Applications*, together with NCSSM modules and software distributed by NCTM, appeared to have strengthened the community. Although only a handful of the Westminster teachers appeared to make heavy use of the ideas and approach promoted at NCSSM, the program did seem to have been influenced almost as much by the Contemporary Precalculus course as by the other reform activities in which the Westminster teachers, particularly Landy Godbold, were involved.

The community of teachers at the Webb School evolved over a period of several years from just one teacher to the entire department. In 1989, Grier Novinger attended a summer workshop at NCSSM and brought home some pilot materials for the book. She had not previously taught precalculus but was given a section to teach in which she pilot tested the materials. For several years, Grier was the only teacher at Webb using these materials or interested in the NCSSM approach. In 1992, Karen Falkenberg and Beverly Johnson joined Webb's mathematics department. Grier commented on the change their arrival made: "Until the department changed [through the hiring of new faculty], I didn't have a whole lot of support. I didn't have people saying I couldn't do it, but—To me it's so much easier when you have someone else to interact with, to bounce ideas off of. I didn't have that before."

The three teachers' areas of expertise complemented each other well, and they worked together to further implement computers and real-world applications into both Honors and College Preparatory Precalculus. The 1993-94 school year was the first time that *Contemporary Precalculus Through Applications* was used for both precalculus courses. That year, all sections of precalculus were taught by either Grier or Karen, and they worked closely together throughout the year to plan and teach these courses. Grier, who became department chair in 1994-95, encouraged the other teachers to attend NCSSM conferences and, as of 1995, all of Webb's mathematics teachers had attended at least one NCSSM winter conference.

The first community of teachers of precalculus at Eisenhower High School had consisted of Tom Seidenberg and Rick Jennings. They embarked on the new course more or less alone, stimulated by not only the NCSSM materials but other developments in mathematics education, including proposals for the greater use of small group instruction and the activities of groups such as a network of mathematics teachers who had attended Woodrow Wilson summer institutes. With the approval of all but one of the department faculty members, Tom and Rick tried out the precalculus materials in 1987-88 when they were still under development and continued to use the available materials thereafter. At the NCSSM winter conference in 1988, Tom and Rick gave a talk they entitled "Introduction to College Mathematics: If We Can Do It, Anybody Can," which referred to their experiences as public high school teachers attempting to implement an innovative course. Both spoke at subsequent winter conferences, and Rick later served on the advisory board of NCSSM's NSF-supported Lead Teacher Project.

As Tom and Rick developed more experience with the course, first Collette Heffner and then Verne Bakker joined the fledgling local community. The teachers at Eisenhower found that it was invaluable to have a colleague who was also teaching the course. As Collette observed, "The first year that we taught [the course] here, Tom Seidenberg and Rick team taught. And I think that that's a nice approach, because it gives you someone else to work with." She had looked forward to the possibility of team teaching the NCSSM calculus materials with Verne, with whom she was paired in the NCSSM Lead Teacher Project, but ended up teaching it by herself in 1994-95. Interviewed about his experience at Eisenhower, Tom twice referred to the "hand holding" that he and Rick had done as they worked with new technology and struggled to understand the new approach to functions and data analysis.

By 1995, all of the mathematics faculty from 1987-88, except for Rick and one other person, had left and been replaced by new teachers. Collette, Rick, and Verne, the teachers of precalculus, seemed to have a reasonably consistent view of how the courses should be taught. The rest of the faculty, however, did not seem to share that view. The faculty appeared rather sharply divided between those who believed strongly in the *Contemporary Precalculus Through Applications* approach to precalculus and those who did not. The local community at Eisenhower never extended to the entire department and, as of 1995, seemed in danger of disintegrating.

Before Paul Myers became the department chair at Woodward Academy, the usual practice was for all sections of a particular course such as precalculus to be taught by as few teachers as possible. Paul made a concerted effort to change this tradition and involve as many different teachers in a particular course as possible. In 1992-93 and 1993-94, Paul taught all three sections of the Data, Models, and Predictions (DMP) course himself because the text was new, and he wanted to get a sense of how the course might be organized before asking other teachers to teach it. He had a long-standing interest in modeling and the use of technology, so he was anxious to see how these ideas might play out in a course. Because he was the only one teaching the course, he was in many ways a community of one. However, there were other teachers in the school with interests in these areas with whom he could discuss particular ideas, questions, or problems.

In 1994-95, Paul decided to involve two other teachers in teaching DMP. Paul, Jeff Floyd, and Mike Wylder each taught one section of the course. Three sections of the course met at different times, but it was not possible for the teachers to visit each other's classrooms except on rare occasions. They met often, but informally, in the halls or during planning periods to discuss certain topics or approaches they might take. Much of the conversation took place between Paul and Mike, because Paul taught DMP first period and Mike taught it second period. Mike would often stop by Paul's room between those periods to see what topics Paul had covered that day and what problems or questions had come up in the discussion. Mike was a relatively inexperienced teacher, so he tended to seek frequent input and support from both Paul and Jeff.

Although there was no mandate that the three teachers had to cover the same material in the same order, they tended to try to stay on similar schedules of content coverage, and they made an effort to use similar forms of assessment. Paul generally took the lead in suggesting pacing and sequencing of content because he had taught the course before. The three teachers shared responsibility for constructing assessments, writing programs for the Texas Instruments calculators, and exploring technology.

An activity of a community nature that occurred at Woodward, but did not necessarily revolve around the DMP course, was a series of technology workshops. In 1994-95, Paul offered the faculty an opportunity to come in before school one day a week for a quarter to become familiar with the computer software available in their lab. The sessions covered such topics as using word processing packages to write tests, using electronic mail, and using specific mathematical programs in instruction. The workshops were attended by a majority of the faculty members.

Although we have described the local communities at the schools we studied as centered on the Contemporary Precalculus course, in most cases the course was just a stimulus for a like-minded group of mathematics teachers who might already have been so inclined to work together. It provided an inducement to collaborate and gave a focus to their work.

Teachers' Connections to the National Community

Teachers at each school we visited were involved in the activities of the larger, more amorphous, national community. This community was a composite—a collection of interlocking clusters of teachers. A key nucleus emerged from those teachers who had attended the Woodrow Wilson Summer Institutes for Teachers at Princeton University. Through publications, small grants, and workshops during the year, this program ultimately reached thousands of teachers across the nation. One estimate is that "the number of students affected by the program was well over 2.5 million in 1991" (Wick et al. 1994, p. 232). Several teachers we spoke to gave credit to the Woodrow Wilson program for setting the climate and nurturing the people who subsequently attempted to bring applied mathematics into the high school curriculum. Henry Pollak's key role as advisor, academic director, and then director or codirector was influential in seeing that topics such as data analysis, modeling, and precalculus mathematics were chosen for the institutes. He encouraged the practice of including high school teachers among the Woodrow Wilson institute faculty and helped forge the connection between the Woodrow Wilson leadership program and NCSSM precalculus reform activities. NCSSM faculty who participated in Woodrow Wilson activities, in addition to Dan Teague, who was codirector much of the time from 1984 to 1993, included Jo Ann Lutz, Gloria Barrett, and Kevin Bartkovich.

Many of the teachers we interviewed cited the Woodrow Wilson program as influencing their teaching. For example, Paul Myers said that his attendance at the 1987 Woodrow Wilson institute "kicked [his thinking about teaching] into high gear." Jerry Carnes said that the Woodrow Wilson institute he attended "totally changed me, totally changed the way I think about mathematics, and in essence, changed our entire curriculum at the school." At every school we visited, except Webb, teachers had attended or taught at Woodrow Wilson institutes. Landy Godbold, for example, attended in 1984, along with Dan Teague, and took over the computing responsibility after Dan moved from computing director to codirector. Rick Jennings was a "Woody" in 1988, when he learned about the Contemporary Precalculus course (then termed Introduction to College Mathematics, but rather different from the subsequent ICM course at Eisenhower). The NCSSM-Woodrow Wilson connection was important in sustaining change. As Henry Pollak put it, "The Woody network helped people not to feel alone even if they were lone voices in their own schools."

A similar cluster of teachers formed around the summer Conferences on Secondary School Mathematics and Technology at the Phillips Exeter Academy. These conferences brought together teachers interested in exploring how they might use technology in teaching mathematics. Many of the teachers attending NCSSM winter conferences reported having attended Exeter conferences as well. The Exeter and NCSSM conferences seemed mutually supportive. Grier Novinger said that several teachers from Webb went to an Exeter conference and that was where they had met some of the NCSSM faculty, who were teaching seminars there. Paul Myers met Kevin Bartkovich at Exeter one summer when Kevin was teaching a seminar in discrete mathematics.

A program conducted by NSF to recognize and bring together the winners in each state each year of the Presidential Award for Excellence in Teaching led to the formation of an association of prominent mathematics teachers, the Council of Presidential Awardees in Mathematics, that overlapped the other clusters in the national community. As one indication of that overlap, Henry Pollak estimated that roughly a fifth of the presidential awardees had attended Woodrow Wilson institutes.[35] Awardees among the teachers we interviewed included

---

[35]See also Wick et al. 1994, p. 229.

Helen Compton, Landy Godbold, John Goebel, Rick Jennings, Tom Seidenberg, and Dan Teague. Many of these had been influenced in their instructional approaches by the Contemporary Precalculus and related materials before receiving presidential awards. In 1993, for example, when Rick Jennings was the awardee for the state of Washington, he was cited for his work at Eisenhower on discrete mathematics (West 1993). The Council of Presidential Awardees was not just an honor society; its members were active in working with other teachers. Grier Novinger attended an NSF-sponsored workshop on geometry taught by presidential awardees. She remarked, "I had been to a National Science Foundation summer workshop for 24 geometry teachers in Arizona, and that was really what spurred me on to realize that I needed to make a lot of changes in my class. That was teachers teaching teachers."

The validity of Grier's comment was reflected in NCSSM teachers' work. Involvement in a variety of professional development activities showed the teachers at NCSSM that classroom teachers who were actually "in the trenches" had more credibility with other classroom teachers than university professors or other academic personnel. This realization strengthened their conviction that high school teachers could and should write curriculum materials and invest in the professional development of their peers. It is more than coincidence that the NCSSM, Woodrow Wilson, Exeter, and presidential awardee programs, as the core of the national community of mathematics teachers seeking reform, all involved teachers as leaders and not simply participants.

Additional programs that brought new teachers into the national community were a series of calculator courses and workshops conducted around the country by members of the Ohio State University mathematics department. We spoke with several attendees at NCSSM winter conferences who said they had participated in activities sponsored by Ohio State. Some had been to summer conferences at Phillips Exeter or were involved in projects conducted by COMAP. The high level of involvement of teachers we interviewed and observed in various curriculum reform activities, each with a community of interest around it, made it impossible for us to trace the influences of any particular community.

In addition to participating in activities of the clusters already identified, numerous teachers in our study were also involved in other curriculum reform activities as of the mid-1990s. Several teachers from Woodward, including Paul Myers and Jeff Floyd, were working with teachers at other schools to write a high school mathematics curriculum for honors students. The project was funded by the Keck Foundation. As of 1995, materials had been produced for 9th and 11th grade courses.

Teachers from Westminster and Eisenhower participated in the NSF-funded ARISE project, which wrote innovative thematic units for grades 9 to 11 mathematics. Landy Godbold was a principal investigator for ARISE; Rick Jennings was on the author team; and Dan Teague was a member of the advisory panel.

In 1991, Karen Falkenberg at Webb became involved in Engineering Concepts for the High School Classroom, a summer course for high school science and mathematics teachers. The course, motivated by a concern to improve the preparation of prospective engineers, was developed by Dartmouth College's Thayer School of Engineering. Karen was asked to return in the summer of 1992 as a lead teacher facilitator and was the national director the following two years.

NCSSM teachers were involved in a variety of curriculum reform activities, including The College Board's Pacesetter Mathematics program to create a new 12th grade mathematics course, committees to advise The College Board on AP Calculus and AP Statistics, and a program that took American teachers to Japan to see how teachers there were using technology.

Asked if she saw the same people participating in conferences and workshops sponsored by NCSSM, Woodrow Wilson, Exeter, Ohio State, and so forth, Karen Whitehead, formerly of NCSSM, said, "Absolutely." She went on to suggest what had brought them together:

> There's probably a group maybe this big [makes small circle with fingers] with lots of fringes [moves fingers out]. So, lots of these people also have colleagues or meet other people. But I think it is amazing . . . This network consists of people who really are interested in lifelong learning. They really care about continuing to change, and to do better. There is not a satisfaction in them that, "My classroom is fine. Leave it. I'm going to keep doing what I'm doing."

Although the national community of mathematics teachers working to change secondary mathematics courses and teaching was not well-defined, it seemed to provide a powerful source of energy and support for reform. Several of the teachers we interviewed played prominent roles in national professional associations promoting reform such as NCTM, the Mathematical Association of America, and the Mathematical Sciences Education Board. Their meetings and publications provided opportunities for the teachers to exchange ideas with a wider audience. The teachers in our study seemed more strongly influenced, however, by others in the national community than by the activities of these associations.

In the Study of Exemplary Mathematics Programs (Driscoll 1987), several themes were found to be associated with excellence in precollege mathematics instruction. Effective leadership was one; another was careful decisions about the curriculum and its implementation. "But the most important theme of all involves the teachers and their teaching . . . It is impossible to overstress the value of teacher collegiality to the quality of a mathematics program" (Driscoll 1987, p. 2). Seeking to raise the quality of their programs, the teachers in our study demonstrated clearly that one's colleagues might be down the hall, but they might as easily be in a neighboring state or across the country.

The Perception of Change

"Change is a journey, not a blueprint. (Change is non-linear, loaded with uncertainty, and sometimes perverse.)"—Michael Fullan (1993, p. 24)

"Changing the curriculum also involves changing individuals."—Hilda Taba (1962, p. 455)

As noted earlier, this study, which began as a study of innovation in curriculum, evolved to include as well the study of change in teachers and their teaching. As the study progressed, we came to see change in a different light. There are many metaphors for change: blueprint, journey, chaos, pendulum, entropy, growth. We found that none of these satisfactorily captured the nature of the diverse changes we observed. A blueprint deals with goals, not reality; a journey implies a clear path and some destination; and so on. The changes we identified were neither regular nor chaotic. They did not always seem guided by rational or even predictable processes, but neither were they random. Moreover, the changes we observed were not always the same as those the teachers saw in themselves and their work. Teaching a different mathematics differently did not necessarily feel the way it looked to an observer. Real change, in a curriculum or in a teacher, operates at too many levels to be captured in neat formulations. Real change is messy and problematic. In this chapter, we attempt to capture some of the manifold facets of the changes we and the teachers saw by offering contrasting poles of some dimensions along which change appeared.

Changing Curricula

The "lived" curriculum of the student is not the one to be found in a national syllabus or a textbook (Kilpatrick and Davis 1993). Similarly, the Contemporary Precalculus course developed at NCSSM did not follow a plan laid out in advance. The textbook NCSSM teachers published managed to capture much of what the course once was, but the content of the course continued to evolve everywhere it was taught, especially at NCSSM itself. Begun in large part as a response to the challenge of making mathematics more relevant to students' lives and interests, the course adapted to changing times and circumstances. In particular, developments in technology—computer software, powerful calculators, and easily handled apparatus for measuring physical phenomena—allowed students to have more firsthand experience in gathering and manipulating data. Teachers increasingly saw the value of problems and investigations that might take more than a lesson segment to complete but that seemed to pay off in deeper understanding. When the book was published, the teachers at NCSSM were thrilled. They finally had their own textbook to use after years of teaching the course from photocopied materials. That excitement diminished, however, as they found that the book no longer fit the course they were teaching. Karen Whitehead, a teacher who came to NCSSM after the book was written, could see the change: "I think that if you asked them now, the way they're teaching precalculus at this school is not the way they wrote the book. They're way past that."

Teachers at NCSSM freely spoke of the flux they saw in the course. Many ideas seemed to be a response to activities tried out in class. For example, Gloria Barrett found that the availability of spreadsheets on the computer allowed students to look at exponential functions from a different angle. She began by experimenting with spreadsheets to investigate how various functions could be developed recursively. She found that, by moving the work on iteration before the work on exponential functions, she could not only contrast linear and exponential functions in a new way but also illustrate two forms of representation. The students would ultimately use spreadsheets to come up with a closed-form representation as well as a recursive one:

> What I'm really excited about is [that] before we finish this unit, we'll go back to the problem we could only do recursively, and by data analysis—not by algebra but by data analysis—we will come up with the closed-form function. They'll be able to see—The pieces are going to fit. And we've never taught that before. That's something that developed.

These teachers felt a responsibility to their students to stay abreast of developments in technology and frequently used new features of the technology as springboards to problems and techniques that might eventually become incorporated into the course. Gloria notes: "We really work hard to stay on the cutting edge [of technology] in terms of not holding the kids back because we don't know something." The course was continually being reexamined, not in light of a grand plan but rather in a kind of tinkering to see what worked best.

When work by Claude Lévi-Strauss on primitive thinking, with its image of the thinker as *bricoleur* (tinkerer, putterer), was first published in English in 1966 (Lévi-Strauss 1962; also see Berry and Irvine 1986), various members of the Association for the Teaching of Mathematics in England began to apply the image of bricolage to mathematics teaching. David Wheeler (1980), in particular, spoke of mathematics educators as bricoleurs who took what they could from their observations, bending it to suit their specific needs, while using their experience of the various problems they faced to sharpen their attention to potentially useful material produced by others.

Others have subsequently argued for the image of teaching as bricolage (e.g., Huberman 1993b) to claim that teachers are more independent artisans than members of a collectivity. In this view, the department rather than the school would be the reasonable place to look for a community of artisans to emerge. Our observations suggested that the mathematics department in the schools we visited was indeed the locus of innovation connected to precalculus. The rest of the school appeared largely indifferent to, or unconcerned by, the reforms we were tracking. We found that the communities of artisans concerned with precalculus did not, except at NCSSM, span the entire department; but we also found rather less independence among all these artisans when it came to course content and structure than one might have expected from reading the literature on teacher autonomy. Somehow these bands of mathematics teachers at each school had managed to become interdependent artisans.

Structure Versus Function. The new math reform efforts introduced the term *structure* into the vocabulary of the school mathematics curriculum. It meant that, in addition to the language of sets, which could be used to express mathematical ideas more formally and abstractly, students would also come to understand the various abstract structures of modern mathematics—groups, rings, fields, vector spaces—and their connection to the numbers, geometric figures, and algebraic expressions the students were encountering in their mathematics classes. Mathematical structures, resting on a foundation of axioms, would be the scaffolding on which their learning would be built, definition by definition, theorem by theorem.

The shift to building mathematical learning on a base of applications was a shift away from structure. In the Contemporary Precalculus course, students did not ordinarily prove theorems. Instead, they learned to operate with mathematical ideas to solve realistic problems. The movement was from structure to function: not merely *function* as a mathematical object, although that was a central notion, but *function* in the sense of the use to which mathematics can be put in one's life and work.

John Goebel expressed the shift in terms of practical consequences:

> We also back off from some of the formality of proofs. We do very little of that in [the precalculus course]. I think that's where before—we were teaching to the top 5 percent we thought were going to be math majors, and the rest of them could have cared less about the structure of mathematics. I would much rather have policymakers in the White House who know enough mathematics to make good decisions than knowing there is a structure to mathematics. The mathematics is too important for policymaking and decisionmaking to have it bogged down in the structure.

The movement to what might be termed a more functionalist school of mathematics is related to a movement in mathematics itself that has been developing over the past half-century. Against the formalist portrayal of mathematics as the science of rigorous proof, the view has emerged that mathematics is a conjecture-laden activity of fallible human beings: "In recent years, a reaction against formalism has been growing. In recent mathematical research, there is a turn toward the concrete and the applicable. In texts and treatises, there is more respect for examples, less strictness in formal exposition" (Davis and Hersh 1981).

School and university pedagogy has been slower than mathematics itself to adapt to this reaction. Part of the reason is the usual inertia in the educational system. Teachers are natural conservators of established practices. Another part is a nagging concern mathematics teachers feel that young learners need structure. They need the base of the subject laid down before they can fly. Tom

Seidenberg expressed the struggle many mathematics teachers were having with Contemporary Precalculus and other innovative courses as they attempted to balance the push toward an applications-driven curriculum against the pull of more traditional programs:

> There's got to be some happy medium. I'm concerned that if there's *no* structure, you know, if it's all problem solving—I'm not sure that's good. I mean, I just have this funny feeling that there has to be some basis, some certain level of mathematical ability that the kids need to have. You know, when have they learned mathematics? And what is it that they've learned, and that we're going to now call mathematics, or precalculus?

The problem of defining their subject was a serious one for these mathematics teachers. Some had reconciled themselves to omitting certain standard topics from the curriculum in the belief that the new content was a well-justified replacement. Others felt so strongly about the standard topics that they included them in their course anyway.

Like several other teachers using the *Contemporary Precalculus Through Applications* textbook, Verne Bakker said he did not do anything with proof in the course he taught. Asked if that bothered him, he replied simply, "No." His colleague Rick Jennings felt the same way, arguing that there were enough other topics in the course that dealt with "the thought analysis that goes into a proof." In contrast, Karen Falkenberg said that she and Grier Novinger used a second book to provide students with the practice in verifying trigonometric identities that they thought the students needed in developing logical reasoning ability.

For critics in university mathematics departments, too many important mathematical topics were omitted or mishandled in textbooks such as *Contemporary Precalculus Through Applications* that were attempting to shift from structure to function—or, as mathematicians might say it, from pure to applied mathematics. The so-called Steve Test requiring the dropping from the curriculum of concepts that could not be introduced with an application seemed to rule out much of the mathematics students needed to know.

Richard Askey (1995) of the University of Wisconsin-Madison complained about a first-year algebra textbook that used a graphing calculator to solve quadratic equations, thus missing a chance to teach some important mathematics, namely, the derivation of the quadratic formula. He went on to say:

> Another example of missing material is in the precalculus book written by the staff of the North Carolina School of Science and Mathematics. It does not contain either the binomial theorem or the geometric series, although it has at least four problems which contain formulas obtained by summing a geometric series. These are applied problems, and formulas are used to solve each of them. Instead of deriving the formulas with the geometric series as part of the derivations, the formulas are given, as if they appeared magically out of thin air, and then used to obtain answers to the original problems. The fact that these formu-

las are not derived, either in the text or by the students, bothers me very much. The fact that otherwise reasonable people support this bothers me even more (Askey 1995, p. 81).

Askey argued, furthermore, that complex numbers ought not to be skipped in precalculus mathematics, and that trigonometric identities ought to be used for purposes beyond applications. Other mathematicians might disagree. The precalculus course cannot do everything, they would say, and the benefits of studying the topics in the Contemporary Precalculus course outweighed a treatment of the binomial theorem, geometric series, or complex numbers—each of which might easily appear in preceding or subsequent courses. Some of the teachers we interviewed, however, agreed with Askey and supplemented their courses by adding material of their own or from other textbooks so that "missing" topics might be included. Askey's most telling point seemed to be that certain theorems and derivations were essential to one's mathematical development and ought not to be omitted. In other words, the move toward applications and modeling in the Contemporary Precalculus course had been too extreme.

In a gentler critique of the reform movement in mathematics education, the past president of the Mathematical Association of America, Deborah Tepper Haimo (1995) argued that although it is important that students experiment with mathematical objects and pose conjectures about them, that is not enough. Students need to see the deductive side of mathematics. Not all conjectures are valid. All students need to see the limitations of what experimentation can provide and should see at least an intuitive outline of a proof, if not a rigorous one. The ablest students should be expected to prove conjectures and validate them. Applications are only part of the story, and not necessarily the part that appeals to everyone:

> While the need to make the subject more relevant may be inescapable, it should not be so all encompassing that the essence of mathematics as a major discipline in its own right is totally lost. It is a field with problems that attract and fascinate some with no interest in the applied or the relevant (Haimo 1995, p. 106).

Finding an appropriate balance between structure and function is a perennial problem in mathematics education that has become more acute as applications have become easier to manage through technology. Each teacher of Contemporary Precalculus sought his or her own balance point somewhere along the line; for most of these teachers, that point continued to shift as they and their students pursued an evolving vision of what the course should be.

Coverage Versus Depth. The movement from a structurally oriented toward a more functional precalculus mathematics was accompanied by a change in how the curriculum syllabus was viewed: from less emphasis on an ordered list of topics toward more emphasis on an interlocking set of themes. When the teachers at NCSSM began to take apart the precalculus course and rebuild it on a foundation of applications and modeling, they found themselves abandoning the view that the course had to follow a well-defined sequence. They—like teachers we observed and interviewed at several other sites—felt free to add, drop, or rearrange units, depending on how those units seemed to fit their changing conception of the course.

In the traditional view of the mathematics curriculum, the teacher and the student face a long list of topics or objectives. The goal is to get through them all by the end of the year. For example, the North Carolina Department of Public Instruction presents teachers of first-year algebra with 81 objectives on which students are to be examined at the end of the year (Hancock 1994). With 81 objectives and something short of 180 days of actual instruction available to them, teachers feel compelled to arrange their lessons so that their students are tackling an average of one objective every two days. Such a portrayal of the mathematics curriculum as an ordered sequence of individual topics to be treated one at a time leaves students and teachers alike feeling that there is little freedom for maneuvering or adaptation. On the other hand, it provides organization and can give them a sense of fulfillment when all the topics have been treated.

As another example of the traditional view, consider the syllabus for AP Calculus. It attempts to capture all the important topics included in calculus courses at universities. To do that, it must reflect the contents of major calculus textbooks, which, to be widely adopted and used, cannot stray too far from a norm that has emerged over the decades. Although updated from year to year, the AP syllabus must change slowly because calculus courses and the corpus of available textbooks change slowly. Gradually, a more or less agreed-upon sequence of well-defined topics can be found in AP Calculus classes because those topics are on the AP Calculus syllabus. They are on the syllabus, in turn, because it must reflect a canonical curriculum. Thus, the calculus curriculum becomes ossified.

The view taken in the Contemporary Precalculus course, in contrast, seemed to be that there was no specific body of material that needed to be "covered" in the sense of teaching a set of fundamental concepts and principles. What was needed were activities that would enable students to work intelligently with data and functions to solve problems. John Goebel, for example, spoke about changing from being a teacher who could not see applications for the mathematics he was teaching to being someone who decided about each topic "that if there weren't good applications, then it probably isn't appropriate to be teaching to these kids." A view like that allowed topics from a crowded syllabus to be dropped in favor of fewer themes treated in greater depth. As the *Contemporary Precalculus Through Applications* textbook put it in the last sentence of the preface, "The course has been constructed with the philosophy that the quality of learning is more important than completing a syllabus" (Barrett et al. 1992, p. xi).

We were struck by how often teachers told us about changes they had made in the units they were currently teaching from what they had done previously. They felt comfortable modifying, reordering, or dropping units that did not seem to work as they should and adding others. Collette Heffner spoke of how she and the other teachers had treated the modeling section in different ways; for example, in the 1992-93 school year, they had moved it to the end of the year as a final project for the students rather than treating it in class. Landy Godbold told us that he incorporated a unit on statistics into his version of the course that was not in the *Contemporary Precalculus Through Applications* textbook.

The spirit of experimentation in the construction of the Contemporary Precalculus course meant that every topic was open to question. Nothing appeared sacred apart from a commitment to engage the students in learning and doing significant mathematics. Functions, data analysis, modeling, and matrices seemed to be the key themes for the course, but even they were questioned and transformed. They were not topics to be covered but themes to be explored.

According to Jo Ann Lutz, early in the history of NCSSM, the Department of Mathematics and Computer Science decided not to offer courses, such as differential equations and linear algebra, that "extended into the college program." They decided that three semesters of calculus would be as far as they would go. Instead, they would offer courses that were "less on the direct path to a technical degree," such as discrete mathematics, modeling, number theory, and abstract algebra. That decision took the department out of the game of accelerating students as far as possible toward and into college mathematics—which is probably the most common curriculum strategy used in U.S. programs for students gifted in mathematics. Instead, the faculty was free to offer a greater variety of courses.

At the level of individual courses, however, Jo Ann said: "I do not think we ever decided as a group to cover material more in depth and therefore to omit some topics. I think that is what most of us believe, but that belief has not been formalized. It certainly has happened in precalculus and calculus, but not as part of a conscious curriculum-wide decision."

The decision to opt for depth rather than coverage is one that not all teachers are free to take or would feel happy doing even if they were free. Lists are comfortable things. You can check off topics as you cover them and get a sense of accomplishment. Proposals for innovation can be dismissed by saying, "I can't cover everything in the syllabus as it is." Moreover, as Lew Romagnano observed in his study of an attempt to change ninth grade general mathematics, a movement from topic-driven to problem-driven mathematics instruction means that "when working on problems, several different topics may be salient to different students at different times" (Romagnano 1994, p. 86). Gone is the simplicity and security of tackling one discrete topic at a time. New difficulties confront the teacher who attempts to help a class of students cope simultaneously with diverse topics.

The teachers we observed and interviewed seemed well aware that a thorough treatment of applications and modeling took much, much more class time, per problem, than was typical in school mathematics courses. It also implied that their assessment practices needed to change to reflect these extended problems. Some were not bothered by these demands; others were. Those who seemed to be bothered appeared to cope by moving more quickly through application problems, giving students hints on how to proceed, or modifying assignments.

One of the curious phenomena we saw was that, at some schools, the *Contemporary Precalculus Through Applications* textbook was deemed too thin, literally, for a precalculus course. It had to be supplemented with other materials, including other textbooks. At other schools, however, the teachers found the book more than ample for their courses. Rick Jennings said, "We don't get through the entire book in a year with our precalc kids." Jo Ann Lutz saw the difference as connected with how much teachers could see to do with the units from the NCSSM perspective:

> It takes our teachers all year to complete the book as it is . . . And yet, people who just pick up the book and use it, I think, think it's too short. They get done in less than a year. When I talk to people that have just sort of picked it off the shelf and use it and have never been to any of our programs or anything, they get done fast. They get done in time to put in the extra things that they think that their calculus teacher wants them to teach, like sequences and series, which really aren't in there at all, or complex numbers—those kinds of topics. I'd put those in too. But we can't because we know too much. The teachers here just have too many things they want to say about every topic they come to. So they can't get through fast.

Although there were always topics such as complex numbers or vector analysis that teachers might have wished could be added to the course, those most imbued with the NCSSM spirit seemed to recognize that if data analysis and modeling activities were to be treated in depth, some valuable precalculus topics would always have to be omitted or shortchanged. Speaking to other teachers at an NCSSM winter conference, Helen Compton told of whispering "trigonometry" in her students' ears as they walked out the door at the end of the year; by which she meant that she had not been able to bring closure to the topic the way others might have liked.

To critics such as Askey, this view of the curriculum was a kind of abdication of the high school mathematics teacher's responsibility to prepare students going to college for the study of calculus and more advanced topics. To these teachers, however, the best preparation they could give their students was to enable them to see usefulness in the mathematics they were learning and to be both prepared and disposed to use it. Dan Teague expressed his view as a desire that students reflect on what they knew and not simply remember it. He wanted his students to be able to view the world mathematically even when they did not always possess specific skills. The skills would come if the students were confident in their mastery of mathematics. Skills by themselves, however, did not provide that mastery.

Some teachers, in fact, did not see the fundamental curriculum issue as either structure versus function or as coverage versus depth but as, in Jeff Floyd's words, "How important are mechanical skills?" Jeff said that it did not bother him to see a student who could not, for example, add rational expressions whose denominators must be factored, but he acknowledged that he might feel differently if he were teaching AP Calculus. We talked with other teachers who worried whether the students in their precalculus courses were acquiring the mechanical skills they would need later.[36] There is no necessary conflict between a curriculum that emphasizes depth over coverage and one that emphasizes skills, but many mathematics teachers would prefer to teach skills as part of a logically organized syllabus rather than on an as-needed basis in the context of problems being solved.

We saw most of the precalculus teachers in our study as attempting to find some middle ground between depth and coverage in the course curriculum. Michael Huberman (1993a, p. 227) found the same phenomenon in a study of 160 secondary teachers in Geneva, Switzerland. The teachers were caught between covering the entire syllabus at all costs, curtailing it for the so-called weaker students, or modifying it by dropping some topics and elaborating others. Many of the teachers in Huberman's sample saw external constraints such as examinations, lists of objectives, and levels of student achievement as sharply reducing their freedom to adapt the curriculum to their students and to their own preferences. Much depends, of course, on how one views these constraints.

Constraint Versus Opportunity. Quite a few of the teachers we interviewed and observed obviously felt the pressure of external constraints. The specter of the AP Calculus weighed heavily on them. If they were not themselves teaching AP Calculus, they worried that their colleagues who were teaching the course would find the graduates of their precalculus course ill prepared. Students would sometimes report back to them that they felt poorly prepared for a standard calculus course. The AP Calculus course is a high-status course in most schools; its presence in a school allows the principal to boast that the school has a superior mathematics program even if he or she knows little else about the program. Parents and principals alike want students to graduate with AP credit. Teachers, therefore, find it difficult to argue that alternative versions of precalculus and calculus are superior. When *Contemporary Precalculus Through Applications* was used in a course that was not followed by AP Calculus, teachers felt fewer constraints than when it was. To keep these constraints from building, schools would use the textbook with students who were not destined for calculus, or they would use it in a second precalculus course for students who had, presumably, mastered the standard prerequisites for calculus.

High school precalculus, however defined, does not as yet have a high-stakes examination attached to it. That circumstance was both an advantage and a drawback to the Contemporary Precalculus innovation. The absence of an examination left teachers freer to formulate their own syllabus, but it also meant that the Contemporary Precalculus course was vulnerable to replacement by

---

[36]The question of striking an appropriate balance between skill and understanding in the mathematics curriculum is a persistent one (see, e.g., Brownell 1956).

other courses. Indeed, the teachers at Woodward said that when the AP Statistics course becomes available in 1997, AP Calculus and AP Statistics would likely become full-year courses in their honors track. That would entail reconfiguring their course structure, and *Contemporary Precalculus Through Applications* would probably then be dropped as a primary textbook.

Another constraint felt by many precalculus teachers we spoke to was the textbooks used in other courses. In the schools we visited, there was departmentwide pressure to adopt a series of textbooks for the secondary mathematics program rather than using books from different authors and publishers. The University of Chicago series was popular for this purpose because, while not as radical as the *Contemporary Precalculus Through Applications* textbook was perceived to be, it offered a curriculum that was seen as moving in more or less the same direction. Many teachers and students felt more comfortable with other textbooks and textbook series than with the NCSSM book because the other books followed the typical routine to be found in school mathematics: brief exposition of concepts, procedures, and worked examples, followed by exercises fitting the pattern laid down by the examples. Students had met this routine in their previous mathematics courses, and teachers were accustomed to it as well.

It is impossible simply to drop an innovative course into a school curriculum without causing ripples that extend throughout the program. Whether those ripples produce barriers to be overcome or openings to exploit is largely a matter of how they are perceived. Some of the teachers we interviewed saw constraints that kept them from making the changes in precalculus that they wanted to make; others, in much the same situation, saw opportunities to make changes anyway. The difference seemed to be one of perception. If teachers had adopted an orientation toward teaching mathematics that led toward applications and modeling, they found ways to change. Dan Teague considered it a question of attitude toward one's teaching:

> I think the attitude came before the book. I think writing the book was, in a way, trying to express the attitude. In order to teach this stuff, how would you have to do it? We thought about all kinds of ways to try to say it. I mean, you don't want to go into the course and just teach it the way you taught Dolciani. We would be very disappointed. I think it's much more the attitude than the book. If you're teaching Dolciani with a good approach, I think that's better than teaching our course with a bad approach.

Reforming precalculus in the direction they sought gave these teachers opportunities to grow. They found not only that their students were learning mathematics in a different way but that they themselves were confronted with the challenge of learning a new and different mathematics. Learning mathematics together helped build the communities that were so important for sustaining the approach to teaching mathematics that they had adopted.

Changing Teachers

Alone Versus Together. No teacher working alone would likely have constructed the precalculus course that emerged from the NCSSM Department of Mathematics and Computer Science. Not only was a division of labor required to write and try out so many novel units and problems within a few years, but moral support from colleagues was needed to keep the course from regressing toward more conventional treatments of mathematics. We heard repeatedly from NCSSM teachers how much effort had to be put into developing the course and how much they had relied on one another. At most schools, they would have had to confront barriers that kept them from collaborating. At NCSSM, a climate of collaboration had emerged, driven in part by the school's mission to reach out to other teachers and therefore to work together in pursuing that, and in part by the departmental dynamic brought about by Steve Davis of funding projects through grants that required virtually everyone's participation at some level. Support for these operations in the form of money to buy time (although never enough time) gave these teachers an unusual freedom to innovate. But they could not have done much with the extra time if they had not been prepared to use that freedom. John Goebel said it well:

> I think it's a matter of having colleagues here, all of whom are willing to try something and share their successes and be supportive. I'm not sure that I would have done any of this [curriculum development work] in my old school where we had so little time to discuss anything with each other. In most schools, teachers work in their own little rooms. They do all these wonderful things maybe, but they rarely share them with anybody. The idea of sharing materials is really foreign to most teachers, especially in public schools. There are a lot of reasons for that. I don't really think it's selfishness. It's a matter of never having a common time they can meet. After teaching five or six classes, the last thing they want to do is have a meeting.

The mathematics teachers at NCSSM worked hard to maintain collegiality. None of them had come from a department in which collaboration was the norm. Jo Ann Lutz pointed out that there was nothing particularly special about the teachers when they first came. Although the teachers hired into the department were all considered good teachers, most had been teaching in nearby schools, and none was nationally known at the time. She said, "We grew into what we are. I think that underlying everything we do is the belief that almost anyone can grow into what you want them to be, if you help them, like [Steve Davis] helped us." Keeping that belief alive required that newcomers to the faculty be given special attention and socialized into the group. In the mid-1990s, the department lacked the funds to hire teachers with extensive experience. Jo Ann did not regard that as handicapping the department; on the contrary, it helped keep the team together: "We have to hire less experienced teachers, but that hasn't hurt us at all. We do have to spend a lot of time and tremendous energy bringing people along. But when you hire a hotshot, they come in and do what they want to do. You don't have a program, you have a teacher."

The department's cohesiveness helped it resist outside pressures that a single teacher might not have been able to withstand. Once the Contemporary Precalculus course had become established at NCSSM, students taking AP Calculus there began to question the topics taught and to ask why they could not use their calculators or work on open-ended questions. The teachers saw the point, and AP Calculus was dropped. The teachers then arranged for students to study on their own and take the examination. There was considerable overlap between the evolving Contemporary Calculus course at NCSSM and Calculus AB, but students taking the Calculus BC examination had to work quite hard to prepare for the examination. Nonetheless, the department was able to resist pressure from some of the school's trustees, parents, and administrators who were concerned about the lack of attention in class to the AP Calculus syllabus.

The department, in the words of one member, had "never had a good relationship" with the school director. The first director, Chuck Eilber, had taught mathematics and seemed to understand the direction the teachers were taking. At the beginning, he helped considerably simply by leaving them alone to do what they wanted. He eventually became more conservative about change, however, at least in the eyes of some of the mathematics faculty, and clashed with them over various issues. The next director, John Frederick, learned from experience how the department worked:

> There is a group mentality in the math department that is unusual for math departments: the nature of the individuals, their level of preparation, their complementarity, their way of choosing new staff members, the way as individuals they allow this collaboration to occur and result in a curriculum and in a textbook are unusual for a typical high school.

When he had arrived as director, he was unaware of the department's expectations regarding royalties from the book. The department faculty felt that all the royalties should be returned to the department. The NCSSM Board of Trustees had instituted a new patent and copyright policy in 1989, however, that would have given the department only 60 percent of the royalties. Some department members thought that the director and the board did not sufficiently appreciate the effort and sacrifices they had made in writing the textbook. They thought they had an understanding with the previous director that, because their work had begun before the policy had been approved, it did not apply to them. The department hired a lawyer and went to the board. Ultimately, the director recommended a compromise division of royalties that was accepted by both sides.

Some NCSSM teachers also expressed concerns about how their teaching had been evaluated when administrators first began visiting their classes. One teacher who was teaching by "going with the flow, using discovery, investigating, and those kinds of things" said that administrators were applying traditional standards in their evaluations. The teacher said that the mathematics faculty members had long ago mastered teaching in traditional ways and were discouraged that their efforts to change had not been appreciated. If they had not had one another's support, they might have given up.[37] Similarly, they were enabled to resist complaints from those students who felt they were being used as guinea pigs when the course was in draft form.[38] They could band together to explain their new program to questioning parents.

In contrast with NCSSM, teachers at the other schools we visited often worked alone much of the time. At each of those schools, one teacher had found out about the Contemporary Precalculus course first and then brought it back to the school. Whether other teachers then became involved with the course depended on who else, if anyone, was teaching it and how interested they were. For example, Grier Novinger was alone in teaching the Contemporary Precalculus course at Webb until interested colleagues were hired; and Paul Myers, at Woodward, taught the course himself for years before turning it over to others. Team teaching, which might help teachers get started with an innovative curriculum, was used, at least initially, at Eisenhower but nowhere else. Teachers conferred occasionally, usually between classes or during free periods, but in general they told us that they had little time or opportunity for collaboration. The local communities that developed at these schools were communities of teachers who—despite a shared vision of precalculus—planned, taught, and reflected on their courses more or less in isolation.

The kind and level of collegiality we observed at NCSSM was present at no other school we visited, yet it seemed to be something toward which some other teachers aspired, especially once they had seen it up close. When Karen Whitehead left the NCSSM faculty, she obtained a position teaching mathematics at a similar, but newly established, state residential academy for talented students. She was able to use the *Contemporary Precalculus Through Applications* textbook with a class of students who were not destined for AP Calculus, but she had no success in persuading her colleagues to adopt the book for the regular precalculus course. Among her colleagues, she found only one visiting teacher who taught as she did; the others seemed uninterested in moving away from a rather traditional approach. She had found it easy to influence other teachers to try other ways of teaching mathematics while at NCSSM. In her new school, she was alone, and she found it difficult to make her lone voice heard:

---

[37]The NCSSM administration contended that, far from being an impediment, it had facilitated the department's objectives. As an example, the director noted that he had volunteered, with the department's agreement, to negotiate their precalculus book contract with Janson. Later, he did the same with their calculus book contract. He reported that the department faculty were "very pleased" with the terms of both contracts, having gained all their requests and some concessions they had not expected.

[38]One NCSSM teacher claimed that the "guinea pig" phenomenon characterized the tryout of the new calculus course more than the new precalculus course.

Most people would say I'm in this tremendous school. I'm in a school just like this. But I am an island. The words that I used to be able to say, they just don't—When they come out, they don't sound as strong. They don't convince people. The [others at the school] don't ask me questions because they know I'm going to go on a tirade of what should be done and is not being done.

Neither their preparation to teach nor the circumstances of their professional lives ordinarily prepare teachers to work collaboratively to develop innovative curriculum materials and approaches to their teaching. Mathematics teachers have no tradition of professional development that leads them to teach each other mathematics, to learn together, and to work together to change their teaching practices. They tend to conduct their professional lives as individuals rather than in groups. The flip side of the autonomy teachers may feel when they close the classroom door is the isolation they may also feel. Individualism, autonomy, and isolation are recurrent themes in the literature on teaching (Hargreaves 1993, Huberman 1993b, and Little 1990). Efforts to encourage collaborative work among teachers appear to go against the grain of social organization and workplace conditions in most schools. The teachers in the NCSSM Department of Mathematics and Computer Science managed to create a culture in which collaboration was a way of life. Whether that culture could be maintained was a question they faced every day.

Up There Versus Out There. Some efforts to change curricula and teaching are mandated. They come from above and are meant to be carried out by teachers in the trenches below. The innovative precalculus course we studied was an attempt to shift the locus of change from "up there" to "out there," where the teachers are. The method of innovation, like the method of teaching, moved away from telling toward inviting, suggesting, and supporting. A current cliché to express proposed reforms in mathematics teaching says that teachers should stop being "the sage on the stage" and become "the guide at the side." NCSSM teachers had not only abandoned lecturing to their students from the front of the classroom, they also worked with other teachers in ways that allowed the teachers to explore mathematics on their own. They worked beside these teachers rather than from a position of authority above them.

As we observed NCSSM teachers in action at conferences and workshops, we were struck by their great tolerance for teachers who were having trouble with the mathematics, the technology, or NCSSM teachers' ideas about teaching. They wanted to show these teachers what they were doing and to discuss the instructional problems they were struggling with themselves, but they were not insistent. If a teacher found something useful, fine; if not, fine. Either way, it was the teacher's own decision. No one was pushed beyond what was comfortable.

American education is beset by people with something to sell. Teachers' mailboxes overflow with catalogs, brochures, and other come-ons. Teachers attending conventions find not only the exhibition space occupied by purveyors of instructional materials but also the meeting sessions filled with speakers touting their latest patented nostrum. NCSSM teachers, too, had something to sell—a course, a book, an approach to mathematics and to teaching—but in their work with other teachers, they acted like colleagues and not merchants. They seemed unconcerned about converting others to a cause or making a sale. They were neither evangelists nor hucksters. Rather, they wanted to exchange ideas.

The communities of teachers with whom NCSSM teachers worked had not been formed under coercion. The communities, local to national, comprised self-selected professionals who were seeking change. These teachers were less impressed by advanced degrees than by firsthand knowledge and experience. They returned to the conferences and workshops because they found them valuable and liked the atmosphere. Collette Heffner described her experience one summer: "One of the things that I really liked about the workshops that we had when I was back there [at NCSSM] for the two weeks is that they didn't just stand in front of us and tell us things. They actually taught us lessons in the manner that they would teach their classes."

NCSSM teachers earned their colleagues' respect because of what they had to offer and how they offered it, not because of their names. Although several members of the NCSSM mathematics and computer science faculty, as well as others in the community centered at NCSSM, had become nationally known figures in mathematics education, the atmosphere at conferences and workshops was marked by a high degree of social equality. One-upmanship was rare.

The same lack of pretension that was an asset in dealing with their colleagues, however, had been a handicap when NCSSM teachers initially sought to get their ideas heard outside the communities in which they worked. Until their work was known, they found their proposals discounted when they spoke to foundation officials. At meetings where they were outnumbered by university professors, they were ignored. As mere high school teachers, they lacked status. Although some of these teachers possessed doctorates, they did not ordinarily advertise that fact. They did find, however, that in certain situations, when others learned about their degrees, they began to pay more attention. Thus, they got respect from outsiders through their credentials and reputation, whereas within the NCSSM community, the respect they received was based on their accomplishments.

The status differential between high school teachers and university professors operated in apparently conflicting ways. On the one hand, it was a hurdle for these teachers to surmount so they could get support for the work they wanted to do. Recall that Steve Davis wanted the materials they produced to have a professional sheen and not look as though high school teachers had produced them. On the other hand, other teachers gravitated to NCSSM conferences and workshops precisely because they were conducted by high school teachers and not professors. By working "out there" with other teachers at their own campus, around the state, and across the nation, NCSSM faculty were able to reach their colleagues in ways denied to educators who did not face teenagers every day in the mathematics classroom. We had halfway expected to find that many conference participants would dismiss much of what the NCSSM teachers said or offered because their students were so special or because they were teaching in such a privileged school. The only teachers who expressed such views, however, were teachers who had decided, for whatever reason, not to adopt the approach to mathematics teaching NCSSM teachers were practicing. The rest saw no reason why that approach might not work for them.

The phrase "out there" as a dimension of change can also be applied to how NCSSM teachers located themselves and their colleagues. They considered themselves very far "out there" in their teaching, out on the edge of experimentation; and they recognized that other teachers might have just begun, or begun to consider, changing their practice. For NCSSM teachers, the issue was not how far "out there" you were, it was whether you were reflecting on your own teaching and trying to change it. They knew the exploration process was a personal one, and they attempted not to interfere, but to help.

Low Risk Versus High Risk. Teaching in a less traditional mode, in which teachers move away from prescribed lesson plans and full-frontal instruction, is one of the most difficult aspects of pedagogy for teachers to master during their career (Huberman 1993a, pp. 229-31). We saw every teacher in our study as being somewhere on a personal quest with respect to less traditional teaching. Some, the diehards, found good reasons not to move very far in the directions implied by the innovative precalculus curriculum. They might have enjoyed whatever contact they had with the NCSSM faculty, course, or materials; but only bits and pieces of the curriculum could be seen in their classes and almost nothing of the approach to teaching. Others, the skeptics, had made more changes but remained doubtful as to the wisdom of implementing "an NCSSM approach." Still others, the true believers, were doing everything they could to reproduce in their precalculus course something of what they had seen it might become.

We saw the teachers' willingness to take risks in changing their mathematics teaching as a key factor in determining their status as diehard, skeptic, or true believer. How willing they were to change, in turn, seemed to depend at least in part on contact with other teachers, particularly those at NCSSM, who were attempting less traditional teaching. It seemed no accident that at none of the schools we contacted had mathematics teachers simply picked the *Contemporary Precalculus Through Applications* textbook off the shelf and started using it. Observing that every teacher in his department had gone to an NCSSM winter conference, Jerry Carnes stressed the advantage of hearing from NCSSM faculty members what they meant by what was in the textbook. He saw the preparation they got through such contacts as essential:

> Can you imagine a traditional teacher who's been teaching things very traditionally and all of the sudden picks up this textbook? It must be absolutely the worst thing. You wouldn't understand what is there. You would do a lousy job.

There's no way you could do a decent job with that. Unless you just happen to have that background somewhere along the way, there's no way you would understand what's going on with it.

We tested Jerry's conjecture by showing the textbook to some relatively inexperienced mathematics teachers who had not had contact with NCSSM and found that, for the most part, they perceived it as lacking important topics, boring, difficult to read, and unattractive. They did not recommend that it be adopted as a primary text for a precalculus course. Similarly, we noted that in one state, the book was not initially included on the list of state-approved textbooks by a selection committee composed of mathematics teachers until the state supervisor of mathematics pushed for its inclusion.

It seems that to use a textbook as unconventional as *Contemporary Precalculus Through Applications* requires both knowledge and courage. To open up one's class so that students are pursuing problems whose outcomes cannot be easily foreseen is a hazardous business. Engaging in genuine data analysis and mathematical modeling feels risky to teachers and students alike. A teacher is not likely to promote such engagement if he or she feels uncomfortable with the mathematics or with letting students struggle. The teacher's willingness to take the risk of teaching differently is also a function of his or her perception of the teaching situation. What constraints apply? What opportunities are available? A movement to include modeling and data analysis, after all, does not resolve all questions of curriculum and instruction. In fact, it raises new questions. "Doing mathematics requires a certain autonomy of thought and risk-taking on the part of the doers" (Romagnano 1994, p. 101). What happens when students refuse to work independently? When they balk at taking risks in their learning? What if one's colleagues, the students' parents, or the school's administrators decline to take their own gamble of supporting what David Cohen (1988) terms "adventurous teaching."

Teachers may feel that they are venturing out on the ledge when they appear not to have moved very far. Cohen tells of Mrs. Oublier, a second grade teacher who believes she has revolutionized her mathematics teaching but whose innovations "have been filtered through a very traditional approach to instruction" (Cohen 1990, p. 327). It is important to understand that change, for Mrs. Oublier, has occurred despite an observer's inability to see much of it. She has taken the risk of changing her teaching, drastically in her view. And for her, it has been rewarding and refreshing.

The teachers who developed the Contemporary Precalculus course, along with their like-minded colleagues, were exhilarated by the risks they were taking. As they saw it, their teaching was more effective. Their students were not only learning more, but were learning differently. There was more involvement with and interest in doing mathematics. These teachers saw risk taking as paying off for them and their students. Many teachers, however, are not drawn to take such risks. If they risk a new venture, they may not consider it successful or worth sustaining. Risk takers among teachers range from the bold to the timid. In any event, just walking into the classroom each day is itself a risky business. No wonder changing teaching can feel so scary.

Making Change

When one looks at the school mathematics curriculum globally and "from above," it often seems like a giant ocean, with tidal waves of reform sweeping across political boundaries and giving a sense of great movement and mutation while structures at the bottom keep life there relatively stable. Our case study concerned the ocean-bed curriculum.

We saw a group of teachers in a very special school and in a new and vigorous department undertake to reform their curriculum and their teaching practice. They chose to begin with precalculus, a course near the top of the college-preparatory curriculum, but without the baggage of an entrenched syllabus or a mandated examination. It was a course that, in most high schools, drew teachers with the most experience in teaching and best preparation in mathematics and students who had survived the winnowing process of algebra and geometry instruction. The course content was full of potential applications. An influx of data analysis and modeling activities, moreover, was in tune with current trends in mathematics itself and with national proposals for reform in school mathematics. There were ample opportunities in the course to explore the potential of new, inexpensive technology; display and manipulate data; graph functions; and perform complex calculations.

Even granted these advantages, the Contemporary Precalculus course took immense effort to develop and implement. Funds were needed to buy time for the teachers to write; arrangements had to be made for them to confer and plan revisions; software needed to be developed; a new orientation toward mathematics and toward pedagogy had to be constructed virtually from the ground up. Spreading the word to other teachers called for the production of a textbook, which itself became a demanding process, not only in the stresses of cutting and polishing the text for publication but also in the demands of producing ancillary materials that would enable schools to adopt the book.

Whatever successes these teachers had in producing this innovation did not translate easily to other sites. Students resisted participation in a transformed course that did not fit their expectations for mathematics instruction. Unsympathetic colleagues withdrew their support. Skeptical administrators and parents questioned why precalculus needed to change. What seemed to enable the course to survive, and in some cases its spirit to permeate other parts of the curriculum, were the various communities of interest that developed around the course. These communities equipped teachers intellectually and sustained them emotionally to undertake the challenges of an adventurous pedagogy.

In this case study report, we have tried to portray what one kind of real teacher change looks like. It looks different from the inside than from the outside. It feels risky and needs support in the form of leaders and colleagues. It proceeds incrementally, does not follow well-trodden paths, makes heavy demands, and jeopardizes established pedagogical conventions in the classroom and school. It does not follow simple accounts of social engineering or policy implementation. Rather, it takes the same sort of cross-country ramble that these teachers wanted their students to take through mathematics.

This was a study of a course-based reform initiated by teachers. It counterpoints other studies of attempts to exhort or mandate reform. The teachers who engaged in this reform did so of their own volition. They changed as much or as little as they chose. Consequently, this study may have little to say to those who seek reforms that "scale up," in the sense of being easily duplicated nationwide. The teachers engaged in this reform represent a tiny fraction of the teaching force. They were helped and encouraged by reform proposals trumpeted by professional organizations and spurred by large-scale projects, but their work had at best a modest impact on the broader educational scene. The changes these teachers undertook were small scale and personal, demanding desire, courage, and vision.

For Americans, success is numerical: More is better. A curriculum project is not successful if it reaches only a handful of teachers and students. Reform efforts count for little if they cannot go national. By those standards, this innovation might be judged a failure. But the teachers in our study were after a different prize. They sought to be more effective in making mathematics intelligible, applicable, and relevant to their students. They did not claim to have the answers to questions of curriculum or teaching. Instead, they understood that to teach, just as to do mathematics, means to encounter open questions continually.

References

Askey, R. 1995. Review of algebra. *American Mathematical Monthly* 102: 78-81.

Banner, L. 1987. *A passionate preference: The story of the North Carolina School of the Arts*. Winston-Salem: North Carolina School of the Arts Foundation.

Barrett, G. B., K. G. Bartkovich, H. L. Compton, S. Davis, D. Doyle, J. A. Goebel, L. D. Gould, J. L. Graves, J. A. Lutz, and D. J. Teague. 1992. *Contemporary precalculus through applications: Functions, data analysis and matrices*. Rev. ed. Dedham, MA: Janson Publications.

Berry, J. W., and S. H. Irvine. 1986. Bricolage: Savages do it daily. In *Practical intelligence: Nature and origins of competence in the everyday world*, ed. R. J. Sternberg and R. K. Wagner, 271-306. Cambridge: Cambridge University Press.

Brownell, W. A. 1956. Meaning and skill: Maintaining the balance. *Arithmetic Teacher* 3: 129-36.

Cohen, D. K. 1988. Teaching practice: Plus que ça change . . . In *Contributing to educational change: Perspectives on research and practice*, ed. P. W. Jackson, 27-84. Berkeley, CA: McCutchan.

———. 1990. A revolution in one classroom: The case of Mrs. Oublier. *Educational Evaluation and Policy Analysis* 12: 327-45.

Davis, P. J., and R. Hersh. 1981. *The mathematical experience*. Boston: Birkhäuser.

Davis, S., and P. Frothingham. 1985. A special school in North Carolina. *Yearbook of the National Council of Teachers of Mathematics 1985: 184-88*. Reston, VA: National Council of Teachers of Mathematics.

Davidson, N., ed. 1990. *Cooperative learning in mathematics: A handbook for teachers*. Menlo Park, CA: Addison-Wesley.

Douglas, R. G., ed. 1986. Toward a lean and lively calculus. *MAA Notes* 6. Washington, DC: Mathematical Association of America.

Driscoll, M. 1987. *Stories of excellence: Ten case studies from a study of exemplary mathematics programs*. Reston, VA: National Council of Teachers of Mathematics.

Eighth annual MCM winners announced. 1992. *Consortium: The Newsletter of the Consortium for Mathematics and Its Applications* 43: 2.

Fullan, M. 1993. *Change forces: Probing the depths of educational reform*. London: Falmer.

Hancock, L. 1994. *Coping with a mandate: Effects of a revised end-of-course test for first-year algebra*. Ph.D. diss., University of Georgia, Athens.

Hargreaves, A. 1993. Individualism and individuality: Reinterpreting the teacher culture. In *Teachers' work: Individuals, colleagues, and contexts*, ed. J. W. Little and M. W. McLaughlin, 51-76. New York: Teachers College Press.

Haimo, D. T. 1995. Experimentation and conjecture are not enough. *American Mathematical Monthly* 102: 102-12.

Hesse-Biber, S., T. S. Kinder, P. R. Dupuis, A. Dupuis, and E. Tornabene. 1993. HyperRESEARCH [computer software]. Randolph, MA: ResearchWare.

Huberman, M. 1993a. *The lives of teachers*. Trans. J. Neufeld. New York: Teachers College Press. (Original work published 1989,)

————. 1993b. The model of the independent artisan in teachers' professional relations. In *Teachers' work: Individuals, colleagues, and contexts*, ed. J. W. Little and M. W. McLaughlin, 11-50. New York: Teachers College Press.

Hughes-Hallett, D., A. M. Gleason, D. E. Flath, S. P. Gordon, D. A. Lomen, D. Lovelock, W. G. McCallum, B. G. Osgood, A. Pasquale, J. Tecosky-Feldman, J. B. Thrash, K. R. Thrash, T. W. Tucker, and O. K. Bretscher. 1994. *Calculus*. New York: Wiley.

Kilpatrick, J., and R. B. Davis. 1993. Computers and curriculum change in mathematics. In *Learning from computers: Mathematics education and technology*, ed. C. Keitel and K. Ruthven, 201-21. NATO ASI Series F: Computer and System Sciences, vol. 121. Berlin: Springer.

Lampert, M. 1990. When the problem is not the question and the solution is not the answer: Mathematical knowing and teaching. *American Educational Research Journal* 27: 29-63.

Lévi-Strauss, C. 1962. *La pensée sauvage [The savage mind]*. Paris: Plon.

Little, J. W. 1990. Conditions of professional development in secondary schools. In *The contexts of teaching in secondary schools: Teachers' realities*, ed. M. W. McLaughlin, J. E. Talbert, and N. Bascia, 187-223. New York: Teachers College Press.

National Council of Teachers of Mathematics (NCTM). 1989. *Curriculum and evaluation standards for school mathematics*. Reston, VA.

North Carolina School of Science and Mathematics (NCSSM). 1990. *The North Carolina School of Science and Mathematics: 1980-90, A decade of achievement* [brochure]. Durham, NC.

————. 1994. *The North Carolina School of Science and Mathematics: A profile, 1994-1995* [brochure]. Durham, NC.

North Carolina School of Science and Mathematics (NCSSM) Department of Mathematics and Computer Science. 1988a. New topics for school mathematics: Vol. 1, Geometric probability [materials and software]. Reston, VA: National Council of Teachers of Mathematics.

————. 1988b. New topics for school mathematics: Vol. 2, Data analysis [materials and software]. Reston, VA: National Council of Teachers of Mathematics.

————. 1988c. New topics for school mathematics: Vol. 3, Matrices [materials and software]. Reston, VA: National Council of Teachers of Mathematics.

———. 1993a. *Assessment resource: Contemporary precalculus through applications.* Dedham, MA: Janson.

———. 1993b. *Graphing calculator lab manual (TI-81 & Casio 7700G): Contemporary precalculus through applications.* Dedham, MA: Janson.

———. 1993c. *Instructor's guide: Contemporary precalculus through applications.* Dedham, MA: Janson.

———. 1993d. *Supplementary resource: Contemporary precalculus through applications.* Dedham, MA: Janson.

———. In press. *Contemporary calculus through applications.* Dedham, MA: Janson.

Organisation for Economic Co-operation and Development. 1993. *Science and mathematics education in the United States: Eight innovations.* Paris.

Pollak, H. O. 1970. Applications of mathematics. *69th Yearbook of the National Society for the Study of Education,* Part 1: 311-34. Chicago: University of Chicago Press.

Polya, G. 1945. *How to solve it: A new aspect of mathematical method.* Princeton: Princeton University Press.

———. 1981. *Mathematical discovery: On understanding, learning and teaching problem solving.* Combined ed. New York: Wiley.

Romagnano, L. 1994. *Wrestling with change: The dilemmas of teaching real mathematics.* Portsmouth, NH: Heinemann.

Stake, R. E. 1986. *Quieting reform: Social science and social action in an urban youth program.* Urbana: University of Illinois Press.

———. 1995. *The art of case study research.* Thousand Oaks, CA: Sage.

Steen, L. A., ed. 1987. Calculus for a new century: A pump, not a filter. *MAA Notes* 8. Washington, DC: Mathematical Association of America.

Taba, H. 1962. *Curriculum development: Theory and practice.* New York: Harcourt, Brace & World.

West, P. 1993. Educators seek to connect numbers to real world. *Education Week* (April 28): 1, 12-13.

Wheeler, D. 1980. A mathematics educator looks at mathematical abilities. Paper presented in June at a conference on mathematical abilities at University of Georgia, Athens.

Wick, C. A., S. K. Westegaard, and C. Q. Wilson. 1994. An agent for change: The Woodrow Wilson National Fellowship Program. *Yearbook of the National Council of Teachers of Mathematics 1994: 227-33.* Reston, VA: National Council of Teachers of Mathematics.

Appendix A:
Table of Contents for Contemporary Precalculus
Through Applications

From *Contemporary Precalculus Through Applications* by Barrett, et. al., the North Carolina School of Science and Mathematics. Copyright 1992 by Janson Publications, Inc., Dedham, Massachusetts, 02026, USA. Reprinted by permission of the publisher.

Appendix B: Questionnaire Survey and Responses

[Questionnaire not in original format.]

NCSSM Precalculus Curriculum Project Questionnaire

School name: _____
Address: _____

School Information
Our school is a _____ (public/private) institution where approximately
students are enrolled in grades _____ .

Information about Precalculus Classes
Number of Precalculus classes taught this year: _____

How many Precalculus classes use *Contemporary Precalculus with Applications* as their
primary textbook? _____

How many Precalculus classes use *Contemporary Precalculus with Applications* as a
secondary or supplementary textbook? _____

How many years has *Contemporary Precalculus with Applications* been used at your
school? _____

Information about Precalculus Teachers
How many teachers at your school use *Contemporary Precalculus with Applications*?
_____

What teachers at your school have attended workshops at the North Carolina School of
Science and Mathematics in Durham? _____

Why did you choose to use *Contemporary Precalculus with Applications*?
_____

Please indicate permission for a member of our research team to contact one of your precalculus teachers for further information by filling out the following:

Name: _____

Telephone number: _____

Best time(s) to call: _____

Survey Responses

Survey forms were sent to 131 schools, and 84 responded (64% response rate). Of the schools that responded, 43 were public and 41 private. The response rate was somewhat better in the East and South than elsewhere, and private schools responded at a much higher rate than public schools. (The purchasers were classified as public or private according to the names provided by the publisher; information from the respondents indicated that this classification was almost completely accurate.)

The private schools responding to the survey tended to be much smaller than the public schools. (Enrollment was based on number of students per grade owing to the great variation in number of grades per school.)

Distribution of Schools by Enrollment and Type		
Number of students per grade	Public	Private
1-75	5	20
76-150	2	13
151-225	10	4
226-300	8	3
301 or more	15	0

Note: Four schools did not report the number of students enrolled.

Most of the schools responding used the *Contemporary Precalculus Through Applications* book as a primary textbook in their precalculus courses, but usage as a secondary or supplementary text was common.

Number of Schools Using *Contemporary Precalculus Through Applications* as a Primary or Secondary Textbook			
Primary only	Secondary only	Both	Neither
57	27	10	10

Small schools were more likely to use the textbook as the primary text in precalculus, which was clearly related to the much heavier use by the private schools as a primary text. The fewer years that the textbook had been used in a school, the more likely it was to be the primary text.

Number of Schools Using *Contemporary Precalculus Through Applications* by Enrollment, Governance, and Years Used		
Enrollment (students/grade)	As primary textbook	As secondary textbook
1-75	20	6
76-150	11	4
151-225	9	5
226-300	8	4
301 or more	8	6
Governance		
Public	26	18
Private	31	9
Years using text		
1	21	9
2	20	10
3 or more	16	8

Note: Some schools used the book as a primary text for some courses and as a secondary text for others.

There was no association, by school, between use of the text and attendance at one of the NCSSM workshops.

Distribution of Schools by Workshop Attendance and Use of *Contemporary Precalculus Through Applications*		
Attendance at NCSSM workshops	Used as primary or secondary text	Not used as text
Teachers have attended	37 (45%)	6 (7%)
Teachers have not attended	37 (45%)	3 (4%)

Appendix C:
Case Study Team

Jeremy Kilpatrick is a Regents Professor of Mathematics Education at the University of Georgia. Jeremy was the primary researcher at the Eisenhower High School site and was one of the primary researchers at the North Carolina School of Science and Mathematics. Jeremy's recent professional activities include work with the International Commission on Mathematical Instruction, the Bacomet Group, the National Council of Teachers of Mathematics, and the College Board. Jeremy also serves as a board member or reviewer for a number of national and international journals.

Lynn Hancock is an assistant professor of mathematics education in the Mathematical Sciences Department at Appalachian State University in Boone, NC. At the time of this study, she was a doctoral student at the University of Georgia, and Jeremy Kilpatrick was her major professor. Lynn was one of the primary researchers at both the North Carolina School of Science and Mathematics and the Westminster Schools. Before coming to the University of Georgia, Lynn taught high school mathematics for many years in North and South Carolina. Her research interests include the impact of assessment on instruction.

Denise Spangler Mewborn is an assistant professor in the Mathematics Education Department at the University of Georgia. At the time of the study, she was a doctoral student at the University of Georgia, and Jeremy Kilpatrick was her major professor. Denise was the primary researcher at the Woodward Academy site. She is a former elementary school teacher with research interests in teachers' sense-making in the mathematics classroom and the role of multimedia technology in teacher education.

Lynn Stallings is an assistant professor of mathematics education in the Department of Middle and Secondary Education and Instructional Technology at Georgia State University in Atlanta, GA. At the time of this study, she was a doctoral student at the University of Georgia, and Jeremy Kilpatrick was her major professor. Lynn was the primary researcher at the Webb School site. Prior to pursuing graduate work, Lynn was a high school mathematics teacher. Her research interests include teacher use of communication and computing technologies.

Chapter 3

The Urban Mathematics Collaborative Project
A Study of Teacher, Community, and Reform

Norman L. Webb
Daniel J. Heck
William F. Tate

University of Wisconsin—
Madison

Contents

3

The Urban Mathematics Collaborative Project:
A Study of Teacher, Community, and Reform

The Urban Mathematics Collaborative Project in Perspective

In a three-year junior high school of 800 students—grades 7 (age 12), 8 (age 13), and 9 (age 14)—located south of the automobile bypass around Memphis, Tennessee, a mathematics teacher in a large classroom moves among her students as they work in groups of four. Her students sit at clusters of desks, and the walls are covered with numerical charts and mathematics slogans. Over the past 10 years, this teacher has made a significant transformation from conducting a very controlled classroom dominated by her lecturing. Now her students are much more actively engaged in learning mathematics by working and interacting with each other. She spends much less time standing in front of her classroom telling her students what to do. Her students are doing projects and writing about their thinking, individually and—especially—in groups. She has become very active professionally in writing grants to fund projects, attending professional meetings and institutes, making presentations at national meetings, and field testing a new mathematics curriculum.

In another Memphis school, north of the central city, two teachers in an elementary school have, during the past five years, grown to devote much more of their classroom time to teaching mathematics than to the other 11 subjects they are responsible for teaching. They too have become much more engaged in professional activities. East of the central city and the bypass, in a private K-12 girls school, a seventh grade mathematics teacher of 15 years is more connected with teachers in the public schools than she has ever been before. Her students are doing projects for the first time—recording data, posting statistics on the wall, and using mathematics to make decisions. What these teachers all shared, along with approximately 400 of their colleagues, is their participation in the Memphis Urban Mathematics Collaborative.

Memphis is, in many ways, a typical large urban area in the United States. With a population of more than 650,000, Memphis is a major center for commerce on the shores of the Mississippi River. At one time, Memphis was an important point of departure for settlers who were moving into the American West. Now the city, home to the headquarters of international courier service Federal Express, has a population that is nearly evenly divided between two major racial groups, black and white. In 1993, the Memphis city school district enrolled nearly 100,000 students, over 75 percent of whom were black. A large percentage of students lived in poverty, with nearly 60 percent of the city's students eligible for free or reduced-cost lunch programs. The Memphis city schools operated an Optional Schools program that designated certain school buildings to house specialized educational programs devoted to the arts, science, mathematics, and other academic content areas—part of an effort to produce more racially integrated school populations. Of the nearly 6,000 teachers employed by the district, 350 taught mathematics in high school and middle grades; an additional 3,359 taught subjects that included mathematics in elementary schools.

The Collaboratives

The teachers depicted in the preceding vignettes and many other mathematics teachers in Memphis and around the country were participants in a mathematics collaborative that had been funded by the Ford Foundation in the wake of its 1984 decision to pursue the development of such organizations through its Urban Mathematics Collaborative (UMC) project. The 16 sites that were ultimately founded were scattered across the United States and included some of the largest school districts in the nation. The project was developed on the basic principle that well-connected and professionally active teachers would be able to improve upon the poor mathematics performance of inner-city students. The central goal of the UMC project was to empower teachers of mathematics by increasing the interaction among teachers, between teachers and other professional users of mathematics in business and higher education, and between teachers and others involved in mathematics education reform. Mining the UMC project and its artifacts produces rich material for studying education reform. The UMC story touches on many of the important aspects of schooling—administration, policy, finance, curriculum, teachers, students, learning, politics, unions, and power. The story entails plots and subplots, protagonists, struggles, and triumphs and failures. Much can be gained by understanding how waves from the infusion of $6 million by the Ford Foundation into urban areas over a five-year period beginning in 1985 were still being felt in many of the sites eight years later. The total amount of external money supplied by the Ford Foundation was dwarfed by the multimillion-dollar budgets of the large city school districts and the many thousands of mathematics teachers who had the potential to become active in a collaborative.

What we can learn from studying collaboration and collaboratives centers on teachers—their beliefs, their craft, their knowledge, their working conditions, and their common experiences. We can learn what motivated sustained change within schools. We can learn about the nature of powerful forces within a large district and community that worked to conserve the status quo and worked against reform. We can learn about bridging the communication between those in K-12 education and those in business and higher education. We can learn about the dedication of a large number of teachers who—with creativity and vigor—often worked in less than ideal physical conditions with an increasing number of students who lived in poverty, daily experienced crime on the streets, often spoke a foreign language, and had learned to approach school and mathematics with a sense of hopelessness.

One amazing finding was that nearly 10 years after the beginning of the project, and nearly five years after the last check was written by the Ford Foundation, nearly all of the sites were still active in some way. None of the active collaboratives were operating with large budgets, and most were existing on exceedingly modest grants. What sustained the activities of these collaboratives was their capability to appeal to the willingness of teachers to participate. Volunteerism was important for the survival of these organizations. Understanding how collaboratives motivated teachers to contribute their time and professional energy is central to understanding how the project worked toward reform. The collaboratives each employed a variety of approaches to reform and to the professional development of teachers—all of which had to appeal to teachers' interests and needs. Collaboratives successfully generated a willingness among individual teachers to work very hard toward change. The magnitude of the problem and the number of teachers necessary to reach far exceeded the available resources.

An important strategy developed in the project across the sites was to devote energy and resources to developing teachers into leaders. Collaboratives invested in teachers so that they would become knowledgeable about mathematics, about pedagogy, about leadership, and about other areas needed to influence and work with others. The collaboratives and activities sponsored by the collaboratives have been proving grounds for teachers to gain experience in being leaders. Many of these teachers became valuable resources for districts as presenters, as writers and directors of grants, and as examples of innovative practitioners. These teachers who became leaders maintained their classroom responsibilities.

Flexibility was critical for the operation of collaboratives. Teacher-leaders, with their individual and collective interests and talents, gave the collaboratives flexibility. Collaboratives also gained flexibility through being administered by organizations and agencies on the fringe of the school districts and not by the school districts themselves. Collaboratives had to be flexible adjusting to different levels of teachers' participation and to the changing commitment of teachers over time. Only occasionally did a teacher sustain an intense involvement in a collaborative over a number of years. Changes in school and district administration, state and national goals for education, and the direction of the broader education reform movement also demanded flexibility in the collaboratives.

Change was attributed to collaboratives. Teachers associated their participation in the collaboratives with changes in their professional outlook, in what they taught, and in how and why they taught what they did. The collaboratives were able to garner a variety of forces that appeared to be important in the change process of these teachers. Many teachers found themselves questioning the status quo in their classrooms, schools, and professional lives in relation to participation in a collaborative. The collaboratives were responsible for exposing teachers to new ideas, the latest thinking about how students learn mathematics, and what mathematics should be taught in the curriculum. The collaboratives not only exposed teachers to these new ideas, they provided teachers with explicit examples of how others were applying these ideas. Collaborative activities offered teachers opportunities to see new ideas modeled and to practice them in model situations. The collaboratives encouraged teachers to work together and support each other through a change process. Teachers were not left alone to make a change, but had collegial help that provided confirmation or support in problem solving. Finally, the collaboratives provided an environment for teachers to experiment with new ideas, take some risks, and constructively reflect on the outcomes and merits of anticipated and unanticipated changes.

The Case Study

The research that forms the basis of this case study benefited greatly from the UMC Documentation Project, which chronicled the UMC project from 1985 to 1990. The present case study's first phase consisted of an updated examination of a collaborative site, the Cleveland Collaborative for Mathematics Education (C²ME) (Romberg and Webb 1992). The report on C²ME provided valuable insights into what questions the researchers should be asking about the UMC project and individual sites.

During the summer and early fall of 1993, collaborative administrators at each site participated in a 30- to 80-minute interview over the telephone for the second phase of the case study. The information gathered during the second phase formed the basis of the *Case Study of Urban Mathematics Collaboratives: Status Report* (Heck, Webb, and Martin 1994). The *Status Report* updated the researchers' knowledge of the UMC project and the collaborative sites, providing a basis for the selection of five collaborative sites for in-depth examination during the third phase of the case study.

The researchers made two field visits to each of the five selected collaborative sites in 1993 and 1994 for phase 3 of the case study. Initial visits consisted of interviews with a variety of collaborative participants in order to identify important questions to be explored and important sources of information to be consulted at each site. Second visits allowed the researchers to follow those leads. The second visits consisted mainly of a greater number of interviews with collaborative participants, observations of collaborative activities, visits to the schools of collaborative teachers, and observations of their classrooms.

Analysis of the information gathered during phases 2 and 3 was performed using an evolving coding system to identify and explore patterns and themes in the data. The researchers were greatly aided by the HyperRESEARCH (Hesse-Biber et al. 1993) qualitative analysis software for coding and compiling information by themes and patterns. Meetings of the researchers occurred weekly to bimonthly during the data collection and analysis phases of the study. During those meetings, the researchers developed and modified the coding scheme as needed, discussed the emergence of patterns and themes from the data, and created matrices for verification of themes within and across sites.

The final phase of the case study consisted of writing. Finding a way to write the case study that addressed all of the interesting and relevant themes, preserved the reality of participants' experiences, and conveyed a compelling story was a challenge that the researchers relished. Writing the case study, sharing our work with one another, and seeking feedback from others provided the most critical filters of analysis and the best tests of our ideas and conclusions. The many people who assisted us, especially the teachers and other collaborative participants, deserve much credit for this case study.

Overview and Origins of the Urban Mathematics Collaborative Project

In 1984 the Ford Foundation initiated the Urban Mathematics Collaborative Project in an effort to improve the quality of mathematics education in urban schools and to enhance the professional lives of mathematics teachers in urban areas. Each of the 16 "collaboratives" that were ultimately founded consisted of mathematics teachers in an urban area who worked in concert with representatives from business, higher education, and school districts to identify and meet the professional needs of local teachers of mathematics. A chronology of the UMC project is given in table 1.

Ford Foundation guidelines for the operation and program of each collaborative were broad: Each of the collaboratives funded by the project was required to serve secondary mathematics teachers, to secure the support—both financial and otherwise—of local businesses and institutions of higher education, and to operate in conjunction with its constituent school districts, seeking district support but avoiding district control. Aside from these general requirements, the collaboratives were expected to develop their own models of organization and activity (Webb and Romberg 1994).[1]

---

[1]See also the UMC project reports to the Ford Foundation for 1986 through 1990 (Webb et al. 1988, Webb et al. 1989, Webb et al. 1990, and Webb et al. 1991).

Table 1.   **Chronology of Events of the Urban Mathematics Collaborative Project**

1983   *Nation at Risk* document released

1984   Ford Foundation initiates the Urban Mathematics Collaborative (UMC) project; Barbara Scott Nelson, project monitor
        Documentation Project established at University of Wisconsin-Madison (February)

1985   Seven collaboratives established:
        • Cleveland Collaborative for Mathematics Education (February)
        • Los Angeles Urban Mathematics/Science Collaborative (February)*
        • Philadelphia Math Science Collaborative (February)*
        • San Francisco Mathematics Collaborative (February)
        • Twin Cities Urban Mathematics Collaborative (February)
        • Pittsburgh Mathematics Collaborative (September)
        • The Durham Mathematics Council (August)
        Technical Assistance Project—Education Development Center, Inc. (EDC), established in Newton, MA (September)
        First UMC annual meeting

1986   Four collaboratives established:
        • St. Louis Urban Mathematics Collaborative (April)*
        • San Diego Urban Mathematics Collaborative (April)
        • Memphis Urban Mathematics Collaborative (September)
        • New Orleans Mathematics Collaborative (September)

1987   Outreach Project established at EDC

1989   Three replication grants awarded by Outreach Project:
        • Dayton, OH*
        • Columbus, GA
        • Milwaukee, WI
        UMC Standing Committee formed (May)
        First Teacher Leadership Workshop sponsored by EDC
        National Council of Teachers of Mathematics (NCTM) *Curriculum and Evaluation Standards for School Mathematics* published

1990   Transfer of UMC project management from Ford Foundation to EDC
        Eleven permanence grants issued by Ford Foundation
        Final year of data gathering by Documentation Project

*Some later changes in names are reflected in Figure 1.

Table 1.   **Continued**

1991	Final documentation report published
	Site case studies published
	NCTM *Professional Standards for Teaching Mathematics* published
	Three UMC position papers published:
	Equity and Mathematics Reform
	Teacher Professionalism
	Assessment
	One replication grant awarded by Outreach Project, Worcester, MA
	Classroom Assessment in Mathematics funded by National Science Foundation
1992	Last replication grant awarded by Outreach Project, New York
	Approximately 2,175 teachers active in UMC
	Last annual meeting
1993	Initial data gathering for UMC project case study
	Fourteen collaboratives still operating, approximately 2,000 teachers active
1994	*Reforming Mathematics Education in America's Cities, The Urban Mathematics Collaborative Project*, based on Documentation Project, published
	Leadership in Urban Mathematics Reform Project established by EDC at six UMC sites
	UMC national governing board reorganized and incorporated in PA
1995	UMC project case study published
	NCTM *Assessment Standards for School Mathematics* published

In February 1985, the Ford Foundation awarded grants to establish the first five urban mathematics collaboratives in Cleveland, Ohio; Minneapolis-St. Paul, Minnesota (cited hereafter as the Twin Cities collaborative); Los Angeles, California; Philadelphia, Pennsylvania; and San Francisco, California. By September 1986, six more grants were awarded and collaboratives were established in Memphis, Tennessee; New Orleans, Louisiana; Pittsburgh, Pennsylvania; St. Louis, Missouri; San Diego, California; and Durham, North Carolina (Webb et al. 1988).

The Ford Foundation also funded the Documentation Project at the University of Wisconsin-Madison and the Technical Assistance Project at the Education Development Center, Inc. (EDC), in Newton, Massachusetts. The Documentation Project chronicled the development of the UMC project for its first five years. The Technical Assistance Project helped the collaboratives to identify and use local resources and to keep abreast of trends in the field of mathematics education (Webb et al. 1988). In addition, the Technical Assistance Project created an infrastructure that allowed the collaboratives to form a national network for sharing ideas and experiences (Webb et al. 1991).

Finally, the Ford Foundation started an Outreach Project in 1987 to expand the UMC project. Administered by EDC, the Outreach Project identified additional urban sites at which the foundation could use relatively small grants to establish new collaboratives. Between 1989 and 1992, five "replication" collaboratives had been established and linked to the national network of existing collaboratives. The new collaboratives were located in Columbus, Georgia; Dayton, Ohio; Milwaukee, Wisconsin; New York, New York; and Worcester, Massachusetts. In comparison with the 11 original collaboratives, there is relatively little information about the development of the five replication collaboratives. For an overview of the locations of the sites and associated activities, see figure 1.

By 1990, the Ford Foundation had approved "permanence grants" for each of the 11 original collaboratives. The collaboratives' permanence proposals were intended to define a plan under which collaboratives would use a final financial contribution from the Ford Foundation. The proposals were also expected to show how the collaboratives planned to shift their funding sources from the foundation to local supporters and maintain a permanent organizational structure. At the same time, the data gathering of the Documentation Project ended, and the collection of information about the collaboratives since 1990 has been much less detailed and systematic than it had been in earlier years.

Each project began with the premise that developing collegiality among professional mathematicians and teachers can reduce teachers' sense of isolation, foster their professional enthusiasm, expose them to a vast array of new developments and trends in mathematics, and encourage innovation in classroom teaching. Considered individually, each collaborative was a unique, locally controlled project. But viewed as components of a wide-reaching national network, each comprised an efficient, cost-effective, and comprehensive field experiment that served to enhance the knowledge and professionalism of participating teachers while functioning as a testing ground for new modes of thought and fresh approaches to larger issues of professional enrichment and subject area expertise. The Urban Mathematics Collaborative project served as a case study of teachers involved in the paradox of reform: While experiencing both external pressure and internal desire to increase problem solving, reasoning, and inquiry by students, the collaborative teachers existed in large inert school systems where a high percentage of students struggled with basic computational skills.

The collaboratives varied in their structure and activities, but each in some way brought together teachers, school district administrators, mathematics users from business, and college and university educators. People from business and higher education served on committees and governing boards, provided summer

Figure 1.   **The Urban Mathematics Collaborative Project**

Original Sites

① **Cleveland Collaborative for Mathematics Education**
Cleveland, OH

② **Durham Collaborative: The Durham Mathematics Council**
Durham, NC

③ **Los Angeles Urban Mathematics/Science/ Technology Collaborative Professional Links With Urban Schools (+PLUS+)**
Los Angeles, CA

④ **Memphis Urban Mathematics Collaborative**
Memphis, TN

⑤ **New Orleans Mathematics Collaborative**
New Orleans, LA

⑥ **Philadelphia K-12 Math/ Science Teachers' Network (formerly the Philadelphia Math Science Collaborative)**
Philadelphia, PA

⑦ **Pittsburgh Mathematics Collaborative**
Pittsburgh, PA

⑧ **Urban Mathematics Collaborative of St. Louis**
St. Louis, MO

⑨ **San Diego Urban Mathematics Collaborative**
San Diego, CA

⑩ **San Francisco Mathematics Collaborative**
San Francisco, CA

⑪ **Twin Cities Urban Mathematics Collaborative**
Minneapolis-Saint Paul, MN

Replication Sites

☐1 **Columbus Regional Mathematics Collaborative**
Columbus, GA

☐2 **Alliance for Education (formerly the Dayton-Montgomery County Public Education Fund Mathematics Collaborative)**
Dayton, OH

☐3 **Milwaukee Metropolitan Mathematics Collaborative**
Milwaukee, WI

☐4 **New York City Urban Mathematics Collaborative**
New York, NY

☐5 **Greater Worcester Urban Mathematics Collaborative**
Worcester, MA

Funding Agency

△1 **Ford Foundation**
New York, NY

Documentation Project

△2 **Wisconsin Center for Education Research**
Madison, WI

Technical Assistance and Outreach Project

△3 **Education Development Center, Inc.**
Newton, MA

internships, made presentations, served as members of small focus groups with teachers, conducted site visits, visited schools, and applied their knowledge and resources to support teachers. Those from business and higher education both donated their time and garnered benefits by striving to influence their future students and workforce and through gaining a greater appreciation of secondary schooling.

Collaboratives were purposefully formed outside the structure of school districts in order not to become overburdened by large bureaucracies. Existing tangentially allowed collaboratives to circumvent the layers of communication within the urban school districts and, in some cases, forge new relationships between mathematics teachers and the district administration. Mathematics teachers and their efforts to address problems in mathematics education became more visible. Administrators used what mathematics teachers had achieved as models for other content areas. Participating teachers became less isolated from other teachers and from the larger community. They felt more valued by the community, gained support from other teachers, engaged in group problem solving, expanded their knowledge of the latest curriculum materials and resources, and grew in leadership skills.

Collaboratives in Close Focus

Collaboratives as Learning Systems for Teachers

This case study is dedicated to understanding and describing the evolution of the urban mathematics collaboratives, both collectively and individually. Each of the collaboratives maintained as a core goal the creation and support of opportunities to learn for mathematics teachers. Collaborative teachers as a group were concerned about the effectiveness and appropriateness of their pedagogy as they sought to improve student learning in mathematics. Often at their own initiative, beyond school hours, and at their own expense, they attended mathematics meetings and conferences, took in-service courses, studied for university qualifications, communicated with other teachers, and read professional journals to learn new ideas for teaching mathematics to students. These efforts to grow professionally were reflected in each of the UMCs. Several intriguing questions arose. First, how did the collaboratives create opportunities to learn? Second, what was the nature of these opportunities to learn and what learning occurred? Finally, how do such opportunities to learn differ from past notions of in-service and professional development for teachers?

Volunteerism as an Entry Point for Professional Growth. Volunteering was the common entry approach for most teachers participating in collaborative activities. Only 1 of the original 11 collaboratives defined mandated district in-service sessions as collaborative-sponsored activities. The volunteerism of the collaborative teachers was arguably the most important requisite of their professional growth. To understand why volunteerism was important, we need only recall more traditional approaches to teacher development. Generally, mathematics staff development provides teachers with an opportunity to hear about a curriculum idea or pedagogical strategy. It is often compulsory. This approach to staff development is associated with a top-down training model in which experts introduce teachers to methods of cooperative learning, portfolio assessment, or some other teaching strategy (Little 1993). This model of staff development relies on the transmission of information to teachers as passive receivers. It does not account for the individual or institutional realities of the teachers' lives. It does not allow teachers to inquire and create, to make decisions, or to transform ideas as active learners.

Typically, many teachers feel a sense of frustration that, even after attending a mathematics in-service session, for example, they are unable to use a new curriculum activity or teaching method to improve the learning of their students. Unfortunately, it is very common for teachers to find themselves teaching in the same ways they always have, perhaps incorporating some of the new curriculum ideas but adapting them to fit traditional teaching behavior (Porter 1989 and Stodolsky 1988). Many teachers are cognizant of this pattern but feel bound, unable to escape this cycle. Ultimately, teachers develop a cynical view toward new initiatives and opt not to volunteer for further professional development (Bell and Gilbert 1994). Thus, more traditional approaches to professional development often fail to influence the pedagogy, practice, and lives of teachers.

In contrast, the collaboratives were composed of many volunteers who knew, understood, and experienced the life of urban school teachers. The volunteerism of the collaborative participants provided important components in the development of a learning system for mathematics teachers—namely leadership, a wealth of opportunities for professional interaction, and a teacher-generated knowledge base built on the realities of teachers' classrooms, schools, and districts.

Building on Teachers' Realities. A strength of volunteerism in the learning system is that it allows teachers, the main learners, to create professional development opportunities that are consistent with their needs and realities. What volunteerism does not provide is an assurance that the learning system will reach all mathematics teachers within a district. Despite this weakness, the urban mathematics collaboratives continued to provide willing teachers with an outlet to pose and resolve their own questions about curriculum, pedagogy, and school leadership. Denise Haverstrom, a collaborative participant from higher education, described how the Philadelphia collaborative attempted to build on the reality of teachers' experiences:

> We developed a retreat and then a teacher leadership institute . . . This was to be teacher-driven, to . . . empower teachers, because so much of the activity that is taught down at the school district does not necessarily address what the teachers feel are their real needs . . . the importance of the leadership is not only empowering the teachers to feel more confident, but also to really take an

active role in . . . the individual school where that teacher may work. Moreover, teachers were preparing to meet the entire district's needs as far as math/science teacher enhancement is concerned. And also, the teachers themselves identified the kinds of content knowledge that they felt was important.

The collaboratives served as vehicles to identify and respond to teachers' needs. This meant supporting teachers' efforts to develop and resolve questions that were meaningful to their everyday practice. This kind of learning and teacher empowerment was manifested in the following ways:

- **Collaborative teachers participated in professional development activities designed to address needs they often identified for themselves.** For example, in the Milwaukee collaborative's Algebra Network, teachers' needs and realities were the focal point of activity and learning. Winston York, a high school teacher, commented:

  > The teachers involved basically determine what their needs are . . . . To me, it's more responsive to the teachers' needs . . . When you look at a districtwide type of in-service, someone has the idea and says, "Okay, we're going to have an in-service on this [topic]." The teachers may be sitting out there saying, "This [other topic] is what we really need" . . . The Algebra Network is more flexible in that sense . . . I am still very committed, in an organization like this, it's teacher empowerment, teacher-run . . . I think a teacher-led organization is the focus . . . And I agree 100 percent with that. I think it's very necessary. I would caution against giving away your empowerment in a sense.

- **Teachers' opportunity to learn extended beyond formal professional development activities into peer observation, counsel, and communication.** Trevor Updike, a high school teacher in Los Angeles, offered the following comments, which trace new opportunities to learn in his experience as a collaborative teacher, from formal activities to departmental dynamics to daily interactions with colleagues:

  > The Exeter Conference, which was part of the +PLUS+ thing[2] . . . there were about eight of us from the LA area, we went together and were from different high schools, so we've become a network . . . We're not so isolated anymore either, because of the networking that we have with the other high schools around. When you go places, you see them. We can call them on the phone and talk to them.

  > Writing a grant to become a +PLUS+ school, we had to work together. We've continued to work together in different areas as part of our +PLUS+ program. We became more of a team . . . We shared all our ideas and worked together on almost all of our things . . . Not everyone participated in that, but it brought us together as a department. We were able to com-

---

[2]Professional Links With Urban Schools, commonly used to refer to the Los Angeles mathematics collaborative.

municate with each other a lot more . . . +PLUS+ gave us some incentive to work a little while together with money, and then we have to continue that process on our own.

You take a risk [in your classroom teaching], and it's nice to have a friend that's also . . . willing to take that same kind of a risk. You kind of bounce ideas off of each other.

- **Teachers' opportunity to learn "new" mathematics curriculum and different pedagogical strategies resulted in teacher-initiated policy formation and recommendations.** Collaborative teachers in Memphis who developed and successfully lobbied for a new course in the high school curriculum demonstrated how learning contributes to teachers' initiative in the realm of policy. What began as a collaborative project to improve the success of students enrolled in Advanced Placement (AP) Calculus throughout the district resulted in a policy formulation, as Tracy Xistris, a collaborative teacher leader, recalled:

We took [our results] to the board and said, "This is what we think ought to happen." And . . . some things did come out of it. One of the things is that Introduction to College Math course that we have [at this school] and we have in a couple other [schools] . . . and even the school system is starting to try to push that. It's a legitimate course now. We were kind of responsible for that, that's why I kind of like that course. But for some students who are not ready to do especially what calculus is right now, that formal stuff, they need something else that would better prepare them for any kind of mathematics that they wanted to go into. So we suggested that.

To build such influence on their professional lives and school practice required the collaborative volunteers to move beyond the traditionally defined boundaries of the teaching profession. This move to reform the roles of mathematics teachers required the collaborative volunteers to develop enduring strategies that supported individual and collective learning and change.

Teachers' Opportunity to Learn. As volunteers, teachers who participated in collaborative activities could not be held as a captive audience, nor could they be forced to change. Ideas regarding new ways of teaching mathematics and different emphases on what content should be included in the mathematics curriculum had to be advanced with sensitivity to the institutional realities of teachers, and with compelling arguments that inspired teachers' desire to make change. One underlying rationale for collaboration as a process of reform was for mathematics users from business and higher education to help provide that argument, but within the context of the classroom and school experiences of teachers. Moreover, opportunities for teachers' learning had to attend to the knowledge teachers required to make changes and the confidence demanded to deviate from existing practice. An important task that collaboratives assumed was to create opportunities for teachers to learn, in ways that accounted for all of these teacher-related factors.

Powerful forces were imposed on teachers to maintain the status quo. Constraints the collaboratives and the reform movement in mathematics education faced included the already overloaded schedules of teachers; district and state mandates, especially standardized testing, impeding changes toward the new recommendations; large classes with many hard-to-reach students; and a dearth of resources. In the face of these obstacles, the collaborative endeavor demanded the creation of both a variety of opportunities for teachers to learn and means to sustain this learning over an extended period of time.

All Collaboratives Were Learning Systems. Organizational learning in the collaboratives was operatively defined as the capacity or processes within the collaborative to maintain and improve teacher knowledge and performance, based on a combination of teacher experience and collective information gathered by the participants. Given this operational definition, each of the collaboratives, as well as the collaborative network, functioned as a learning system. That is, all had formal and informal processes for the acquisition, sharing, creation, and utilization of knowledge about mathematics curriculum, pedagogical strategies, and assessment practices. Learning within the collaboratives, moreover, extended to broader notions about mathematics education and the teaching profession. Collaborative participants communicated—both informally and formally, within and across collaborative sites—and assimilated language, norms, procedures, and classroom-based outcomes, beginning with the initial socialization of participants and continuing throughout participation.

The collaboratives often utilized very traditional methods for acquiring knowledge about mathematics education. For example, collaborative teachers attended local, regional, and national mathematics meetings. However, the collaboratives were nontraditional in that the teachers were expected to bring information back to the collaborative and to share with colleagues. Furthermore, teachers who employed new ideas in their practice also shared experiences, successes, pitfalls, and transformations of those ideas with other participants. Most of the collaboratives organized formal opportunities for teachers to share this information. Philadelphia organized extensive content and leadership institutes. Memphis and Columbus developed mathematics camps and retreats. Los Angeles provided mini-conferences to expand the base of learning. All of the sites sponsored workshops in which teachers presented mathematics to colleagues. The collaboratives supported teachers' creating learning opportunities for teachers, as the Saturday workshop series sponsored by the Los Angeles collaborative exemplified.

During the 1993-94 school year, the Los Angeles collaborative's Teacher Council assumed full responsibility for managing and conducting the workshop series independent of the collaborative's host agency. The vast majority of the workshop series—which included as many as eight workshops per year—was designed and led by collaborative teachers with attention to classroom realities. George Hillman, an active teacher-leader, described the workshops:

The model +PLUS+ workshops' use is that we provide training or we provide exposure to certain ideas, but the teacher is then to go back to the classroom and test those ideas. That it isn't just "show and tell." It is show and tell, and then go use, apply it. Then come back and critique it at the subsequent workshop. Modify or absorb new ideas or incorporate new ideas; go back again and field test whatever it is you are learning that is specific to the workshop.

Not all learning opportunities were confined to formal settings, however. What distinguished the collaboratives from more traditional staff development opportunities was the relationships that developed among participants, as Hillman described:

And if I have a problem, if I have a difficulty, I know someone I can call for help. Because I get calls from other people. [That] never would have existed before. And if we said, "Hey, can you guys make a recommendation for a good algebra 1 book?," we could call 20 people, you know, in a day and get an opinion, and a valued opinion. We would have been out on our own otherwise. If anything this has done, it would be the opening of communication and contact among our people.

Pearl Quigley, the Los Angeles collaborative's director, viewed the workshops and the informal relationships similarly:

Our workshop model . . . is one of the most powerful things we've developed because it created a mechanism . . . an outlet for teachers who had special opportunities through the math collaborative to share with others what they had learned . . . Teachers have told me over and over again what makes the +PLUS+ workshop so unique is that, "I just don't go learn something new, I am required to use it, and I have to bring back the results to my peers before the next session and the next month" . . . It parallels the fact that learning is a process.

For many collaborative teachers, participation in the collaboratives was largely about new and expanded professional relationships. The nature and extent of those relationships revealed the depth and breadth of the collaboratives' impact.

Understanding the Relationships. Relationships among teachers in the collaboratives extended beyond the traditional boundaries of the school building. Teachers across schools and districts met and discussed issues relevant to school mathematics reform. Such substantive relationships persisted most observably among collaborative teachers across school sites, rather than within buildings. It was somewhat rare to discover more than a few teachers within one building who were very active in the collaborative. School mathematics departments, the traditional organizing feature of mathematics teachers within the school building, continued to serve mainly in an administrative capacity. In notable, if infrequent, cases, some school mathematics departments developed a substantial instructional focus in relation to teachers' participation in the collaboratives.

Still, the teachers who organized themselves around instructional goals and learning, mainly active collaborative teachers, represented only part of the mathematics teaching corps of the school. Thus, the collaboratives served as extended "departments" in which teachers and others were organized chiefly around mathematics instruction.

Generally speaking, the development of relationships among collaborative teachers involved three steps. First, a core group of teachers and collaborative staff organized around some idea. The core group usually was composed of original collaborative participants and a few other committed teachers with less collaborative experience. These groups utilized and extended the collaborative infrastructure as they created potential activities and programs focused on the original idea. The Memphis, Milwaukee, and Philadelphia collaboratives, for instance, utilized teachers who had attended leadership institutes at EDC to develop and lead regional and local leadership institutes.

The next step was to establish channels for communicating ideas and information about programs and events. Most collaboratives created mechanisms to communicate with their target population of teachers, which varied across sites by district, grade level, and content area. Most collaboratives used periodic newsletters or special mailings to target teachers and to solicit their support and participation. Lydia Oates of the Memphis collaborative remarked, "It seems that every secondary mathematics teacher . . . in the Memphis city schools was automatically on the collaborative's mail-out, so they were getting information about special projects." Other communication efforts were also employed to keep teachers informed about collaborative activities and, more importantly, to expand teacher participation. Collaborative staff, including teachers, telephoned individuals and visited different schools to garner greater participation. However, the most widely used strategy for garnering participation was informal communication among teachers. In combination, these efforts generated interest in collaborative activities.

Once a teacher demonstrated interest in a collaborative activity, he or she was often greeted with individual encouragement for further involvement by collaborative teachers and staff. This step was of great importance: first, because many teachers expressed reservations about joining an organization that required substantial time and devotion to mathematics education; and second, because new teacher participants should not feel excluded from deeper levels of involvement. Oates, an elementary teacher, remarked:

> I have to say that as far as mathematics, I am a person of great math anxiety . . . so when a person from the math collaborative approached me about working for extra hours for the collaborative, I was a little bit unsure whether it was something that I wanted to get involved in because I really, at that time, even though we were a collaborative school, I really didn't understand what the collaborative was all about. And in getting a much closer look at it and getting involved with it, I was encouraged . . . and invited to be a presenter at a work-

shop in an area that I don't really feel that I had expertise. But the more I worked at it, the more I got involved . . . the more comfortable I began to feel in that world, dealing with my colleagues or with my peers, teaching them and instructing them. So I think the collaborative has had a big influence on me personally in terms of just overall confidence . . . and then I think I've grown a lot professionally since I've been involved with the activities.

Kyle Keys, a higher education representative who served on the board of directors for the Columbus collaborative, had witnessed the same phenomenon of teacher growth and viewed it as an important renewable resource: "We have some teachers that started out with us the first year that are still with us. That have matured in their leadership style and knowledge and they are now teaching other, younger teachers how to be leaders within their schools. It's a good expansion model."

Solid teacher participation, new relationships among teachers and other professionals, and strong communications were cornerstones of the collaborative enterprise. Equally important was mathematics, ultimately the substance of participation, relationships, and communication.

Building a Common Language of Mathematics. A basic premise of sustaining any learning organization is the development and adoption of a language that can serve as the building block of the organization's purpose. Collaborative participants were engaged in a process of building, adopting, and promoting a common language of school mathematics. When we think of school mathematics as a language, we often associate it with the discipline of mathematics (Ernest 1991). On a surface level this assumption appears reasonable. Teachers create opportunities for students to think about and manipulate symbols of mathematics. However, the mathematics in schooling is not necessarily what mathematicians, scientists, business people, musicians, or designers do. Instead, school mathematics curriculum evolves as a process of reconceptualization and reformulation of language that borrows from and extends beyond the disciplinary boundaries of mathematics (Popkewitz and Myrdal 1994, and Stanic 1991). Moreover, this process involves excluding language typically associated with mathematics.

Mathematics is used as a part of the language of schooling in which specific patterns and practices are a product. The language practices of schooling give structure to what mathematics students are expected to learn and, at the same time, give organization to the manner in which teachers are expected to produce that learning (Popkewitz 1988).

For example, many teachers and urban school districts have long associated the language of school mathematics with the basic skills movement (NCTM 1980). Teachers have been inundated in past years with curriculum guides that have constructed mathematics as a discrete, disconnected language. Further, many standardized measures of mathematics achievement have been developed and aligned with language derived from the basic skills movement (Berlak 1992). The collaboratives provided teachers with a mechanism to situate these calls for basic mathematics skills into a larger, more contemporary conversation of mathematics education reform. The introduction of the *Curriculum and Evaluation Standards for School Mathematics* (NCTM 1989) and the *Professional Standards for Teaching Mathematics* (NCTM 1991), coupled with the evolution of the collaboratives, advanced a language that transcended the basic skills movement. Brenda Baston, a collaborative participant and district administrator from Georgia, remarked:

> I think the collaborative has helped the teachers to look beyond regional and state mandates. Our state for a long time has settled for what's basic . . . and . . . required. In response, the collaborative has tied itself to the *Standards* . . . Also, the teachers who participate in the collaborative had gotten involved with teachers of other states. That's exciting. It gives you insight into what you should do and gives you excitement to do other things.

Teachers in the collaborative made the language of the National Council of Teachers of Mathematics (NCTM) *Standards* the common language of mathematics education in the district, as Baston explained: "The collaborative is what really brought us the *Standards* . . . Probably we would have found out about them, and I'm sure there were isolated teachers who knew about them. However, as far as a systematic way of utilizing those standards, it was from teachers in the collaborative that we came to know." Lana Lassiter, a high school teacher and collaborative participant in another school district in Georgia, offered a similar thought: "I think that what [the collaborative has] done is put the NCTM *Standards* up on this billboard and say, 'Hey, this is something that you need to concern yourself with.'"

These remarks reveal the role the *Curriculum and Evaluation Standards for School Mathematics* and the *Professional Standards for Teaching Mathematics* played in creating a language for change among collaborative participants. Many of the collaborative teachers used the language of the *Standards* to free themselves from the pedagogical and curriculum obligations attached to outdated mandates and the basic skills movement. These efforts were nurtured by teachers sharing strategies and practices grounded in the language of the *Standards* with their collaborative colleagues. This process of sharing strategies and ideas helped to create a language and way of knowing mathematics that was very different from the language of basic skills. The language of the *Standards*, moreover, gave teachers a greater control over their pedagogy. Nadine Nelson, an elementary teacher in Georgia, stated:

> I also think [other teachers in the collaborative] help to give you reasons for [mathematics reform] so that if you had to speak with an administrator or if you had to speak with a parent, and give them a rationale for doing something—like we play cards in the classrooms and play different games—I have a sheet of the different standards that they meet so that if I have a parent say, "Well, why are they doing that?" they will have a rationale for it.

The language of the *Standards* and the larger mathematics education reform played a similar role in the experience of Jo Lamar, a middle grades teacher in Milwaukee:

> [Participation in the collaborative has] built confidence in me . . . I'm the curricular chair. I find it easier to chair meetings, to come to a consensus and solutions . . . to direct things. And to also challenge people. I have a wealth of topics from the EDC [leadership] experience and from the internship, and things that I've become interested in through those, that I can [use to] challenge my other teachers. And I can say, "How are you doing this and such from the *Standards*?" . . . It was good for textbook selections because we were able to look for certain things that were pointed out to you through the curricular experience.

Many educators and parents associate school mathematics more with the language of basic skills than with the language of the *Standards* and the current mathematics reform. Collaborative teachers were prepared and positioned to communicate the current language and practices of mathematics reform effectively and thus could promote awareness of reform and help advance it.

Communicating and Advancing Agendas. The collaboratives provided teachers with a vehicle to share the notions and language of mathematics reform with school administrators, parents, and other stakeholders associated with the educational process. The sharing of language among collaborative teachers and other decisionmakers in the educational process resulted in a sense of authority in mathematics education among teachers.

For example, Claudia Fagan, the director of mathematics in one school system, noted that collaborative teachers were informed about mathematics reform and proactive in efforts to change mathematics education. She stated:

> There is a strong awareness of possibilities out there among collaborative teachers. In terms of exploring the literature, making me aware of opportunities for teachers to go to staff development . . . I get a lot of calls in that area. I think that is just an outgrowth of knowing that [as a teacher] if I'm interested in something that there is someone that I can approach to help me to get what I really want if I want to be involved in . . . a workshop . . . They are reading and they are saying, "Here is another possibility. Here is something we can do," and they bring that forward.

Collaborative teachers identified a new sense of knowledge, authority, prerogative, and success when they approached administrators with new ideas. A high school teacher in Georgia, Rochelle Robinson, related the following example:

> We had the support of the administration [when we were] getting the TI-81 calculators. I went to see [the district administrator] in that regard. I said, "I want a classroom set of the TI-81 calculators," and she said, "I want you to have it, too." We have the support of the administration . . . It took the collaborative

idea and the support from the administration to get it done and we had both. Without the collaborative we wouldn't have known about support from the administration, because it would not have occurred to us to ask the question.

Why did collaborative teachers feel empowered to approach their principals and district supervisors with new ideas? The answer is complex, though several important components of the explanation can be identified. First, collaborative teachers were willing and increasingly able knowledge seekers. Second, the national UMC network and EDC made a concerted effort to provide leadership training to collaborative teachers, an effort that was adopted formally and informally in all sites. The creation and exploration of strategies for developing and advancing an idea with administrators and others who were instrumental for supporting change was often built into leadership development of teachers. Third, collaborative teachers were willing to experiment with new ideas in their classrooms, and many benefited from the support of similarly engaged or like-minded peers. This combination was likely responsible for moving many teachers to action.

Administrators recognized how teachers' professional growth through the collaboratives made them more effective in efforts to promote change. Brenda Baston, the administrator with whom Robinson worked, commented:

> Rochelle said, "Brenda, we have got to have these [graphing calculators]." She told me exactly what she wanted, how much it would cost . . . Funding is sometimes at our discretion as to where we spend it. She built the best case for it, and then I said "Okay, this is the amount of money," which was not as much as she asked for. And she got on the telephone and she got bids and she got all she wanted.

Suzanne Sanchez explained how the collaborative improved teachers' effectiveness in such requests:

> Part of what the collaborative does to support that is that they help the teachers become knowledgeable about all of those peripheral facts that make a big difference as they come back to the district and present a request, so that they become very knowledgeable . . . The classroom teacher is aware of all of those ramifications, legislative, political, certainly academic.

Moreover, many administrators sought out collaborative teachers for insight and assistance during the reform process. For example, Justine Jordan, a high school teacher and collaborative leader in Georgia, commented that interaction with her principal had been considerably different before and after her participation in the collaborative. Previously, she would expect to request resources such as graphing calculators, then wait for the principal to approve the purchase. She could wait months, even a year, before receiving requested resources, if ever. In contrast, as a collaborative participant, Jordan found a vehicle to borrow graphing calculators and to receive training in a technology-based curriculum. Effective use of new resources became a leveraging tool for Jordan to influence her principal's decisionmaking: "A lot of times when we can show an administrator here that the kids are using something, the purse strings loosen up a little quicker." She later elaborated:

> When I come in, like with the graphing calculators, and I show them [to administrators] and say, "Look, this is what my kids can do with this. Come up here. You'll see these kids using this." And then he turns around and says, "Well, how can *you* manage to have those for us so that we don't have to send them back?" That's one way a lot of schools have gotten graphing calculators and other materials as well . . . It gives you more credibility. And when you say, "This isn't particularly working, but this here does work," it gives you more credibility. It's not that you just read it or that it's some off-the-wall idea that you've come up with. You've either done it yourself or you've seen it work somewhere else.

In many collaborative sites, the teachers accepted the challenge to change their mathematics programs with the support of administrators, rather than in reaction to requests or mandates from administrators. This shift in authority was built on the demonstrated ability of collaborative teachers to introduce innovations in classrooms on their own initiative, an ability that teachers largely attributed to ongoing professional growth and peer support within the collaboratives. Thus, the newfound responsibilities many collaborative teachers took on was a product of the learning opportunities made available within the collaborative and of the recognition by administrators of the professional growth of teachers. The communication and advancement of an agenda for mathematics education reform by teachers, then, depended on the strength and credibility of the learning system.

Collaborative Teachers and Partnerships. Partnerships with business, higher education, school districts, and other organizations played an important role in creating and substantiating opportunities to learn for teachers. Furthermore, the nature of partnerships in part determined the extent to which teacher learning translated into educational change. Although these partnerships differed considerably among the collaboratives, several general observations about partnerships are pertinent.

First, teachers played a significant role in giving direction to the partnerships. Most often, teachers interacted with business and higher education partners on collaborative boards and committees. Kyle Keys, a representative from higher education in the Columbus collaborative, described his experience: "The [teachers] on the board . . . play an aggressive position. They do not sit back and let the business and higher education run things. They are in there saying their bit and raising objections or support . . . whatever they need to do."

A business representative in Memphis, Harold Kristoff, viewed the interaction of teachers and collaborative members from business and higher education similarly, with representatives from all sectors contributing to important discussions. Asked what the most important outcomes of the collaborative had been, Kristoff responded:

> Communication, I think, would be put right up at the absolute top . . . Open,
> free communication is the thing . . . between college teachers, high school
> teachers, elementary teachers, business people. People who have been involved
> in math and have gone in different directions are now coming back together.
> It's interesting to see that they [all] still have good ideas.

Second, the most successful collaborative partnerships with business and
higher education occurred when all partners had been given or had assumed very
specific roles. The cultural mismatch between business operation and the education enterprise, and to a lesser extent between K-12 education and higher education, at times was an obstacle in developing good working relationships between
partners from the different sectors. However, the collaboratives that overcame
this cultural mismatch did so by delineating specific roles for all participants.
Critical to any collaborative partnership was recognition that each partner contributed uniquely and importantly to the collaboration. The words of Shane
Unnan, an business internship sponsor in the Milwaukee collaborative, depicted
this requirement admirably: "I'm an engineer—a manager of quality and the operations group. My training is not in teaching or curriculum development. That's where
the teachers' expertise comes in. I can show them science and math applications and
it's really the teachers' roles, as I view it, to really develop the curriculum."

In addition to internships, many other partnerships developed in the collaboratives. For example, Temple University, Christian Brothers College, University
of Memphis, Columbus College, and Marquette University provided substantial
in-kind support to the collaboratives (e.g., office space, clerical support, fiscal
agency, hosting of collaborative functions) as well as administrative support—in
some cases salaried positions for collaborative directors. Benjamin Edwards, a
higher education representative in Memphis, witnessed an unprecedented
growth in collaboration between mathematics teachers and colleges and universities: "The intense relationship of mathematics [teachers] with the colleges is a
direct offshoot of MUMC [the Memphis Urban Mathematics Collaborative]. If
they are comfortable to say, "Could you give us a classroom?" or something like
that . . . this has all been [due to] the collaborative."

Partners from higher education often provided expertise in grant writing and
in knowledge of current issues in mathematics education reform. In collaboration with an informed and active corps of mathematics teachers through the collaboratives, powerful partnerships were formed. Nowhere was this phenomenon
more evident than in Philadelphia, where the collaborative's Teacher Congress
was viewed as an ideal partner for university efforts in mathematics education
reform. Evan Irving, a partner from higher education, described plans to involve
teachers in his institution's effort to improve the mathematics background of
incoming students:

> The idea is to get juniors in high school, to give them . . . the placement test
> that an incoming freshman would take and diagnose their potential for
> college at that time. And that if they are not up to speed, make recommen-
> dations for math courses that they can take in their senior year. . . So the con-

gress is a natural organization to plug into this, because we already have these highly motivated people . . . These same congress people could do the teaching in high school. They could teach these courses in the senior year.

Denise Haverstrom, a representative from another institution, recognized similar potential for partnerships between the collaborative and her school. In a pending effort to reform the core mathematics and science curriculum, she said, "The teachers from the congress are committee members . . . I brought them in to be curriculum committee members of each of our math and science committees."

Effective partnerships also developed within the context of specific activities. A program in Memphis illustrates the effective delineation of duties by collaborative partners. Tracy Xistris, a teacher-leader, described a unique partnership the collaborative utilized to increase the effectiveness of its calculus program. "We had people from the research department at [a local hospital], the biomedical research people, because they knew research better than we did. And they took all our data [from the program] and they made some correlations and they gave us some suggestions."

A final observation on partnerships concerned the relationships that had formed across collaborative sites. The Outreach Program of the Urban Mathematics Collaborative Project which developed the five replication sites provided a meaningful example. Each of the replication sites benefited from contact with the original sites. The original collaboratives provided a form of technical assistance with ideas and frameworks for administration and programs. Milwaukee's Algebra Network grew out of an idea gleaned from another collaborative, as Yvonne Alberts, the network's director, recalled, "One of the things that I found out that certain [collaboratives] have done . . . is they had an algebra network and a geometry network. We met . . . and we were brainstorming things that we could do as a large metropolitan collaborative. From that [emerged] the idea of networks."

However, there was reduced fiscal support for partnerships across the national network after the Ford Foundation funding cycle ended in 1990. Collaborative participants sought to sustain partnerships across sites in the intervening period, due to their perceived value. Informal contact among collaborative staff and directors remained somewhat prominent, though it was less regular since the UMC's annual national meetings ended in 1992. The collaboratives have since formed a new national governing board charged with coordinating projects across sites. Additionally, two national projects emerged that benefited from cross-site collaboration. The Classroom Assessment in Mathematics (CAM) Project has served as a focal point for discussion about assessment issues in mathematics education across several UMC sites. A new national assessment program built on the foundation of CAM is currently being implemented. A second program, the Leadership in Urban Mathematics Reform Project, also involved several collaborative sites. The purpose of this project was to continue the development of mathematics teacher leaders in urban school districts that began with the leadership institutes of the UMC project. EDC, a partner of the collaborative enterprise since 1985, played a key role in supporting the governing board and administering both projects.

Changing Conceptions of School Mathematics

Mathematics Education From the Beginning of UMC to the Present. The history of the thinking that led to the development by the Ford Foundation of the Urban Mathematics Collaborative Project and the work of the collaboratives spanned nearly 15 years. Over that time, mathematics education in the United States experienced a dramatic period of reform, energy, and effort.

The Urban Mathematics Collaborative Project arose when the mathematics education community and users of mathematics were on the brink of significant change that in 1985 had not been crystallized. The UMC project was conceptualized with a particular perspective toward mathematics, namely, that the mathematics that students will be needing and using in their lives will be different from that needed by previous generations. Moreover, it would be important for students in inner-city schools, through the guidance of their teachers, to participate actively in the latest thinking in mathematics as defined by the mathematics education community, business, industry, and higher education. However, in structuring the project to develop 11 collaboratives, the Ford Foundation did not present a central perspective or position, but allowed individual collaboratives to define and develop their own.

In the 10 years that followed the grant to the Cleveland Education Foundation to establish the first collaborative, the United States saw widespread reform in mathematics education. No one in 1985 could possibly have predicted the attention and the magnitude of the reform in the teaching of mathematics in U.S. schools that has been fueled by the NCTM *Curriculum and Evaluation Standards for School Mathematics* and *Professional Standards for Teaching Mathematics*.[3] All of the collaboratives were active in advancing the NCTM *Standards* and had used the *Standards* as a guiding force for structuring and conducting programs. The *Mathematics Framework for California Public Schools* (California Department of Education 1991) had a similar influence on the activities and goals of the California collaboratives (Heck, Webb, and Martin 1994). Many of the collaboratives held events at which teachers and others discussed and analyzed these documents. One significant indication of UMC project's influence on the mathematics education community was the inclusion of people associated with UMC on the 25-member writing team of the third and final standards document on assessment (NCTM 1995). Whereas members of the writing teams of the previous two standards documents had included but one person from a large urban school district, the assessment standards writing team involved people from three of the UMC sites.

---

[3]The *Standards* are analyzed in more detail in Chapter 1 of this volume.

Locally, site visits revealed that collaborative activities and teacher networking through the collaboratives had indeed influenced changes among individual teachers—in what mathematics was taught, how mathematics was taught, how students' knowledge of mathematics was assessed, and what mathematics students learned. Teachers and students experienced an increase in enthusiasm toward mathematics in schools. Some collaborative teachers were instrumental in getting new textbooks and courses approved by administrative bodies, while other teachers were instrumental in making a strong case to administrators leading to inclusion of reform-guided instructional activities in school and district mathematics programs. Significant change among individual teachers in relation to the collaboratives was apparent, along with at least an informed awareness by a significant proportion of mathematics teachers within a district. What was not evident was significant large-scale change in entire mathematics departments within schools or in mathematics programs across a number of schools within a district.

Change in Vision. Much of the language of the current reform effort in mathematics education can be categorized in three areas: how mathematics is taught, what mathematics is taught, and to whom mathematics is taught. The NCTM *Standards* (NCTM 1989 and NCTM 1991), *Everybody Counts* (MSEB 1989), the California *Framework* document (California Department of Education 1991), and other reform documents addressed each of these areas at length. A vision of mathematics education reform certainly existed in the United States in the early 1990s, though interpretations of that vision varied within the mathematics education community and over time.

One of the growing strengths of the UMC project was its commitment to reform of mathematics education. Nearly all teachers and others interviewed for this case study viewed their participation in the collaboratives as an effort to *change* mathematics education, not to preserve, to improve, or even to perfect what currently existed. The collaboratives' efforts to change mathematics education depended on a vision of how, what, and to whom mathematics should be taught in schools. The collaboratives' vision depended in turn on the current state of mathematics education at the classroom, school, district, state, and national levels as its point of departure.

Teachers maintain a considerable amount of autonomy in selecting what mathematics topics to include in their courses. There is no mechanism—nor is there interest in having a central mechanism—that mandates precisely what a teacher should teach in a course. Although teachers have autonomy in deciding what they teach, there are a number of forces that inhibit variation from the norm—standardized testing, pressure to prepare students for the next course in the sequence, the rigidity of textbooks, established tradition, and a variety of state and district curriculum guides. Moreover, there is evidence that mathematics teachers' choices about how and what they teach are more heavily influenced by how and what they were taught while in high school than by any professional

development experience (Romberg and Middleton 1994). Under such conditions, any large-scale change in the content of courses and how that content is
taught is very difficult.

All of the collaboratives in some way supported activities to make
mathematics teachers aware of what changes were needed in the mathematics
curriculum and why such changes were necessary. For example, the collaborative coordinators and directors gave presentations about the NCTM *Standards*
while the documents were in draft form. Some of these presentations were
orchestrated by the technical assistance project operated through EDC. Some of
the collaboratives had specific activities in which groups of teachers reviewed
drafts of the *Standards* and sent their comments and concerns to the writing
team. District mathematics supervisors from collaboratives were afforded the
opportunity as a group to discuss and think about means of implementing ideas
from the *Standards*.

As a result of the activities of the national UMC network, at least one person at each site had a fairly extensive knowledge of the *Standards* and had given
some thought to their dissemination prior to the release of the first document at
the 1989 meeting of NCTM in Orlando, Florida. By this time individual collaboratives had conducted a series of activities and had gotten a number of teachers
to participate in some way in evening meetings, industrial site visits, summer
institutes, workshops, and other activities. The *Standards* appeared at an oppor-
tune time—collaborative directors and coordinators were seeking more focus
after a productive phase promoting involvement and awareness. In addition to
efforts by the national UMC network to inform collaborative leaders about the
Standards, collaboratives used funds to send individual teachers to national and
regional meetings of NCTM. Thus, teachers participating in collaboratives
became aware of the changes NCTM was advancing. In Memphis, the mathematics curriculum director talked about the *Standards* at every workshop she
gave to teachers. In Los Angeles, +PLUS+ teachers who organized the collaborative's annual workshop series selected topics based on the recommendations
from the California *Framework* and the NCTM *Standards*. In Milwaukee, the
Algebra Network included discussion of the NCTM *Standards* on its agenda.

It is difficult to ascertain whether teachers who had a reform-oriented understanding of mathematics and learning were the ones who chose to participate in
a collaborative, or if participation in a collaborative helped to change what
teachers were thinking. Some combination of the two was probably most accurate. In any case, both collaborative participants and nonparticipants identified
many teachers in their areas who were interested and active in the current reform
movement and knowledgeable about the NCTM *Standards*.

Established networks that the collaboratives had formed among teachers,
business, and higher education served to acquaint others, in addition to the
teachers, with the *Standards*. When a part-time instructor at a university was
hired to serve as a collaborative coordinator and a link with the university, she

was given a copy of the *Standards* before she was hired and told to read it. This was the first time this future coordinator had seen the NCTM document.

Alex Barnaby and Ian Kent, business partners in Columbus and Milwaukee respectively, each identified knowledge of the NCTM *Standards* with their involvement in collaboratives. Barnaby, in fact, attended an NCTM regional meeting with a delegation from the Columbus collaborative.

Having university and business partners read and know the *Standards*, however, could not breathe life into them as well as could teachers who had tried ideas from the *Standards* in their classroom. George Jackson, a mathematics professor and collaborative administrator, reported,

> We . . . have people from the [university mathematics] department come in and present an hour's lecture on the topic that is supposed to be for elementary teachers. At the end of the hour, the report back to us was "nothing was learned." We had these middle grades teachers discuss the same material using NCTM activity-type work, and it was a different experience.

Frances Ilps, the collaborative coordinator, added, "The teachers thought they came out with a gold mine." Jackson explained, "We moved away from having the university person be a workshop leader to selecting middle grades teachers to be the leaders, and there was a tremendous improvement. So we discovered that having peer teachers be leaders is an essential part of making them successful." The strength of the collaboratives' vision of mathematics education therefore depended not only on the quality of the ideas behind it, but also on who advanced the vision and how it was articulated. Collaborative teachers nearly unanimously preferred efforts grounded in teacher leadership, teacher decision-making, and the active participation of teachers. Most collaborative activities evolved to incorporate these components.

Important to an understanding of the collaboratives as a living example of reform is the nature and pervasiveness of the vision of mathematics education that had emerged 10 years after the collaboratives' inception. Although the collaboratives' vision of how, what, and to whom mathematics should be taught was not fully solidified within or across sites, certain themes were consistently present—especially among those teachers most actively involved in collaborative leadership and activities. In particular, teachers' classroom practices included an emphasis on applications of mathematics to realistic situations, and especially those that used technology; hands-on, cooperative learning; and learning how to communicate mathematically. More emphasis was given to mathematics as a language for communicating about and solving problems. Finally, a broader variety and greater number of students were viewed as capable of learning mathematics. A critical facet of the collaborative experience for teachers was that this vision of mathematics education should transform rhetoric into reality in the classroom, in professional development, in interactions with colleagues, and in teachers' knowledge and beliefs.

Teachers' Knowledge and Beliefs: The Challenge of Change. Mathematics education reform in the 1990s is identified with a fairly coherent and widespread vision of change typified in the NCTM *Standards* and other reform documents. Nationally and locally, the UMC project had tied itself to that vision of change that addressed to varying extents what mathematics programs, classroom instruction, and student learning could and should be. Reformers and researchers have, however, recognized that communication of a new vision of mathematics education, or even widespread awareness of that vision, does not necessarily produce change that permeates all or any of these areas. Substantive change in the mathematics programs of schools, the classroom instruction of teachers, and the mathematical learning experience of students relies heavily on individual teachers' underlying beliefs about mathematics and learning (Cohen 1991; Fey 1979; Lichtenstein, McLaughlin, and Knudsen 1991; Simon and Schifter 1991; Thompson 1984; Wilson 1990; Wiske and Levinson 1993; and Woodrow 1992).

Teachers, then, were both the objects and agents of change in the collaboratives' endeavor. They were the objects of change in that the UMC project was founded on the conviction that a concerted effort to improve the lot of urban mathematics teachers would lead to improved mathematics learning for their students. They were the agents of change in that the project also recognized that teachers should become problem solvers and decisionmakers, proactive contributors to a new vision and a new reality of mathematics in urban schools. Teachers were the major focus of efforts to improve mathematics learning in the urban schools in which they worked. Individual teachers' knowledge and beliefs were, therefore, especially important to the UMC project as a reform effort in mathematics education. In order to have significant influence on teachers' commitment to and involvement in mathematics education reform, the collaboratives devoted effort toward changing teachers' knowledge and beliefs about students, about mathematics, and about pedagogy (Pitman 1994).

The UMC project never intended to mandate change in mathematics programs or teachers' classroom practices. Collaboratives instead relied on their capacity to change teachers' knowledge and beliefs about mathematics education and to enable teachers to act on their knowledge and beliefs. Several structural and organizational parameters of the collaboratives dictated the necessity of this approach to change. First, participation of teachers in collaborative governance, programs, and activities was almost entirely voluntary. Therefore, collaboratives had to take into account teachers' existing and changing knowledge and beliefs, and support changes consistent with teachers' level of comfort with reform. Second, the collaboratives as organizations existed independently of their constituent school districts and had no official authority to require specific changes in classroom practices or department, school, and district policies. Lacking this authority, the collaboratives relied on the strength of teachers' convictions about necessary and feasible change. Only then could knowledge and beliefs be translated into new classroom practices, professional involvement, and lobbying efforts for change in policies. Third, the collaboratives prided themselves on the fact that their agenda for change was grounded in the expressed needs and desires of member teachers. Espousing labels such as "grassroots," "teacher-driven," and "teacher-led," the collaboratives' agenda for change depended on their capacity to develop and harness teachers' knowledge and beliefs about students, mathematics, pedagogy, and the teaching profession. Ideas, programs, and suggestions could be strategically introduced, but the collaboratives' vision ultimately depended on the teachers' willingness to articulate, develop, and support it in action.

Many collaborative participants recognized the centrality of this fact. Pearl Quigley, the Los Angeles collaborative director, said: "So many projects stop or stand still or dry up and go away because resources end, but there is a driving force going on here . . . there's something so valuable about experiences these people have had that they are willing to get their own resources to keep it alive." Denise Haverstrom, a higher education representative in Philadelphia, noted, "The teachers of the Philadelphia school district were the driving force and needed a little direction here and there. But it's their baby." George Jackson, a higher education representative in Memphis, similarly stated, "The success that we have had is due to it being teacher-led instead of administrative-led. Every activity was initiated by teachers and teacher groups . . . This is teachers for teachers rather than trying to satisfy an administrative demand."

All of the collaboratives had, to varying extents, adopted the NCTM *Standards* as a foundation for desirable reform in mathematics education. The three California collaboratives viewed their state's *Framework* (California Department of Education 1991) similarly. Furthermore, the Woodrow Wilson Institutes, the activities of the North Carolina School of Science and Mathematics, the Interactive Mathematics Program (IMP), the Geometric Supposer software (Yerushalmy and Schwartz 1985), *Discovering Geometry* textbook (Serra 1989), and many other concurrent developments in mathematics education had substantially contributed to the collaboratives' agendas for change and the programs offered. Many collaborative teachers doubted that they would have had much, if any, exposure to these other developments in the absence of the collaboratives. Collaborative teachers, though, had not been pressured to comply with the goals or visions of any of these contributors. Rather, teachers' knowledge and beliefs had given these contributors their prominence in the history of the collaboratives. The collaboratives' impact as living examples of reform and innovation lay in the processes by which teachers' knowledge and beliefs change and become manifest in mathematics programs, classroom practices, and student learning.

Teachers' Knowledge and Beliefs About Students. An underlying assumption during the development and founding of the Urban Mathematics Collaborative Project was that students' mathematical learning could be enhanced through a sustained effort focused on change among teachers (Nelson 1994). Given the centrality of students and teachers to one another in the educational enterprise (e.g., Bird and Little 1986, and Little and McLaughlin 1991), teachers' knowledge and beliefs about students played a vital role in the change process in which the collaboratives were engaged.

In 1991, the Urban Mathematics Collaborative Project issued a direct statement about students in its position paper on equity: "The Urban Mathematics Collaboratives (UMC) believe that *all* students can and must learn mathematics" (EDC 1991, p.1). Individual collaborative teachers' interpretations of equity and the notion that all students can and must learn mathematics varied. Conservatively, collaborative teachers as a group decidedly believed that *more* students can learn more mathematics than has traditionally been the case; many collaborative teachers included *all* students in this conviction. Eva Harrison, a Memphis middle school teacher had viewed a goal of the collaborative as "restructuring our math curriculum and . . . how we present materials to the students, taking into consideration the total population basically, not just one set, you know . . . I think we're dealing with equality . . . I think that was one of the reasons I bought into . . . the collaborative." Alex Barnaby, a representative from business who is actively involved in the Columbus collaborative, offered a similar thought:

> One of my concerns is that . . . the emphasis of the math collaborative really reach the mass of the population and that we don't find ourselves doing a better and better job of educating the top 15 percent of the math students in the country [and] the other 85 percent are kind of left drifting. We're trying to move the country into a very technical math- and science-oriented environment.

Collaborative teachers frame their beliefs about student equity in mathematics education in two ways. First, all students deserve access to challenging and appropriate mathematics instruction in school. Second, the full potential of all students should be developed, rather than settling for a minimum level of mathematics understanding in each student.

Many collaborative teachers believed that providing the mathematics instruction that students need and deserve involved developing and adopting courses that suited students' varying needs and ensuring that students were prepared and encouraged to study mathematics throughout their school years. Two Los Angeles high school mathematics departments had utilized the resources and collegial support of their collaborative to alter their course offerings and teaching assignments, to work directly with students, and to lobby their counseling departments and administration in efforts to ensure that all students had access to the highest quality mathematics courses and were encouraged to take them. Collaborative teachers in Memphis developed a new course designed specifically to provide a challenging option to students who did not enroll in AP Calculus and who would otherwise receive no mathematics instruction. The Columbus collaborative was actively engaged in a program to provide challenging and appropriate instruction in algebra to all high school students in its constituent districts, in accordance with a recent state mandate that all students had to complete first-year algebra in order to graduate from high school.

A subset of collaborative teachers viewed access slightly differently, believing that a single challenging core curriculum, unlike the existing course sequence, was appropriate for all students. Scott Truesdale, a Los Angeles teacher who taught the reform-oriented IMP—which he had learned about, experimented with, and become fully involved with through the collaborative—believed that a project-based, integrated mathematics curriculum best serves the needs of all students, including gifted students and those with learning disabilities. Still others viewed access to quality mathematics in part as an integration of challenging content from many areas of mathematics into all courses. A member of the Columbus collaborative, Derrick Duncan, said:

> After I became involved in the collaborative, I perceived more the need to teach a higher level of mathematics to all of them . . . than to have general math students and algebra students . . . The more I became involved in the collaborative, the more I realized that algebra and geometry and some of those things were accessible to all students, even if not in a formal course.

Although Truesdale and Duncan exhibited certain differences with regard to their beliefs about students, two common ideas about students' access to mathematics are apparent—that all students are capable of success with challenging and appropriate mathematical content, and that teachers and schools must provide and support access for all students.

A related notion of equity, that the full potential of all students should be developed, follows closely the idea that challenging mathematics is accessible to all. Vincent Valentine, a high school teacher in Georgia, was introduced to graphing calculators through the Columbus collaborative. He reflected on his recent experience with traditionally low achievers:

> Here we are in high school and we . . . don't want to be doing eighth grade math . . . If you'll take that kid and get him out of the boredom of the eighth grade math, let him use a calculator, he's excited, it's different. Most of those kids [have] been doing the same thing for years . . . He winds up being a pretty good student when he can use a calculator, so why not let him use it. You're not going to teach him his math facts in high school . . . But you might be able to slip in linear equations.

Many collaborative teachers similarly found that students who might not have been reached by traditional mathematics instruction were capable of learning and understanding challenging mathematics when given the appropriate opportunity. Eva Harrison, a Memphis middle school teacher, learned of a variety of new instructional techniques by hearing about them and engaging in them at collaborative activities and experimenting with many of them in her classroom. She marveled at the results she was having with students. Her experiments led her to some intriguing conclusions about the varieties of ways that students could appropriately show their understanding: "Sometimes they act [the mathematics] out; they write to me on paper how they got the solution, they use models to show me how. There are all kinds of ways you can do that. Then I have some students who don't want to write it out or act it out. They are going to have a conversation with me and tell me how."

Valentine and Harrison illustrate that collaborative teachers were committed to the idea that, when given the opportunity, more students are capable of successfully learning and understanding challenging and appropriate mathematics. Moreover, their experiences suggest that some teachers adopted or were moving toward the belief that all students can and must learn challenging and appropriate mathematics. Teachers' beliefs about what mathematics is challenging for all students and what mathematics is appropriate for all students, then, are fundamental to the story of the collaboratives.

Teachers' Knowledge and Beliefs About Mathematics. Researchers of mathematics teachers' professional growth surmise that a reconsideration of knowledge and beliefs about mathematics by teachers is fundamental to real change in mathematics programs and students' mathematics learning experiences (Simon and Schifter 1991, and Wiske and Levinson 1993). Studies of individual teachers have shown that utilization of progressive methods and innovative materials in classrooms do not necessarily coincide with reform-oriented beliefs about mathematics among teachers or improved higher order mathematics learning among students (Cohen 1991 and Wilson 1990). Teachers' beliefs about what mathematics is and what constitutes mathematical learning will determine how teachers gauge improvement in their classrooms.

An earlier study of a large sample ($N=490$) of mathematics teachers in districts served by collaboratives found that although conceptions of mathematics varied considerably among teachers, the differences were not discriminated by frequent, occasional, or no participation in the collaboratives (Middleton, et al. 1990). Indeed, among the entire sample of collaborative teachers interviewed and observed for this study, changed beliefs about mathematics were in many cases not apparent or were difficult to identify. For a meaningful number of individuals, however, a reconceptualization of mathematics had been an outcome that they attributed to their involvement with a collaborative. Of particular note, teachers who had participated in activities that involved the learning of mathematics with fellow teachers reported changed knowledge and beliefs in this domain.

A contrast is illustrative. Three teachers in a Los Angeles high school described vastly different beliefs about mathematics. Each had been involved in the collaborative in some capacity, as had nearly all the mathematics teachers in this school. The first, Francisco Hierro, was involved in the initial efforts to organize the collaborative and participated in some of its early activities. Hierro found the activities somewhat valuable, but not fulfilling of his needs as a teacher. He had not participated for several years. He described his view of mathematics as, "the rules . . . that's all it is, really. It's just a bunch of rules. You learn how to use them well . . . you're successful at this." The view of mathematics that Hierro described is characteristic of other "effective" teachers (Remillard 1992). Although teachers may in fact have measurable success in teaching students mathematics based on this view, it remains questionable, especially in light of the current reform movement, whether this brand of mathematics instruction serves the needs of students in an increasingly technological and information-based world.

By contrast, Earl Grant, another mathematics teacher in the school, remained active in the collaborative. His participation had been mainly through the collaborative's workshop series, which stressed experimentation and reflection; also, he and several other Los Angeles teachers had received support from the collaborative for travel to the North Carolina School of Science and Mathematics (NCSSM) for summer institutes. He had as a result learned a great deal about teaching with graphing calculators. Grant offered an intriguing comment about their use in the classroom:

> I think I'm trying to have the kids get a broader picture so that they can use something like a graphing calculator to bring up an answer quickly and discuss from that . . . They can use the calculator to do some of the dirty work. Then we can talk about the significance of the answers or the significance of what they see in the calculator.

Although Grant did not offer a clear explanation of his beliefs about mathematics, he rejected the importance of rules and algorithms ("the dirty work") and emphasized students' discussion of the significance of representations. Nowhere was his emphasis more evident than in formal and informal classroom discussions. Grant constantly rephrased and redirected students' questions to stimulate further thinking and discussion. He consistently allowed several seconds of silence to follow questions and arguments so that students could think and respond to him and to one another with reflective, thoughtful insights.

A third teacher, George Hillman, who had been extremely active in the collaborative as a leader and participant, offered a more explicit explanation of the belief in mathematics he emphasized in his teaching. Hillman described himself as a "chalkboard, paper, and pencil" teacher a few years ago. He related his current beliefs about mathematics while discussing his goals for mathematics students:

> I would like them to be able to think mathematically . . . I want them to be able to look at a problem, and ask themselves questions that lead them toward solving or understanding the problem, and not just mechanically and blindly say, "I need this formula over here when I do that. I don't know why I do it" . . . I want them to be able to think their way through a problem. So, I really try to teach . . . thinking about how they solve the problem rather than memorizing.

Hillman's geometry class was observed while engaged in a brief problem-solving effort. Hillman asked his students to offer several possible ways to solve a particular problem and to determine what information in the problem was useful and what was not. The focus of this and many other parts of the observed class period was on the process of problem solving, rather than exclusively on the solution.

Hillman offered another specific example:

> We're doing probability in the algebra 2 class . . . Rather than memorizing those formulas, I want them to think and make a model of the probability situation, and come up with either the proper permutation/combination answer, or the correct probability under the restrictions of the problem. I haven't taught formulas in there at all. Some purists may dislike that, but we push buttons on the calculator if we want to know a permutation. We understand the background of it, but we push buttons. My concern is—do you know when to use the permutation button and when to use combinations, and then do you know how to use those in terms of probability?

For many teachers, participation in the collaboratives influenced the shift of beliefs about mathematics from what Hierro described to what Grant implied and Hillman illustrated. A past belief in mathematics as computation and algorithms was a common theme among collaborative teachers. Presently, the beliefs of many about mathematics tend to emphasize understanding of mathematical process, application of understanding, communication about processes and understanding, problem solving, and logical thinking. Collaborative teachers described an increased emphasis on the interconnections of mathematical topics, connections of mathematics to other disciplines, and the incorporation of new topics in mathematics into their knowledge and practice.

A middle grades teacher in Georgia, Meg Mobley, explained the shift in beliefs she had experienced, and the strong connection she saw between participation in the collaborative and her change in beliefs:

> I got involved in the collaborative . . . and my big focus before was computation: "Let's get up here and let's add correctly and multiply correctly." I mean, that's what I focused on—if you can do all the skills. So now I'm . . . doing a little bit more and understanding why I'm doing this, and that's kind of a lot from the collaborative . . . Like we do a lot of word problems, but they are stories, you know. I have a series of word problems and at the end, it's a story. The kids have to understand each problem before we can get to the [end]. I just made that up . . . but I kind of got different ideas from people [in the collaborative], what they did . . . My big focus was on computation . . . That's all I worried about, I didn't care if they understood it or not, so the collaborative really changed my mind.

Adam Dooley, a participant in the Milwaukee collaborative's internship program, noted an important shift in his beliefs about mathematics toward an emphasis on realistic applications of mathematics. The internship exposed him to applications of mathematics in business, offered modeling of teaching strategies in a university course, supported classroom experimentation with a unit developed from the internship experience, and advocated long-term reflection on his learning through keeping a journal and interacting with other teachers and university and business participants in the program. The year-long experience had led him to conclude, "I believe more than ever that teachers need to . . . provide more 'hands-on' activities [in mathematics] as well as connect those activities to the real world. Teachers need more than ever to provide students with the skills needed in the real world."

An emphasis on problem solving as a process, rather than a means to an end, also characterized collaborative teachers' changing beliefs about mathematics. Carolyn Graver, a high school teacher in Philadelphia, offered an explicit explanation of her changing beliefs about mathematics. Through her connection to the collaborative, Graver taught the reform-oriented IMP curriculum and participated in a network of IMP teachers who had followed the curriculum together in a learning environment paralleling what their students did in the classroom—cooperatively solving problems, writing about mathematics, and devising multiple means of solving problems. Graver's experience with writing as an instructional tool in IMP led her to the following conclusion:

> [Even if students] don't come up with the right answer, you want to see the process they are going through to solve [problems]. That's what it's all about, it's problem solving and that's a thought process . . . The whole key is to get them to not memorize a formula but to be able to approach a problem and then have different skills to solve a problem. So they can use diagrams, they can use pictures, anything to help them come up with the answer.

Across several of these examples, a single intriguing theme emerged. Many collaborative teachers were shifting to a belief that mathematics is an individually and socially constructed way of understanding and solving problems, as opposed to a static body of knowledge with one best method of solution, one proper answer, and one understanding of each problem. Students were being encouraged in many collaborative teachers' classrooms to use their strengths to find the best ways to understand, to represent, and to make sense of problem situations.

One of the most active leaders in the Columbus collaborative, Justine Jordan, explained her transition to the view of mathematics as an individually and socially constructed way of thinking about the world. In response to a question about the most important mathematical ideas she taught her high school students, she said:

> One is . . . logical thinking . . . To just try to build some kind of logic into their thinking. Before they act, to formally visualize the problem that they're working with and kind of create a rough plan. And then start mapping that plan out. That mathematics is not just "find an equation, plug it in, grind it out." There might not be an equation to plug it in. There may not be an answer. It may be that I have several answers. It could be from here to here. And that there are appropriate times for estimation rather than exact answers. And that there are times when I do estimate and need to overestimate. Or times when I would want to underestimate . . . So many kids think that mathematics is strictly based

on a set of rules, but there are not necessarily rules to be followed. It's more what makes sense for this situation.

Jordan related this conception of mathematics to her experiences with colleagues in the collaborative: "You have a whole group of people, or a network of people, here [at the collaborative] that seems to be creating [a] . . . belief that mathematics is *doing* it, mathematics is not this set of rules that everybody has to fall into."

Many collaborative teachers' beliefs about students and about mathematics changed. They were therefore faced with the challenge of integrating two sets of changing beliefs in their daily work in the classroom. Teachers' knowledge and beliefs about pedagogy elucidate how collaborative teachers met this challenge.

Teachers' Knowledge and Beliefs About Pedagogy. Pedagogy is the critical link between students and content that makes teaching a unique human endeavor (Ayers 1990). Mathematics teachers' knowledge and beliefs about pedagogy have numerous sources and influences, including teachers' extensive experience as students (Bird and Little 1986), their own high school and college mathematics teachers, and their teaching colleagues (Middleton et al. 1990). The act of teaching is a continuous endeavor to improve pedagogy to enable the most effective interaction between students and the content of mathematics (Ayers 1990). Teachers' attempts to improve pedagogy involve the influences both of their own intentions and of innumerable external conditions of the students, classroom, department, school, administration, and community (Bird and Little 1986). The teachers in the UMC project faced some of the most challenging external conditions found in American schools (Nelson 1994). Collaborative teachers also encountered acute internal challenges when rethinking pedagogy to complement their emerging knowledge and beliefs about students and mathematics.

Conceptually, pedagogy cannot simply be equated with teaching methods. Teaching methods refer to *how* a teacher teaches mathematics. Pedagogy encompasses both how a teacher teaches mathematics and *why* a teacher chooses to teach in that way (Wilson 1990).

The collaboratives, to be sure, exposed teachers to many new teaching methods. In particular, many teachers associated their awareness of methods such as cooperative learning, activity-based instruction, and discovery learning with participation in collaboratives. Additionally, many teachers credited the collaboratives with their introduction to classroom uses of calculators, graphing calculators, software, manipulatives, and a variety of other new materials for instruction in mathematics. "I came out of college maybe six or seven years ago," Percy Patterson, a Georgia high school teacher commented, "and I'm not even sure I even knew what cooperative learning was and . . . a lot of the manipulatives that I have seen and that I use come from the collaborative or were inspired by it." Tonya Trier, Patterson's colleague, expressed a similar sentiment: "I think the collaborative has allowed me to see a lot of different manipulative things that are available . . . to use, not just instructional techniques but other physical objects that you can use, things that you can do, activities."

Exposure to these and other new teaching methods came to teachers from many sources. Not only did collaboratives expose teachers to new methods, but textbooks, curriculum guides, district in-service training, and college course work all provided similar exposure. Yet without a reconsideration of why one might choose particular methods in light of the nature of learning and content, new teaching methods—no matter how novel—are nothing more than processes used to teach the same mathematics under a different guise (Cohen 1991, Simon and Schifter 1991, and Wilson 1990). What set the collaboratives apart, in teachers' views, was their attention to both *how* to utilize a new instructional approach and *why* a teacher might choose to use it based on beliefs about students and mathematics. The combination often produced truly new pedagogy.

Scott Truesdale, a Los Angeles high school teacher, described how he came to the collaborative dissatisfied with his pedagogy:

> Ten years ago, I might have still been doing only basic math and high school math . . . trying to get the kids to do their problems and the topic test and to learn to do fractions and division and lots of drill and practice sheets probably. I knew it wasn't doing much good. And then a few years later . . . I quit that. And then you would have seen the kids using manipulatives almost every day in the classroom to do traditional algebra out of a regular algebra book.

Truesdale was then exposed to a new curriculum, IMP, through the presentation of an IMP unit in the California *Framework* (California Department of Education 1991). Sensing its power to improve his pedagogy, he borrowed materials from another teacher and experimented with the unit in his classroom. He was impressed with the results, "The kids did wonderfully with it. Took them longer than it was supposed to, but they loved it. It was great." Convinced of the success of his experiment, Truesdale sought the support of the collaborative: "I went to work on . . . the [collaborative] director. I said . . . 'We've got to get [IMP] . . . How can we get [it]?'" Beginning in 1992, with the support of the collaborative, Truesdale and a colleague began the teacher training for IMP and began teaching the curriculum at their high school.

Teaching IMP offered Truesdale a sense of how to teach a new brand of mathematics. An observed class period revealed students who were engaged in an experiment using a basic pendulum. Students were observing and recording the motion of the pendulum as they varied the length of the pendulum, the weight at the end, and the amplitude of the beginning position of the weight. The students charted and graphed data aggregated from the whole class and began to examine functions that might model the observed motion. Truesdale described why he chose to utilize such activities: "The projects are a concept tree, a place to hang ideas. A place to make connections. A place where kids have to look at kind of a garbled situation and refine out of that and focus on a problem. Then they have to figure out a strategy for solving the problem."

For other teachers, the sequence and significance of events in a pedagogical shift can be quite different, as it was for Rhonda Usher and Sharon Villand, two middle school teachers in Memphis. In their case, dissatisfaction with their existing pedagogy emerged from substantive exposure to and modeling of new pedagogical ideas during Camp Mathagon, a week-long collaborative program. The ensuing shift in their beliefs about pedagogy came about through experimentation during the program and in their classrooms, aided by shared collegial support. Both teachers described a pedagogical orientation prior to their participation in the collaboratives that centered on algorithms and isolated skills. Their instruction was teacher centered and textbook driven. Calculators, cooperative groups, and activities that facilitated discovery learning were absent from their classrooms. The two viewed themselves as "effective" teachers, surmising that their students in fact learned the mathematics that they were attempting to teach them.

Both have changed their pedagogy significantly in the past few years. Villand recognized that previously: "There was not the connection to what it is that [students] see in their everyday life. The connection was not there. The problems were not meaningful. They had to do with things that were totally unrelated to their world." Attending Camp Mathagon made a difference. Villand recalled:

> [Changing my pedagogy] only became important to me when someone slapped me in the face with the fact that it needed to be, and when I realized that the problems that we were working had nothing to do with any thinking, first of all, because it was on the same page as the skill was that we were working on, so they didn't have to think through anything. And most of them were unrelated to what the students did in everyday life . . . Camp Mathagon made me see things so differently . . . I experienced problems at Camp Mathagon where we saw the problems in a group setting. That really made me see how fun math could be if we did things that were more relevant to where the kids were.

Ongoing collegial support helped the two teachers to sustain and reflect on the change as they experimented with it in their classroom. As Usher said, "Of course we had a backup system . . . if something didn't go right with me, I always had that interaction with the sixth grade [teacher] and she with me, and I think that that's an important part."

Collaborative teachers' knowledge and beliefs about students, mathematics, and pedagogy were interwoven with their experiences in the classroom, their professional development activities, their interactions with colleagues, and their involvement in the broader reform movement in mathematics education. Many of these facets of teachers' professional lives changed as a result of participation in the collaboratives. As a result, the teaching and learning that took place in collaborative teachers' classrooms looked considerably different than it did just a few years ago, according to the teachers. Engagement in the learning system of the collaboratives enhanced and focused a number of forces that were driving such changes in mathematics education.

Driving Forces of Change. Important reform documents and the growing body of research literature on mathematics education reform identify a number of societal and educational forces that necessitate an overhaul of mathematics education in the United States. Many collaborative teachers have considerable knowledge of reform documents such as the NCTM *Standards* and the California *Framework*, as well as an awareness of research supporting the current reform movement. More significant to these teachers, and quite possibly to the reform movement, have been the teachers' own professional experiences in and out of the classroom, which have placed the forces identified in the documents and research in a realistic context. Participation in the collaboratives played a substantial role in this regard for many teachers. Such participation allowed long-term exploration of and experimentation with the ideas along with fellow teachers; and it preserved a strong connection to the realities of teachers' daily lives and of the districts, schools, and classrooms. In this way, the forces of change that undergirded the reform documents and research actively contributed to how teachers derived meaning from the language of the reform movement and turned the idea of reform into reality in classrooms, schools, and districts.

Emphases on Topics. Yvonne Alberts, who was very active in the Milwaukee collaborative's Algebra Network, described how her teaching of algebra changed:

> For the second year in a row, we've done spread sheets. I take the graphing calculators in all the time. Every day this week, for example, on our quadratics, we look at it. And we're into projects. We've done a project on fractals and we've done projects on statistics and stuff in my algebra class. I introduce for my kids concretely whenever I can, wherever it's appropriate. Now it might be the graphing calculator . . . or it might be algebra tiles, or it might be my square tiles, or it might be pictures, and statistics kinds of things, [or] human box plots.

A striking and often-cited feature of the NCTM *Standards* is the lists of areas deserving greater and lesser emphasis in the reform-oriented view of mathematics education. Indeed, the *Standards* offer certain justifications for the recommended shifts in emphasis. However, the experience of Alberts demonstrates how participation in a learning system like the collaborative can substantiate and solidify the need for a changed emphasis on topics in mathematics courses and programs. She compared her current and past practices:

> I have a pet peeve . . . It is algorithms and rules . . . Sometimes you've got to have these rules. But more often than not, we want to understand the concept underlying it. We developed area . . . we did a lot of squares . . . that idea that area is squares and perimeter is linear. And we did a lot of work like that before we got to this abstract [formula]. And . . . for example, all these textbooks have the exponent properties—powers, products of powers, powers of powers. It is so silly because those kids want that rule. . . A week later, they forgot the rule because they don't understand that an exponent is just a way of multiplying things repeatedly. So that is what I've gotten away from—these rules in the book—and gotten into the concept underneath . . . So that's how I've changed. I

used to teach all those [rules] and no more . . . I could see the mastery was only
the top . . . good rule pushers. And the other thing that I've found, the good rule
pushers aren't the good problem solvers. They aren't. For example, the classic
[example] is with my geo-boards in geometry. "Given an area of 24, what is the
minimum perimeter? Or given a perimeter 20, what's the maximum area?" The
problem solvers are usually the kids that have not been getting the A's. They're
the ones that are willing to try and fail. Right? And the rule pushers, the algo-
rithm freaks, they put one little rectangle on their geo-board and, "That must be
it." [laughter] . . . What I have seen is that they've done a lot of damage [by]
not letting them experiment . . . And the textbooks still do that.

Participation in the collaborative was an important part of the learning
process that undergirded the change in Alberts's practice. Although many of the
ideas and policies that led to certain changes did not originate with the collaborative, "the urban collaborative has let us revisit these ideas and share them
again." She provided the following example regarding contact with a colleague
in the Algebra Network and in other contexts: "He changed my idea of just
teaching functions . . . from plotting data to getting the best fit line—in freshman
year, we never did that before in the freshman year. But you simplify it with the
best fit line. You don't have to go into the difference of squares and all that
stuff." Change in her practice did not occur without difficulty, but the goals
Alberts sought for her students demanded that she persevere: "Cooperative learning is accepted. It's very hard at the beginning of the year when they haven't had a
teacher who uses it. So they learn, 'Yeah, we're going to work together and solve
problems.' They have to learn that it's okay to make mistakes and learn to experiment and conjecture."

Use of Technology. Development of technology exerts two related,
though distinct, influences on mathematics education. First, the rapidly expanding use of particular types of technology in business and industry demands that
students' learning prepare them to use and understand both existing technology
and new technology, which will doubtless appear during their adulthood.
Second, the development of technology for educational uses—especially
computer software and scientific and graphing calculators—has altered the
necessity, feasibility, and accessibility of various mathematics content in the
school curriculum.

Harvey Jacobs, a high school teacher in Los Angeles, explained how
technology changed his teaching:

The thing that . . . is most different about the way I teach now is . . . really just
in the last few years, the graphing calculator. It lends itself to doing a lot of different kinds of things; there's a lot of experimentation. Mathematics was never
an experimental class before. I mean, it could have been if you had a high-powered computer, and you wanted to lug stuff around. But now the ease with
which you can . . . do things, take off. I find myself much more spontaneous . . .
in that I'll have an idea in the middle of doing something [in class], and . . . it'll
just change what I do for the period.

His introduction to and continued learning about graphing calculators was directly related to participation in the collaborative, beginning with a collaborative-sponsored workshop series devoted to graphing calculators shortly after their introduction into the educational market. More intensive and directed learning followed:

> The +PLUS+ [collaborative] connection led to [NCSSM] which has opened just a huge world to me. We have been using their material . . . since prepublication days. Actually, [we] brought it back one summer and integrated it into the curriculum, used it in a variety of ways. Their emphasis on experimentation, on applications, is something that the department has been moving toward gradually. So now we have, I'd say, at least half of the math department, more than half the math department, actively using graphing calculators.

The potential of technology to change the teaching and learning of mathematics further encouraged teachers at Jacobs' school to explore its use in courses that integrated mathematics with other content areas. This effort provided students with realistic experiences of how mathematics is used in business and industry. For example, a mathematics teacher teamed with a science teacher at the school to offer an integrated course using experiments to generate data and the graphing calculator to perform analysis. Still another teacher conducted portions of his geometry courses without a textbook, a technique he learned from a collaborative teacher at another school. He relied on students to conjecture and develop propositions and conclusions using the Geometric Supposer software (Yerushalmy and Schwartz 1985), which he also learned about through the collaborative. Jacobs himself planned to teach a computer programming course using the TI-82 calculator and expressed excitement about the possibilities for learning based on the positive outcomes he had observed so far:

> Some kids . . . start to look at the stuff that they've been doing for a long time, and they sort of see it in a new light. They really have a visual image of manipulations, they really have a picture of what is happening. I would like to see more kids willing to experiment, willing to try stuff . . . Once you get them started, once you get them over that initial question of comfort level with the calculator, then they are freer. It's going to be interesting, I'm going to . . . teach this . . . computer programming class just using the TI. It'll be fun to see whether the kids . . . just stick to doing the "assignments" . . . or whether they will actually start to [say], "Well, wait a minute, I don't want to do that . . . Why couldn't I do this?" And develop some of their own goals and own projects.

Teachers had to enjoy a level of comfort with new technology before they could use it effectively in the classroom. Olivia Overkamp, a high school teacher in Georgia, stated:

> This is a gradual thing with me. You have to learn how to use the technology, the graphing calculator . . . I was in a workshop this summer . . . the Woodrow Wilson workshop which was sponsored by   the   collaborative and . . . the

Algebra for All Program and . . . we were introduced to the 82s [graphing calculators]. Before that . . . I was introduced to the . . . TI-81 and I learned how to use that. So it's been a gradual thing, but through the various workshops and seminars and so forth . . . I've learned . . . And the teacher has to learn to do these things before she can be brave enough . . . It took me a long time to get up the courage to start using that graphing calculator in front of my kids because kids are so good with technology. You put one in their hand and immediately they are ahead of you . . . I was very shy about it.

Overkamp reported, "The calculators are used in every class." An observation of Overkamp's algebra 1 class revealed a situation in which calculators were always available to students and in which they were encouraged to use them for calculation, visualization, and experimentation.

Paul Randal, an engineer and internship sponsor in the Milwaukee collaborative, understood the importance of teachers' comfort with and knowledge of technology. In his words, what was critical for teachers was "being comfortable with computers and applications . . . understanding what it is that they're doing. Because if they understand what they're doing then they can teach it better . . . If the teacher doesn't understand, the kids will never understand unless they get really lucky."

Teachers' enthusiasm for and experience with technology as a driving force of change in school mathematics continued to spread through their leadership activities. Overkamp was influential in the development and implementation of a collaborative summer mathematics camp for students entering algebra and an accompanying year-round professional development program for algebra teachers. Jacobs belonged to the writing team for an integrated mathematics curriculum that the state of California was developing which focused on problems in the transportation industry. His work in this capacity benefited from certain connections he made through the collaborative, especially that with a local professor with whom he jointly led a collaborative workshop several years before. Furthermore, Jacobs shared his expertise as a teacher-leader on a Woodrow Wilson Institute traveling team, an activity he also linked to his participation in the collaborative.

Pedagogy. Collaborative teachers discovered that in order to teach a different type of mathematics, the means by which they taught mathematics and their reasons for choosing such means also had to change. This realization came in two forms. In some cases, teachers adopted new goals for their students' mathematics learning and found that they had to teach differently in order to address their new goals. In other cases, teachers adopted new teaching strategies and discovered that they and their students began to think effectively and productively about mathematics in ways that they previously had not. For most collaborative teachers, the influence of pedagogy on mathematics teaching and learning in their classrooms was a combination of the two.

The experience of two middle grades teachers, Rhonda Usher and Sharon Villand, who belonged to the collaborative in Memphis, illustrates how ideas about pedagogy can affect the teaching and learning of mathematics in classrooms. Both described themselves as previously effective, but traditional, teachers. Because of changes they made, their classrooms looked much different, as Usher elaborated:

> In my fifth grade room you would have never seen manipulatives; you would have never seen group work, because I didn't do it five years ago. It was only after [I attended the collaborative's] Camp Mathagon that I started it and it's taken me awhile . . . So I didn't buy into the group work until I actually experienced it, and now I'm a firm believer in it. I also use calculators, which you saw today, that I probably wouldn't have five years ago . . . Discovery. For my kids to discover something with manipulatives or the calculators or whatever I use, is something that would be . . . very atypical of me five years ago, because I was in charge. And I told you what to do.

Villand added:

> What you saw today was the lack of a textbook. Five years ago I would have been tied to a textbook or to a worksheet, something that would have emphasized the skill that I was working on, probably from a textbook. You also would have seen a teacher telling rather than a teacher asking; that's a major difference.

Interestingly, the teachers acknowledged that they had some awareness of new pedagogy prior to their experience with the collaborative, but had not chosen to adopt new learning goals or new teaching methods. Their direct interaction with those ideas in the context of the collaborative, Usher related, made a significant difference: "I . . . was at Camp Mathagon . . . and until then I'd never really—of course I've heard about it—but I was there and I experienced it, and then I knew the benefit from the group activities and what I actually learned from the other members of my group . . . We were put in that experience mode and I was willing to try it."

Although Usher and Villand have made substantial changes in their mathematics pedagogy in relation to past experiences in the collaborative, both sensed that their personal changes and the larger reform effort to which they were connected were far from complete. Whereas Villand admitted, "I will show my absolute ignorance . . . I really did not know that all of [these opportunities] were available," the teachers were actively involved in the CAM project, the Maths in Context project, and many other nationally known reform efforts, which influenced their pedagogy. Marilyn Parsons, a Memphis collaborative participant from higher education, viewed teachers' awareness of opportunities as critical: "They have resources outside their own classrooms and their own school. That they get to share resources and share ideas [and] . . . network . . . has been a primary goal of our workshops." Harold Kristoff, a business representative in Memphis, agreed that the presence of the collaborative had been significant in creating new links between teachers and others:

There have been some really good things provided for teachers to attend and do that would not have happened had it not been for the collaborative . . . activities and workshops. If it hadn't been for that original money put into this, nothing would have ever started . . . There's so many relationships and things [that] wouldn't have happened.

Applications of Mathematics. The traditional paradigm of mathematics education assumes that once students have mastered particular sets of rules and procedures, they will be able to apply their skills to appropriate situations. Generally, specific, predetermined applications of mathematics demanding a single answer immediately follow the development of relevant skills in the instructional sequence. A reform-oriented view of mathematics education finds fault with this approach to applications because students cannot be expected to encounter situations demanding applied mathematics in this way outside of school. In fact, most real-world mathematical problems can be solved by many means, and the appropriateness of both means and solutions remains a matter of informed judgment. Furthermore, many realistic problems demand the integration or construction of novel mathematical solutions. School mathematics, in the reform view, must be founded on a realistic view of applications.

In order to incorporate an authentic view of mathematics applications into their activities and strategies, the collaboratives built partnerships with individuals from local businesses, industries, and institutions of research and higher education. Although the development and maintenance of such partnerships was one of the greatest challenges—and, in some cases, chief shortcomings—of the collaborative efforts, exposure both to other professionals who used mathematics and to authentic applications of mathematics was often a powerful force for change. The Milwaukee collaborative was one of the most successful in fostering connections between teachers and businesses through its summer internship program.

In addition to a six-week internship in a local business, participating teachers received training and support to develop a curriculum unit based on their experience, both for use in their classrooms and for broader dissemination. Jo Lamar, a middle grades teacher in a parochial school, worked with several engineers for her internship. She wrote an outstanding unit on blueprinting, which engaged students in measurement, two- and three-dimensional visualization, scale drawing, and proportional reasoning. However, the curriculum unit itself had arguably less impact on Lamar's teaching than her growing understanding of how mathematics is applied creatively and collaboratively in industry, "I was impressed in industry . . . during the internship, at the amount of collaborative effort, group effort on things. It wasn't *a* person's project. And so I've done that [in my teaching]." In addition to her revised emphasis on working cooperatively in mathematics, Lamar also discovered the relevance of communication of mathematics:

> Students should be able to write math, speak math, and communicate it effectively in a standard way. And I didn't have that focus before. I'm constantly saying to children, "What do you mean by that answer?" . . . That I learned from [my internship sponsor]. One of the pleas they made was, "Please, get your kids to communicate the mathematics."

What is meant by communication of mathematics transcends written and oral communication in the context of realistic applications of mathematics. Paul Randall, an engineer who worked with Lamar, characterized her use of blueprinting in this way:

> Jo has had a couple of units since she's been here on . . . blueprinting . . . in the math program that she puts together. Instead of just doing, "This is geometry. Look, there's a square." And everybody goes, "Great, that's a square." They sketch up a part . . . And then they draw it up as if it were going to be made as a blueprint [using] . . . two-dimensional drafting views. And then try to get a feel, three dimensionally, what it is and then go from there. And I know that that's direct from her experience here. And that's good. Students are . . . interested in expressing something three dimensionally. There are very few ways of doing that. Blueprinting or drafting is one of the ways, one of the communication tools.

Moreover, Lamar incorporated into her instruction real mathematical tools used in industry and gave her students genuine experiences with mathematics applications:

> This gave me the courage to [work] . . . in bolder ways on a bigger scale, because I could draw from resources of other people. I could give more than just textbook information to kids. I could give it from the field of engineering; I could bring in tools. You know, I could line things up that were technically authentic. I could quote . . . and it's not just book knowledge. I have taken kids on field trips to [my industry sponsor] to follow through.

An important part of belonging to the collaborative for Lamar was her ongoing interaction with her internship sponsor. Collaborative administrator Liz Navarro encountered the same phenomenon with many teacher-interns: "Many of them still had a connection with the company. They still felt free to call their mentor at the company and ask for equipment to borrow for their classrooms or to bring the students in for field trips or to make arrangements." Although Lamar's participation in the year-long internship program had ended several years before, she maintained a current understanding of how industry was using mathematics and continued to search for new ways to let students experience both specific and general applications of mathematics.

Student Knowledge of Mathematics. A major issue in the current reform movement's conception of school mathematics concerns the idea that mathematics consists of individually and socially constructed knowledge. This view challenges traditional mathematics teaching practice because it replaces the textbook as the center of mathematical authority in the classroom. A constructivist conception of mathematics requires a monumental change in the roles that teachers and students are expected to play in the classroom. Teachers' acceptance of students as mathematicians with unique knowledge and authority to solve problems; to make appropriate decisions; and to draw, communicate, and justify conclusions is paramount to the current mathematics reform movement.

The Philadelphia collaborative brought the IMP curriculum to three high schools in 1993. The collaborative teachers who first taught IMP in the district comprised a corps of volunteers as well as specially selected individuals. They received specific training for teaching the course and continued to meet on a regular basis to share ideas and collaboratively address problems during its implementation. The teachers' frequent contact with one another was cited as a powerful feature of the program by one teacher, Carolyn Graver. Perhaps even more powerful was her realization of how the thoughtful and deliberate use of student knowledge as a basis for instruction changed the mathematics teaching and learning in her classroom:

> They pick up a lot of things ... You'd be surprised [at] some of the things ... This course is saying that everyone does not have to be a mathematical genius to learn mathematics ... I had a girl give a demonstration. I'm looking at it and I'm saying, "Now I don't see how she's doing that." So I asked her to explain it to me. And she did. And you know what, she was absolutely right! [laughter] ... I never would have taught it that way. I certainly would not have taught it that way because ... I would never think of it, because that's not the way I looked at it. So they come up with a lot of ideas. I always say that as long as they can back up what they are doing ... And they all approach it differently.

For Graver, accepting the variety of approaches that students brought to problem solving became essential to the teaching of mathematics: "The whole key is to get them to not memorize a formula but to be able to approach a problem and then have different skills to solve a problem ... You find that kids approach things in all manner of different ways."

Although her comments illustrated that she accepted students as competent and successful problem solvers, Graver cited a greater difficulty with materials that transferred a large amount of authority and autonomy to students and teachers when determining correct or appropriate solutions:

> This book doesn't give you answers. The teachers' edition has no answers in it. We don't have any answers. Sometimes it bothers me. I think to a certain extent they want some of this stuff to be open ended. They don't want ... to say, "Here's an exact answer." They are more concerned about the process the students are going through. Whereas, of course myself as the educator, I would like to make sure that the answer that I have is the correct one.

Graver's focus on correct answers independent of students' construction of knowledge and mathematical authority appears somewhat troubling in the context of the reform movement. However, her instructional methods of handling both students' correct or appropriate solutions and their incorrect or inappropriate solutions remained quite encouraging:

> You have to monitor them. Because some of these kids are really thinking off track, and get it completely wrong. So you have to really monitor them in the groups. That's what the presentations are all about. Because then they get up in front of the classroom, and the kids will see what should have been done or how it could have been done. Or you always try to look for kids who do things differently but are approaching it the right way. So you don't want them to get up there and keep doing the same thing, the same way. You might want three different ways of how students have looked at it. [If students present something that is entirely wrong], what you do is you don't try to embarrass them . . . You let them say what they are going to say. And when they are finished, you talk to the class on the whole, asking them why. You try to get him to see why what he is doing is not right. You don't want to leave it so that the kids think that's the right answer or that's the right way to do it. You have to handle it that way because you don't want them to think that's right, but you ask them questions to maybe get them to thinking that maybe this is not the way that it should be done. So while they are presenting, you are also talking to them. Or even the kids will be asking them questions because you'd be surprised what a lot of the kids know: "No, that's not the way you are supposed to be doing that." You have to get them to not talk that way, but . . . to ask questions about, "Why are you doing that? Well, what did you use? What did you base it on?" . . . Of course you say, "Did anyone else do this differently?" Or, "Did anyone else get a different answer?" And then they'll say, "Yes." And then they'll go up [and present their solution]. Then you have both. Then *he'll* decide, "Well, then mine must be the wrong way, this is the right way."

Carolyn Graver, Harvey Jacobs, Jo Lamar, and the other mathematics teachers described here were developing a new set of roles for teachers and students in classrooms in response to strong forces in education and in society. Their role had become to monitor and support students in the creation and articulation of many strategies and solutions for challenging problems. The students' role had become to experiment with problems, to utilize technology to investigate problems, to apply and communicate their understanding, and to deliberate the appropriateness of different strategies and results. The teaching and learning of mathematics in the classrooms of collaborative teachers had changed.

Mathematics in the Classroom. If all students are to learn mathematics, to become inquisitive and effective problem solvers, and to apply their mathematics learning effectively and appropriately in and out of school, mathematics teaching and learning cannot continue to be the same. For teachers, participation in the collaboratives was an ongoing, interactive process of examining and rethinking goals for their mathematics teaching and ways to achieve those goals in their classrooms. Classroom instruction, mathematics course content, means of assessment, student learning, and teacher and student roles were subjects of scrutiny and objects of change across the collaborative sites and in the national effort. Certainly, a comprehensive reformulation of school mathematics across all of these areas did not occur at either the local or national levels. However, the proven ability of the collaboratives to initiate and support change in each area suggested that the collaborative enterprise represented at least one ingredient of an approach to reform that could address the multiplicity of areas in need of change.

Instruction. One very visible change in mathematics classrooms, which teachers attributed to participation in the collaborative, was more active student engagement in the learning and doing of mathematics-related activities. Traditionally, mathematics instruction in the United States has required students to learn mathematics by listening to teachers' lectures and doing a number of repetitive mathematics exercises to practice and perfect a procedure or concept. In the classrooms of many collaborative teachers, the teaching and learning of mathematics differed significantly from this norm. Specifically, group work and cooperative learning; use of manipulatives and technology; and learning by inquiry, investigation, and discovery were prevalent practices teachers associated with their involvement in the collaboratives.

Justine Jordan, a high school teacher in the Columbus, Georgia, area, explained how her classroom practices had changed:

> I was teaching general math . . . and "drill to kill" doesn't work in the high school level . . . They don't learn . . . [I knew], "Hey, what I'm doing is not working; I've got to go find something else" . . . You sit in your classroom and you say, "There's got to be a better way to do this." And the collaborative is a place to go and look for those better ways.

What Jordan discovered was that not only was the instructional method of "drill to kill" ineffective, but it did not allow her to teach what she ultimately felt that students should know: "We have got to get them to understand concepts." An observation of Jordan's classroom revealed an environment in which students conducted experiments lasting days to weeks, processed data using graphing calculators, and engaged in heated debates about mathematical problems arising in a variety of applied situations. Students in her classes learned about parabolas and tangent lines through experiments conducted by throwing paper balls into trash cans. They examined other functions by recording and analyzing data of change over time, such as the growth rates of dehydrated toys that became dinosaurs when placed in water. Moreover, at the students' request, they also recorded data about how the toys shrank when removed from water, an extension of the activity that Jordan did not plan, but was happy to allow. Her classroom was an environment in which students eagerly experimented and struggled with mathematics to form a collective understanding of their results through discourse with one another and with their teacher.

Jordan credited the collaborative with much of the support she needed to make this shift. In addition to the collaborative's material resources, which she has used to change her classroom practices, she found a great deal of moral and intellectual support through the collaborative's staff and membership. Changing classroom practices is a risk for teachers that demands solid justification and ample opportunity for reflection and modification. Participation in the collaborative made that risk worth taking for Jordan:

> I feel confidence . . . that I'm not going to fall on my face. And well, if I do, it's not the end of the world. I will go troubleshoot it and see what happened. You know, I feel more comfortable getting in front of my kids and saying, "Hey we are going to try this. If it works, it works; if it doesn't, let's see if we can figure out why it didn't work. But I think . . . this should be pretty good. I know that [another teacher] has tried this . . . and it worked . . . with his kids."

A further contribution toward the changes in Jordan's practice was the opportunity to experiment with new strategies in the collaborative's summer camps, "In the camp is when I started, 'Here, you [students] take the chalk. You guys go do this.' You know, I got to experiment with me letting go of the control." The same transfer of control over the learning of mathematics occurred in her classroom:

> Five years ago, I probably would not have stepped back and handed the marker to the kids. I probably would have gone into a fit if a kid got up out of his desk and started coming to the board [to] do the writing. Whereas now, if the kids want to take charge and they want to direct it, then I'm beginning to feel enough confidence that I can pull them back in and I can let go of them and let them have control.

More control and ownership of mathematical learning among Jordan's students allowed them to make conjectures and draw conclusions on their own authority rather than depending on the teacher's prompting or instruction. She described how her students responded to their first exposure to a nonlinear function, which she spontaneously displayed on her overhead graphing calculator one day during a lesson on linear functions:

> Now all of my classes, this year, when we look at something, we have an idea of, "Oh, nothing is being squared. We must be doing this. Oh, something is being squared, we must be . . . " You know, modeling with our hands and our bodies so that we use anything to get it to sink into their minds. And so as soon as they saw the parabola come up, they knew, "Let's don't use $y = mx + b$." But they weren't real sure what to do with it. And those were the first problems that we worked on that were really quadratic, which was today. And I think they made the shift well to it. I was glad to see them not trying to fit it into $y = mx + b$.

Jordan knew that her classroom looked different from others. She understood that she had taken a risk and continued to take risks with her new classroom practices. Change supported by the collaborative, however, proved to be worth the risk:

There are some math teachers that they'll see my class, and my kids at the board and me standing back and I've got five kids hollering at the same time, they would just classify it as totally disruptive, nobody's learning anything and they're just having a good time. Whereas to me the reality of it is those kids were making statements and making conjectures and standing up . . . and saying "I know this is what I need to be doing."

Content. A strong argument can be made that changes in how mathematics is taught and learned significantly influence what mathematics is taught and learned. A traditional lecture-and-drill method of instruction lends itself to a view of mathematics as a fixed set of facts and algorithms to be memorized and mastered. A true inquiry and cooperative learning approach, by contrast, emphasizes mathematics as questioning, experimenting, and making sense of situations individually and collectively. Researchers have shown, however, that mathematics teachers change their practice somewhat and even include new topics in their courses without possessing or communicating a changed view of mathematics (e.g., Heaton 1992, Prawat 1992, Putnam 1992, and Remillard 1992). It was noteworthy, then, that many collaborative teachers who had made shifts in their classroom practices not only built an appropriate system by which to change the content of their teaching, but successfully did so by introducing both new mathematical topics and new ways of thinking about mathematics.

About half of the 24 teachers in the mathematics department in one large Los Angeles high school participated in the collaborative. In addition to a grant proposal for improving mathematics instruction that the department wrote in order to join the collaborative, several of its teachers attended the collaborative's annual workshop series and traveled around the state and nation to learn about and to be trained for teaching new courses. The collaborative directly provided or supported much of the learning that the department's teachers required in order to change the content of their mathematics courses. It supplied the information and moral and technical support for a great deal of the additional learning that continued to take place through the teachers' own initiatives.

Trevor Updike and Alberto Cortez, two of the department's teachers, were responsible for discovering and piloting both a new algebra 2 course (from the Interlocken School of Art in Michigan) and a new precalculus course (from NCSSM). They depended on the collaborative for support throughout the process. Updike related a series of events that began with his attendance as a collaborative member at an NCTM national convention to participation in institutes at the Exeter Academy in Massachusetts, Mount Holyoke College in Connecticut, California State University at Northridge, Ohio State University, NCSSM, and Interlocken. Cortez followed a similar course. He highlighted the support of the collaborative particularly with respect to involvement with NCSSM:

When we got the [NCSSM application] we felt that we just needed a little bit more . . . professional help in how to get our packet more presentable . . . We needed someone who was aware of what's out there. How these people look at these applications and someone who would be more knowledgeable with those kinds of things. So [the collaborative director] was the one we went to right away. She's familiar with all of these things, all across the country. So we went to her expertise, and she was very beneficial in helping us to remember that we had done these other things that we didn't even bring up into the application. She was aware of what we had been doing. She helped to bring out things that we had completely forgotten that we had done. So with her help and some of her input, we were able to get to [NCSSM].

The collaborative's support, as well as the mutual support the two teachers discovered through their joint involvement in the collaborative, led to a very different view of the mathematics content in their courses. For example, Cortez described a typical lesson:

We had another exercise where we're teaching about iteration or recursive routines . . . We do a little lab where we've poured in so much colored dye. It changes the water. What we're trying to [model is] this guy is taking in so much medicine and his body eliminates it after a . . . fixed period of time. What is happening to the real amount in the body? So what we do is develop the fact that he's taking his four milligrams of medicine every six hours. We have them watch that in the calculators. They start putting this in and they start seeing this recursive routine. They put in four and take out four. We put in four more, they take . . . The body eliminates a quart. They put in four more. The body eliminates . . . Well, what happens is that they begin to see a certain saturation point that the body reaches in terms of the amount of medicine in the body. So we try to explain to them that that's really what the doctor's goal is, for them to fight off the infection. The doctor predicts that the body needs so much medicine in there to get healthy, to keep it that way for a certain amount of time. They say that if he keeps taking this amount of medicine, the body is going to stay at that level pretty much. They see this in the calculator. It graphs it and they see the graph.

Well, what we try to do then is have them apply this to a real-life situation. We have them write . . . "You come up with a situation where you're out there and you're the one that gets sick. Tell us about it. You go to the doctor. Give me the prescription that the doctor gives you. How often do you feel that the body eliminates that medicine? How long is your body going to take to reach that saturation point? And then, after you've been on the medicine for so long, how long is it going to take, after you've stopped taking the medicine, for the body to go down to less than one milligram of medicine?" So you can come up with this whole scenario of writing about what they were doing, what caused them to get ill. Then going to the doctor. The doctor giving them this prescription and then they start saying how often they have to take it a day, and how many days it takes them to reach that point, or how many times they have to take that medicine to reach that. How long they're on that medicine and how long it takes them to come down off from that. So again, there's the mathematics all built into that. At the same time, they're realizing that it's something that they can relate to.

The mathematical content in a lesson such as this one is built both around specific mathematical ideas, recursion and iteration, and around a realistic context, medicine. The application and meaning of mathematics are integrated into the content and format of the lesson. Such an approach to the content of mathematics courses was a departure from past experience for both teachers. It allowed them, however, to pursue a type of mathematics in their teaching that past practice did not allow. Updike elaborated, "Of course we didn't have the technology of the calculators. The calculator generates a lot of our understanding now . . . we couldn't do problems that we can do today. Ten years ago, we couldn't have done them. By doing them by hand, you could do them, but it would take forever." Moreover, Updike says, the new approach to content allows the teachers to pursue new goals for their students, "[Ten years ago] the biggest difference was that I always lectured basically, the whole period. I try to avoid that. I try to have [students] become more involved, trying to force them into taking responsibility of learning the material." He later added:

> I want them to be able to use . . . a tool, the calculator, and visualize problems and be able to solve problems with that as an aid. I want them to be able to work together with other individuals, whether they like that individual or not. I want them to be able to share ideas and to exchange ideas with each other . . . I really want them to be able to use the calculator as a tool to help them learn more mathematics and solve more problems. I'm working them toward being problem solvers.

Chris Erlanger, an English teacher at the school, was excited about the new content in the school's mathematics courses and its potential connections to English content. He related one instance in which he and Updike had worked together:

> He had . . . started journals . . . in his math class . . . As an outgrowth of that . . . he had [the students] write a letter to their grandmother explaining conjecture. And he was . . . concerned . . . "Well, how am I going to grade this?" And I said . . . "Don't worry about the grammar. Don't worry about the structure. You read it and see if they got the concept of conjecture across." Then because . . . it so happened that we had a lot of the same students . . . I said, "Give them back . . . and have them give it to me. And I'm going to look at it from another viewpoint . . . did I understand what conjecture was? And I'll take a look at the structure, but I'm not going to heavily grade it." Well, we got to share the idea, and I told him about setting it up, and what were their thought processes, and that was what I was concerned with.

In retrospect, both Updike and Cortez knew that they could not have changed their mathematics courses to the extent they had without their participation in the collaborative. In a discussion of how the new course content met the goals of the California *Framework* and NCTM *Standards*, Updike said, "When [the *Standards* and *Framework*] first came out, I had no idea how I could do it. Now I have an idea on how I can make some of those changes that they're requiring or that they're asking for." Cortez expressed a very similar sentiment when he discussed the importance of the collaborative in providing the teachers with a direct link to the new mathematics courses and their developers: "I don't know that I would have attempted it had it not been for somebody else that's willing to do it with me and had we not gone to the workshops. I don't think anyone could do it without that, to have at least some knowledge of what's going on down the pipeline with this course."

Assessment. Assessment is a rapidly developing field in both research and practice in mathematics education. Considerable emphasis is currently being given to evaluative assessment of student outcomes, reflective assessment of teachers' practice, and formative assessment of mathematics programs. Many collaborative teachers who are interested and immersed in changes involving content and instruction have a particular interest in making changes in their assessment practices.

The influence of external forces regarding assessment is one of the most acute for collaborative teachers (Little and McLaughlin 1991). The knowledge that students will be evaluated and affected by standardized mathematics tests required by districts, states, and higher education places a heavy responsibility on teachers engaged in change. Many collaborative teachers questioned the fairness of standardized tests, which did not measure the same outcomes they viewed as important for students in mathematics, as well as the usefulness of standardized tests as formative evaluation tools for themselves and their mathematics programs. However, they feared that if they did not specifically prepare their students for standardized tests, their students would suffer due to the importance still placed on standardized tests for course placement and college admission.

Some collaborative teachers assessed students under conditions that resembled standardized testing—in some cases, simply to prepare students for standardized tests and in others because the teachers genuinely preferred to assess by testing. Many collaborative teachers, though, were changing their assessment practices in order to provide better summative and formative evaluation for their students and themselves. Referred to as "authentic assessment," "performance assessment," and "portfolio assessment," the variety of changes in assessment practices among collaborative teachers was linked to their beliefs about students, mathematics, pedagogy, and the teaching profession.

Catherine Dorsett was a middle grades teacher in Georgia who had been teaching for 21 years. She traced the roots of her current focus in mathematics education to her graduate coursework about 10 years before at Columbus College, where she worked with the collaborative director and began to rethink her teaching in the hope of incorporating more problem solving into her classroom practice. Early experiments revealed some problems: "Problem solving was kind of like a fun day. And the kids really . . . didn't take it so seriously because they weren't assessed, really, on it like they were assessed on tests. And they knew that those tests continued to count for more . . . of their grade." Dorsett did not want to give up on this change, but was unsure how to address the disequilibrium she felt: "Every math teacher kind of feels like that. They realize that there's something missing here. It's just that it's very difficult to address how to organize for what's missing."

The beginnings of a solution materialized when, through the school district and the collaborative, Dorsett was invited to participate in the CAM project in 1991. Engaging district mathematics supervisors and consultants and classroom teachers in several collaboratives, the CAM project was developed by EDC to provide a mechanism for exploring, creating, and cultivating new techniques for student assessment in school mathematics. After the first CAM meeting, which offered exposure to new assessment ideas, Dorsett began to experiment in her classroom with "more authentic . . . test situations where the kids had to show me and draw this for me, to really get at their understanding." However, she remained dissatisfied with tests as her sole method of assessing students' mathematical understanding. At a later CAM meeting, Dorsett was familiarized with the notion of portfolio assessment by a California teacher who used the technique. She explained the importance of finding a colleague who was employing the new method: "Until you see another teacher . . . say, 'This is what I did' and then show you an example of it from a classroom, it doesn't seem to become real for you. So, it just seemed real to me then. I thought, 'Well, I can do this. I can do this.'"

Initially, Dorsett was uncertain of the logistics of portfolio assessment in her classroom: "At first, I thought, 'This all sounds good. I'll use my authentic assessment techniques,' but I thought, 'This is going to be chaos in the classroom. How am I going to organize for all this?" Moreover, she was unsure of the compatibility of the technique with her own students. She was willing to experiment, though, and began using portfolio assessment with all of her students during the second semester of the 1991-92 academic year. The effects on her teaching and her students' learning were noteworthy.

Upon reflection on her experience as a mathematics student, on her teaching, and on her students' learning in the context of experimentation with portfolio assessment, Dorsett determined that mathematics and various areas of mathematical study are built around a few core concepts that demand continual examination and reflection to promote learning. Portfolios became her primary method of facilitating students' reflection on core concepts. For example, during the 1993-94 school year, Dorsett focused on mathematical patterns of shape, number, and other attributes. She related:

As I reflect on the things that the kids have done, it has really amazed me how [students] now tell me they find patterns in everything. They come up and show me patterns all the time. I don't think that, if it hadn't been for this, they would even have talked like that or even thought about it. They sense that there is some mathematical power in this for them.

She defined mathematical power operatively: "When [students are] given a problem that they feel it is within their realm of thinking to reflect and have some sense of what this problem is about. Even if they don't know the answers, that's not as important as being able to attack the problem on some conceptual level of understanding."

Skill and practice, which Dorsett believed were important, were not absent from her teaching. In terms of class time, she and her students generally spent at least as much time working specifically on skills as they did on understanding. However, an emphasis on her students' effort and mathematical understanding permeated classroom activities and formed the main emphasis of learning. She believed that her students understood that emphasis:

They know that a portfolio is where we do our math thinking . . . They're much more mature in the way that they're talking mathematics. That's one thing I've noticed. They reflect a lot on how they see things. They know that it's not just skill driven and they know that I comment on the way that they think, and so in a lot of their writing . . . in classroom discussions . . . [they] reflect on the way that they see things, or the way that they see things differently. I think that's really important.

Iris Ingram, a district administrator and collaborator on the CAM project, visited Dorsett's classroom and was impressed by the changes she saw:

Catherine . . . asked me to come and see her . . . after we had been to a Classroom Assessment in Math workshop, assess her kids . . . They were making graphs of some materials of some information . . . She was just amazed at how the kids could follow directions and she just assessed them on what they had done together. And they had worked cooperatively.

Ingram also commented on the impression Dorsett's changes made on a visitor from EDC:

[He] came down and visited a day, and we went to the schools. He was just awed with the material they had gleaned from . . . just two two- or three-day workshops . . . They were using them and implementing them into the class. Catherine . . . is just teaching portfolios . . . But she is bringing them back and working through the collaborative to do that.

Reflection on student progress reinforced the changes she made in assessment. In one case, Dorsett recalled that a student wrote, "I think you should definitely do [portfolio assessment next year] because when you do portfolio work, your mind is in a different gear." She added, "I know what he's talking about. There is a different gear that you are in, and whatever that gear is, I think that's where you really learn. It's not the gear when you're sitting and teaching for skill development."

Teacher and Student Roles. If teachers were indeed behaving differently in their classrooms as a result of experiences with the collaboratives (Nelson 1994), then the roles they and their students played in classroom learning were another part of the story of the collaboratives. Descriptions of traditional classroom roles depict teachers who dispense facts and information to students (Cohen 1991, Stake and Easley 1978, and Wilson 1990). By contrast, many collaborative teachers were redefining their classroom roles in light of beliefs about students, mathematics, and pedagogy toward a role that was most often described as a facilitator of learning, rather than a dispenser of knowledge.

Several teachers in a high school in Georgia described similar transformations in their own classroom behavior and their expectations of students. Each was involved in a number of collaborative functions and attributed classroom changes to professional growth within or inspired by the collaborative. During a group interview with several teachers in the department, Tonya Trier, in reference to the teachers' participation in the collaborative, emphasized, "It has acted as the catalyst for all this [change]."

The teachers in this high school who attributed to the collaborative a catalytic role in their professional growth and resulting changes in classroom roles ranged from some with only a few years' teaching experience to one with over 30 years of experience. Percy Patterson, who is in his seventh year, described his expectation of teacher and student roles in the classroom during an individual interview:

> I try to be a facilitator . . . I think a student learns best by doing things. What I try to do as a facilitator is get them stuck at some point or get them to question what's going on or what's happening. And then to get them to work themselves or get themselves unstuck. Whenever they're in the process of getting themselves unstuck or answering a question, that is the most valuable moment of a lesson.

He continued to play other roles in the classroom, "I still lecture and that's still important. But I use it more as a tool . . . I would say, 60 percent of the time, the students are in their groups doing investigations and doing problem solving on their own." He assumed the role of facilitator when students worked in groups:

> I circulate around the classroom . . . monitoring the class . . . I like for them to be as close to the material as possible. I try to remove myself as much as possible from that so that they experience it first hand . . . One thing you'll never see me do in class is give a student an answer. I feel pretty emphatic about that. The students will reveal the answer and will discuss whether it's right or wrong. But if I do it, I really hate myself for it . . . I try not to say that an answer is right. I want the student to understand that they got the answer and they can feel like they can justify the answers they've got . . . If I tell a student the answer, then I've reduced their power as far as their ability to learn.

Patterson's description of the role he expected students to play in the classroom suggests a shift of mathematical authority from teacher to student. Vincent Valentine, a colleague with over 30 years of classroom teaching experience, agreed with this view:

> I would like for [students] to stop and think . . . "Make sure you're right. Be your own authority." . . . I keep trying to tell those students, in algebra 2, "You've had more math than 90 percent of the students at this school. Don't keep looking at the back of the book for the correct answer. You be your own authority . . . You know how to prove it. You know how to check it out. Don't ask me if it's the right answer. You know whether it's the right answer, because you're the authority now."

A similar shift in teacher and student roles occurred in the classroom of Trier, who was in her seventh year. She recalled that her role as a teacher at the beginning of her career was mostly as a lecturer. She had since "gone to almost totally cooperative learning groups." She believed in cooperative learning in mathematics because it helped to meet her goals for students to tackle problems by trying different strategies, sharing ideas, and using their understanding.

> [I expect students] to always look for "why" and not be afraid . . . [I] still have some students that are more hesitant than others, but they are more willing to work a little bit harder and longer on a problem, because one of the rules in my class generally is "three before me." If your whole group decides that they do not know the answer and cannot come up with anything, then I'll come and give you a hint or whatever. But not immediately giving up. It is knowing that you've got the tools.

Brenda Baston, a district administrator who had worked intensively with these teachers, observed, "Over the past five years, the majority of the math department of this high school has changed from a . . . lecture standpoint to a hands-on [standpoint, with students] involved." Classroom observations supported Baston's assertion. In Tonya Trier's geometry class, she designed novel problems involving the area of circles and squares. The students worked together to design strategies for solving the problems. Trier guided a discussion of their strategies, emphasizing students' explanation and justification of their reasoning. In one instance, she refused to accept a student's use of a formula from the textbook until he could justify its use with his own reasoning. Down the corridor in Rochelle Robinson's precalculus class, several students encountered an error message, rather than a usable result, while attempting a new procedure on their graphing calculators. Instead of quickly demonstrating the correct procedure, Robinson used the experience to help students discover how the calculator had interpreted the equations they had entered and what they could do to correct the problem.

These teachers and several others at this school not only participated in, developed, and led numerous collaborative activities, but also relied on one another for collegial support on a daily basis. Teachers at other collaborative sites attested to similarly strengthened professional relationships with colleagues in their schools due to shared professional growth in relation to the collaboratives. New forms of professional involvement with other teachers and the broader community, in part, motivated teachers to undertake many of the described changes. Moreover, collaborative teachers' ongoing learning sustained those changes.

Change in Student Learning. Ultimately, mathematics education reform must be judged on the basis of student learning. Major criticisms of the present state of mathematics education in the United States are that successful mathematics learning is exclusive, especially of women and minorities; and that the nature of mathematics being learned is not appropriate for modern society. In evaluation of the collaboratives within the current reform movement, then, two aspects of student learning are salient: Which students are learning? And what is the nature of students' learning?

As has been emphasized, a major ideological underpinning in documents such as the NCTM *Standards* (NCTM 1989, NCTM 1991, and NCTM 1995); *Everybody Counts* (MSEB 1989); the California *Framework* (California Department of Education 1991); and the UMC position statement on equity is that *all* students can and must learn mathematics that is both challenging and appropriate. The experience of many collaborative teachers has given this ideology a grounding in real school and classroom situations.

Collaborative teachers have not universally unequivocally adopted the vision that all students can learn mathematics. However, a number of collaborative teachers have discovered that changes in their teaching have led to substantial learning among students who had not previously been successful in mathematics courses. This realization has led to the widespread belief among collaborative teachers that more students—in some teachers' cases, all students—*can* learn mathematics.

Eva Harrison, a Memphis middle grades teacher, told interviewers of her success with cooperative learning, which she learned about through collaborative activities, "Because of the grouping, I don't even think I have, this semester, I do not believe I have an F in my room. Something is wrong, isn't it? I don't think so, not for the semester average." This remarkable experience has reinforced the beliefs in a goal of student equity which initially led Harrison to explore the collaborative. In some respects, her experience had proven that goal to be reachable, but she did not believe her own or the district's work toward that goal is complete, "I think it was a long-range goal; I think of goals as being long range anyway, and it's going to take some time. There has been effort put forth and it's spread . . . to help restructure mathematics for all, but . . . because of funds or whatever . . . we haven't met the goal."

Not to be discouraged, Harrison continued to change her own practices in order to approach the goal of equity in mathematics education: "I started the writing in mathematics, I guess about three or four years, and it's because of the collaborative. The collaborative had . . . an equity project . . . that emphasized

. . . journal writing and students writing in mathematics, and that's where I really got off into the writing." She has found that writing allows students who otherwise might not be able to demonstrate their learning to express themselves effectively, "[I] want to see how they're thinking on paper . . . I have some students who wouldn't dare go to the board, they're just shy, so this is a form of having them to talk to me."

In conjunction with a belief that students can learn mathematics, collaborative teachers as a group expressed strong sentiments that more, or all, students *must* learn mathematics. This belief was largely based on their knowledge about the applicability of mathematics to a growing variety of careers and real-world situations. The experience of Scott Truesdale, a high school teacher in Los Angeles, is illustrative. Since he joined his school's mathematics department from the school's special education program, he attended and led collaborative workshops; received several grants directly and indirectly through the collaborative; and taught IMP, which he and another teacher brought to the school by utilizing the collaborative network.

Truesdale commented about his students:

> They don't all have to be rigorous mathematicians because they're going to work on teams with other people. If they're good at something else . . . the art work or the communication or something else, at least if they have a mathematician on their team, they'll know how to use that person because they will understand something about statistics and calculus and algebra and geometry and . . . discrete math.

Interestingly, Danielle Framer, an administrator at another school in Los Angeles, made a similar statement about what students should be doing in the classroom during an independent interview:

> It's all part of solving problems in a logical way. That's the whole purpose of mathematics as far as I am concerned, because most kids are not going to be mathematicians . . . They're going to be something else. But they are going to learn from mathematics a logical way to solve problems and look at options. That should last them for their whole lives. At least I hope it will.

The many changes Truesdale made in his mathematics courses, especially the incorporation of experiments and projects from the IMP curriculum, proved successful with a broad range of students:

> We brought 25 severely learning disabled students into the [IMP] program the first year, into the mainstream classes . . . Numerically, they did as well or better than the mainstream kids . . . At the end of last semester, I had four A's . . . and two of those were severely learning disabled kids out of the resource specialist program. And one kid . . . who read at the second grade level last year . . . is getting an A in college prep math. For three semesters, he has had an A consistently. He comes up with, more often, with insights into what's going on. Because the insights are supposed to come from the kids out of the structured activities. And he comes up more often than any other kid in the class with

insights into what's going on. But if you handed him an algebra book and told him to do it, he wouldn't have a chance to pass.

Victoria Wells, an administrator at Truesdale's school, offered similar reflections on student learning, especially among students who had been hard to reach in the past: "I have gone into their classrooms and I've watched the kids . . . using the manipulatives in the integrated method. It is very exciting to see kids that I know in a traditional class . . . wouldn't even hear what the teacher was saying, let alone be doing math. And here these kids are really trying to figure it out."

The nature of student learning in mathematics, too, changed in the classrooms of many collaborative teachers. Rather than expending energy on memorizing formulas and perfecting the use of algorithms, students were learning to use mathematical concepts, models, and tools to make sense of situations. Suzanne Sanchez, a district administrator in Georgia, observed that teachers who participated in the collaborative taught mathematics differently and that their students thought about mathematics differently:

> I see a change in philosophy on the part of the teachers. There is an openness and willingness to allow students an opportunity to have that hands-on experience—the time to make conjectures and test those and draw conclusions. I see more of a facilitator kind of an approach to teaching as opposed to the authoritarian person in the classroom . . . Comments [of students who have been involved in this approach to mathematics instruction] were so powerful—comments like, "Having the opportunity to have the graphing calculator in the classroom and work with that and see the graph has made more difference in my understanding than I would have expected it to," and "Having the opportunity to communicate about the mathematics that I'm studying and to learn that mathematics within a real-world context has made a difference. It makes it come to life. It makes it fun."

Changes in conceptions of school mathematics in relation to the collaboratives—including how, what, and to whom mathematics is being taught—are more teacher specific rather than schoolwide or districtwide. Although the most significant changes in conceptions of school mathematics in relation to the collaboratives may be described as individual, these changes should not be seen as isolated. The forces that undergirded the described changes among collaborative teachers depended to a large extent on teachers' reduced isolation from one another, from other professionals, and from current and relevant ideas and developments in mathematics education. Moreover, the same vehicle—the collaborative—that allowed teachers to make individual changes, also offered them leadership opportunities by which to share their experiences and utilize their influence to promote change throughout their schools and districts.

Teacher Professionalism

Several stated goals of the UMC project addressed the professional lives of teachers. Envisioned changes included the following goals:

- Teachers would be less isolated from each other and from local, regional, and national mathematics-using and mathematics education communities.

- Teachers would become more intellectually stimulated, have more up-to-date knowledge, and be in a position to participate actively in improvement efforts in their classrooms, schools, and school districts.

- Teachers would develop a broader definition of what constituted professionally relevant ideas and relationships (than the then-current notions of "in-service") and would be more enthusiastic about their professional lives.[4]

Across collaborative sites, many teachers demonstrably expanded their professional contacts with one another and with others, viewed themselves as more current in their knowledge and involvement in improvement efforts, and reformulated their conception of what constituted a career in teaching mathematics.

Teachers' redefined professional lives in the context of the collaboratives were most notably characterized by networking and leadership. Networking encompassed broadened professional relationships with other teachers and other professional stakeholders in mathematics and mathematics education. The language, ideas, and purposes that were shared with others were also components of networking. Leadership connoted the authority and willingness with which teachers became informed about and made decisions regarding changes in their classrooms, schools, districts, and states. The collaboratives offered teachers considerable support for both networking and leadership by providing a center of activity and communication, a link to the broader mathematics education community, and financial and organizational backing for teacher-led change.

A reformed vision of the teaching profession required not only that teachers be engaged in substantially different roles, but also that they view their professional lives differently. What collaborative teachers believed about themselves and the teaching profession gave meaning to the manifestations of networking and leadership that characterized the collaborative venture.

Knowledge and Beliefs About the Teaching Profession. Two factors that have dominated the experience of teachers in the United States for decades are that teachers spend the vast majority of their contracted time in the classroom with students, and that teachers' professional training is largely limited to preservice education for licensure and sporadic in-service training for license renewal (e.g., Bird and Little 1986, Sato and McLaughlin 1992, Talbert and McLaughlin 1994, and Weiss 1995). Researchers, reformers, and practitioners

---

[4]This idea was already recorded in a 1988 memorandum to the Ford Foundation by Barbara S. Nelson (cited in Nelson 1994, p. 21).

have—especially over the past decade—called for a reformed view of teaching that includes more ongoing opportunities for teachers to learn and more time for teachers to meet and plan with one other (e.g., Ayers 1990, Battista 1994, Bird and Little 1986, Clark and Astuto 1994, Goswami and Stillman 1987, NCTM 1991, Sykes 1991, and Wiske and Levinson 1993). Reports on field tested modifications of teachers' time and responsibilities have attested to the power of this new vision of teaching in advancing change in classrooms and schools (e.g., Keedy and Rodgers 1991, Leinwand 1992, Lytle and Fecho 1991, and Roberts 1992). Observers of the UMC project, too, have noted that many participants in the collaboratives have utilized their professional time and have viewed their professional responsibilities differently than prior to their participation in the collaboratives (Bruckerhoff 1991, Little and McLaughlin 1991, McCarthy 1994, Middleton et al. 1990, and Richert 1994). By placing teachers at the center of substantive educational change, the collaboratives had advanced new ideas about the professional duties and responsibilities of teachers. The translation of these ideas into individual teachers' beliefs about the teaching profession represented a critical piece of the story of the collaboratives.

Two themes consistently emerged that characterized collaborative teachers' beliefs about the teaching profession. First, collaborative teachers viewed substantive, ongoing professional growth and learning within the context of a professional community as pivotal to the teaching profession. Second, collaborative teachers viewed themselves as competent educational decisionmakers and authoritative instructional leaders, both in and out of the classroom.

Tonya Trier, a Georgia high school teacher, cited the importance of ongoing professional growth and learning: "I think [the collaborative] has shown me what a difference a teacher can make as far as growing in education. And keeping up with your interests. Changing with the times." She had a number of opportunities for professional growth through the collaborative, including local meetings and conferences as well as attendance at institutes in Massachusetts and a travel grant to visit collaborative schools in Philadelphia. Reflecting on how the collaborative affected her seven-year teaching career, Trier related:

> I think several of the most effective things that come to mind immediately are [the collaborative's] willingness to support the teachers in different things that they want to try, sending them to different conferences and things like that, providing in-service . . . mini-workshops and things to just share and show new hands-on type of equipment, new things being done in other classrooms . . . any type of teaching method that is . . . being used and being used effectively. And being there as a resource when you have a question or when you have a need. And providing anything from videos to posters to manipulatives to seeing what other teachers are doing . . . Those types of things have been tremendously helpful in my classroom, just seeing what other people are doing.

The importance of support for teachers who are changing and the need for ongoing learning by teachers was echoed in statements made by Brenda Baston, an administrator in Trier's district who had been very involved with collaborative governance and activities:

> [Teachers have] felt more professional because it's given them an organization. The collaborative has supported a number of [teachers] in attending regional, national . . . math leadership institutes. [It has] been a force [in the] high school's moving ahead in math. [The collaborative] didn't just give them a copy of the NCTM *Standards,* [it] gave them to the people and then . . . helped them to understand them and helped them to know how to implement them either directly or indirectly through the workshops.

Notably, the collaborative teachers' desire for ongoing growth and learning in mathematics education spilled over into the professional life of Baston, who said, "[The teachers'] involvement in state and national professional organizations has certainly increased. Mine has, too. I was not a member of the National Council of Teachers of Mathematics before they became involved." Encouragement for Baston to join NCTM originated with collaborative teachers.

Yvonne Alberts, a bilingual high school mathematics teacher and collaborative leader in Milwaukee, professed her beliefs about the teaching profession with special emphasis on authoritative instructional decisionmaking. Reflecting on her own decisionmaking in the past few years, Alberts said:

> What's happening to me for algebra is that I'm starting to let go more and more and more and saying, "I'm just going to not spend as much time doing [one topic], and I'm going to spend more time doing [another] . . ." If you want to have algebra for everybody, you have to connect in a lot of ways and let each one find what he understands best. And that's [why] I've tried to expand my repertoire.

Alberts's capacity to expand her teaching repertoire was enhanced by her participation in a number of programs, including, importantly, the collaborative's Algebra Network. Participation in the collaborative was not about learning a single new approach to teaching. Rather, it was about learning the wide variety of available options and making informed decisions:

> We brainstormed issues to set our agendas. We have shared materials . . . If we found interesting problems, we brought those to share. We've addressed . . . textbooks . . . the *Standards* . . . technology in the classroom . . . different ways of teaching algebra . . . computer-intensive algebra to the different textbooks and approaches that you can use. We've looked at manipulatives, algebra gear, and things like that . . . Several times we've had graphing calculator people come in. Problem solving has been in there. I've mentioned spread sheets before. One time we were talking about topics, what we think that we need to do in the middle school. The whole spectrum. What's going to lead to what? What are our expectations? What do we need to emphasize and de-emphasize? . . . [A speaker] told us about Algebra for the 21st Century project. He shared problems with us and we're looking at . . . more project-based, exploratory kind of curriculum in the algebra class . . . In May, since we're going to just dabble

in geometry a little bit, what are our expectations as we get kids through algebra? More kids through algebra? How is geometry going to change? . . . We discussed how do we deal with all of these levels of ability coming into our algebra for everybody. [A presenter] came to talk about grading and assessment issues.

Given the wealth of information about new approaches to teaching mathematics, the connections made through the collaborative with a community of fellow educators had enhanced Alberts's willingness to make instructional decisions:

[One particular colleague] is wonderful. Here is a guy who probably couldn't retire if he wanted to. He comes to our network meetings. And he's such a risk taker. He is such a risk taker. If I take a little risk . . . If I jump off a little cliff, [he] is jumping off a big high one. And I think for all of us, just to go someplace with him coming . . . It helps me professionally a lot because I asked him once . . . "How can you do that? Don't people get upset?" And he said, "Listen. You've got to have broad shoulders." So you know, I think that's another thing the network does is let people have broad shoulders and know that it's okay . . . There are neat people out there doing this stuff and . . . giv[ing] us ideas that we use.

Julie Krauss, a high school teacher in Los Angeles, had been teaching only three years. She offered several strong statements about the mathematics programs of her school, her district, and the state of California. The conviction and authority of her statements demonstrate the authority she believed teachers should assert as educational decisionmakers. During an interview Krauss remarked, "I like what the California state *Framework* says about integrating . . . math more. We're not doing it and we should . . . more discrete math, more statistics across the board." She added, "Just because the textbook says you start at chapter 1 and end at chapter 12 doesn't mean everything in between is important." Describing the extent to which this vision had been achieved in her own school, Krauss contended:

I don't think [mathematics has] changed as much as it needs to at our school . . . I'm frustrated at my school site. There's a group of teachers who want to do it the same old way, who have always done it, and there is a group of us who want to change and go with the new state *Framework* and with what the *Standards* say to do . . . This one group will say, "Well, until we get more textbooks . . . " and I keep saying, "Until you do something different it doesn't matter what textbook you're using."

Krauss's comments are revealing in two respects. One, she expressed a vision of mathematics teaching that she was willing to make happen on her own authority, with the support of the professional community to which she belonged:

I use a geometry book that's written totally around teaching in groups, Michael Serra's *Discovering Geometry* . . . I don't think I would have attempted it if I didn't feel . . . that other people were doing it and that it was an appropriate way to go . . . I just can't stand teaching [from a] traditional book. I'm not going to spend a year pulling teeth to do proofs. First of all, we don't need to do them anymore; second of all, [students] don't want to do them. So why don't we use something that they can . . . at least they'll do, and do some real math with it. I don't think I would have had the guts to say [that] if I hadn't felt that was . . . appropriate . . . I feel myself willing to stand up a little more for what I think is how the department in the state is going in education and how math should go . . . And I'm more willing to stand up for the kids at this point, and the curriculum and what should and shouldn't be in there. I'm more opinionated now.

Krauss's sense of authority stemmed from two sources. In part, she viewed her knowledge of what students needed as a source of authority. Thus, she de-emphasized traditional geometric proofs because both she and her students believed them of less value and usefulness than many other potential topics of study. Also, she grounded her authority in the knowledge that fellow teachers were making similar decisions.

Two, Krauss stated that teachers should be articulating goals for their own profession and for education and that they should be entrusted to work toward those goals: "The school district needs to . . . allow the teachers to have a little more impact and say in what's going on in the [teachers'] training . . . What teachers need is time to sit as just a department or small group and look at the curriculum and look how you change it to meet the needs of the students. " Participation in the collaborative had been singularly important in the development of Krauss's beliefs about teachers as authoritative educational decision-makers. Her main source of information and inspiration about mathematics education was the collaborative, especially through "interaction with other math teachers beyond just my school site. To know that what I've struggled with at my site, they struggled with at theirs. That there is a possibility of change in a school site; that it can happen if enough people believe in it." Krauss also credited the collaborative with "encouragement to be active elsewhere." "I don't think I would have started the master's [degree] program if I hadn't been involved in [the collaborative]. There's a number of [collaborative teachers] taking it also, which is also encouragement in doing summer-type projects, whatever is out there, [and] to try new things in my classroom."

Ongoing professional growth, enhanced decisionmaking, and increased professional authority among collaborative teachers were noted by other participants as well. The Memphis collaborative provided an illustrative example. Donna Green, a district administrator, said simply of collaborative teachers' desire for ongoing professional development, "I think the teachers going back to training to improve what they were doing was brought on by the math collaborative."

New professional relationships and responsibilities emerged from teachers' professional growth. Paula Strauss, a collaborative participant from higher education, noted that mathematics teachers' influence in the school district was slowly but recognizably increasing in relation to the collaborative:

> I think as a result of the collaborative and some of the ideas that the collaborative has given to the city school system, it has enhanced the city schools to go into the area of deregulated schools . . . whereby the teachers are given quite a bit more freedom to do things that they think are necessary to get over knowledge to the students. Whereby before you have a set curriculum—this is what you're supposed to teach, this is how you're supposed to teach it—and the other teachers didn't have too much to say about it . . . I think some of [the change] has to do with the collaboratives' influence, and in that teachers are now a little bit more involved in terms of what's going on in the system.

Claudia Fagan, a Memphis district administrator involved in the mathematics program, confirmed this perception:

> Teachers in the collaborative are very much a part of the district's mathematics program. And I think that participation in the collaborative was . . . kind of the genesis of their really catching on. I think I have seen members of the collaborative group participating in all phases. For example, textbook adoption: They have been very much a part of that, because they realize the importance of having an appropriate program to move forward with the *Standards*, the curriculum and instruction, and also to help teachers to grow professionally . . . They come with more knowledge and more of the leadership, and it is not like having to begin at the ground floor to get the teachers prepared for what we are looking for.

In relation to the Milwaukee collaborative, a similar phenomenon has occurred surrounding the school district's development of a new assessment system. Norman Platt was a district administrator who had played a key role in the Milwaukee collaborative's substantial participation in the CAM project. He viewed the district assessment development and CAM as mutually compatible:

> It isn't enough for me to be knowledgeable in assessment reform. We've got to get teachers out there who are doing it and then help to spread that among other teachers. And so the notion of teacher power, teacher empowerment, is an absolute key . . . What we wanted to do was to get teachers who would become informed and then extend that knowledge base.

CAM was a vehicle by which Platt's vision was approached. He described the importance of the four Milwaukee middle grades teachers who participated to the district's effort: "These people right now practically live and breathe alternative assessment . . . They are also at the core of our assessment reform effort here. They are key members of our math assessment committee." Platt and the district trusted in the authority and decisionmaking skills of these teachers in this very demanding and meaningful enterprise:

Last year the school board abolished our competency tests, and we're in the process with replacing that with a performance assessment that we're calling our proficiency examination . . . Next spring, we're doing it for real. That means the 11th graders have to succeed on that in order to get a diploma. The CAM people are right involved in the development in this. In fact, [they] take leadership roles . . . We operated in small groups and the CAM people were among the group facilitators.

Teachers' beliefs in their individual and collective authority as educational decisionmakers was manifest in the collaboratives. Networking and leadership were the two most notable areas in which collaborative teachers asserted their authority.

Networking. Collaborative teachers considered their teaching colleagues a very significant influence on their teaching (Middleton et al. 1989). This finding is consistent with collaborative teachers' belief in ongoing professional growth in the context of a professional community. Yet it remained surprising considering the reality of isolation, even within schools, of urban mathematics teachers from one another (Nelson 1994). Participation in the collaboratives offered an escape from professional isolation by providing a forum for professional discourse among teachers. Prior to the collaboratives' inception, opportunities, incentives, and purposes for sustained contact with other teachers were reportedly uncommon. For many veteran collaborative teachers, contact with other teachers was not only common, but an essential ingredient of the teaching profession.

The case of two middle school teachers in Georgia, Carmen Yannon, a seventh grade teacher, and Jill Zilinski, an eighth grade teacher, illustrates the power of networking in the context of the collaboratives. The two were involved with the Columbus collaborative for four years. Their participation began quite inauspiciously. "We had no idea what we were getting involved with," Zilinski said, "we were just asked to work with the [collaborative's] math camp." The two teachers spent additional summers working in the collaborative's summer mathematics camp for middle grade girls (called Positive Reinforcement in Mathematics Education—PRIME), serving as joint coordinators for an outreach session of the camp during the summer of 1993. Furthermore, they made substantial changes in their classroom practices and greatly expanded their professional involvement in and out of the collaborative. Growth in their professional networking undergirded all of these changes. Yannon related:

I think that the biggest benefit of [the collaborative] has been that it gives teachers confidence—to get in the classroom and try new ideas. Because before hand, we were never told that there are new ways to teach. Everyone taught the same way . . . Any math classroom you went into, everybody's doing the same thing. But now, on any given day, there are all sorts of activities going on . . . [Through] the rapport with the other math teachers, we have really gained a lot just by going and talking with other math teachers and getting new ideas.

Zilinski added:

> This is my 22nd year of teaching . . . When we first got involved with the col-
> laborative . . . most of the time, I taught as I was taught . . . A lot of the things
> that they introduced us to, we do use, and we call them in for materials every
> now and then . . . I sort of felt like I was living in a cocoon or something when I
> realized all that stuff was out here. But in the past, [our school system] has not
> encouraged us to go to anything . . . and it's just been in the past three years or
> so that we have.

These comments not only reveal how networking can broaden teachers'
knowledge of mathematics education, but also underscore the isolation these
teachers previously felt in their profession. Networking with other mathematics
teachers played an important role in breaking the two teachers' sense of isolation
and changing their perceptions of teaching. Zilinski revealed:

> We worked with and got to know teachers from several different school dis-
> tricts. We would never have been allowed to do that before. We found out that
> the problems that we had in the classrooms were the same ones they had . . . A
> couple of teachers that work [in another school district], they worked at PRIME
> camp the first summer that [we] worked down there. And when we went . . . to
> the Georgia [Council of Teachers of Mathematics] convention, we made a point
> to find them. If we went to different sessions, we shared the ideas that we got
> there . . . They gave us a lot of ideas that they had already used, and they were
> willing to share because they knew it worked in their classroom. I think that's
> important. Sometimes I think we don't share enough with other teachers . . . To
> me, it made me more enthusiastic . . . I just get really tired and frustrated some-
> times. It's a little boost to get to go somewhere and hear some new ideas. It
> needs to be different every now and then. You need a little bit of change every
> now and then to keep it from being the same old thing all the time.

Broader professional involvement changed the way Zilinski viewed her role
in the teaching profession: "I . . . never felt that I had any say-so over what went
on with the curriculum. I was given my books and my materials. I just taught
what they told me to teach . . . I guess I never realized that I could have any say-
so."

The value of networking with other mathematics teachers in the collaborative carried over into the teachers' professional relationships within the district.
"We started meeting with the high school math teachers," Yannon said, " . . . and
they've outlined the skills that they want us to teach and make sure that they get
incorporated before these students get to high school." Zilinski concurred, "I
agree, and I think a lot of that came from the collaborative . . . Since we've been
involved with the collaborative . . . we've gotten together with them at some
things that they wanted us to hear."

Two important outcomes of networking for Yannon and Zilinski were the
authority it gave them to make instructional decisions and the development of
their leadership abilities. Yannon traced the development of authority in instruc-

tional decisionmaking to her participation in the collaborative's mathematics camp.

> [The camps provided] an opportunity to experiment in a small environment . . . They provide the manipulatives and . . . background materials that we need in order to go in and prepare a lesson. They assist us in preparing the lesson. And then we're given the opportunity to implement it in a very small setting. And once we've implemented it in that small setting, it's very easy to come into a regular classroom and experiment with a group of 30 kids.

Armed with the experience and knowledge of shared experimentation, Yannon felt that she made better decisions: "Each classroom has been allowed to purchase their own math manipulatives . . . And basically, our supply list came from stuff that we had used at the collaborative that we knew that we wanted to use within our classroom on a regular basis." The teachers' sense of improved instructional decisionmaking was corroborated by their principal, Walter Westerhoff:

> When we made our . . . purchase . . . I had asked them to develop a list of materials that they had tried and knew would be of benefit to them and their students. We did go ahead and purchase them. At that point, I didn't feel like they were shooting in the dark. I felt that they knew what they were talking about because they had experience . . . When they came and made their . . . presentation with the things that they wanted, I felt comfortable with it. I didn't feel like they were buying things that would not be used . . . They were able to make a very good case for the things that they were interested in having.

Professional networking and leadership have since become regular parts of these teachers' careers. They were instrumental in working with the collaborative director and their district's assistant superintendent for instruction to establish an outreach session of the collaborative's mathematics camp in their district. The teachers' involvement as leaders in the district's mathematics program became commonplace, as Zilinski described,

> When there has been something that has come up in the county . . . about middle school math, they'll call . . . us. Because I think they feel that we've been involved enough with things and we sort of know what's going on now . . . They may ask us our opinion . . . We planned a workshop for elementary teachers.

Both teachers have maintained their involvement with the collaborative, which they perceive as a vital link to new ideas and resources. They consider networking and the sharing of ideas among teachers as critical aspects of the profession. Both have also extended their formal education as a result of their participation in the collaborative. Zilinski admitted that she had felt that her formal professional education was at an end several years before; however, after her involvement with the collaborative she felt compelled to expand her professional involvement. Regarding her enrollment in a mathematics education program at a local college, she said:

I couldn't pass it up . . . It's something that will really help me and I can really do something with it in the classroom . . . I really think that all of this has come about because of our involvement with the collaborative . . . So I really can't say enough about it because we were ignorant . . . I'd never heard of the [NCTM] *Standards*. We'd never heard of any of this before we went [to the collaborative].

Networking gave these two Georgia teachers new knowledge, which advanced them into leadership roles in the collaborative, in their school, in their district, and in the broader mathematics education community. Teacher leadership, in many forms, was a cornerstone of the story of teachers' experience with the collaboratives.

Leadership. Traditional leadership roles for teachers are quite limited (Bird and Little 1986, and Little and McLaughlin 1991). Within a school, a few teachers may advance to positions of department chair, lead teacher, mentor teacher, or potentially a position combining administrative and teaching duties. Within a district, active teachers often hold few leadership positions, though some administrative and instructional leadership positions are considered promotional opportunities for teachers who leave the classroom. Teacher leadership in the context of the collaboratives was not a direct challenge to traditional leadership roles, mainly because the vast majority of teacher-leaders in the collaboratives remained in the classroom (Little and McLaughlin 1991). Some collaborative teachers became department chairs or lead teachers in their schools, and the collaboratives expanded opportunities for teachers to contribute to decisionmaking structures in their districts. Such forms of teacher leadership are important and represent considerable impacts of the collaboratives.

Teacher leadership in the context of the collaboratives, however, connoted far more than promotion to formal positions of authority in schools and districts. Collaborative teachers' voluntary commitment to professional growth was a form of leadership. Their service as designers and conductors of professional activities was a form of leadership. Their accepted responsibility to maintain the relevance and currency of their professional knowledge was a form of leadership. Their willingness to share ideas, experiences, reflections, and opportunities with colleagues in both formal and informal relationships was a form of leadership. Teacher leadership in the collaboratives, by these definitions, was not limited to the few who rose to the top. Rather, leadership was a fulfilling component of the teaching profession for all, and it did not necessitate a surrender of classroom teaching.

The case of a mathematics department in Los Angeles is instructive. Maria Naranjo, the department chairperson and an active collaborative participant, attributed her desire and confidence to seek a leadership position to her involvement with the collaborative. Several years ago, Naranjo moved from a school that was very professionally involved to one that was not—a situation she said "depressed me immensely because I enjoy being active. I kind of like to be around people who are doing things. Maybe I wasn't always the initiator, but I liked a place where people did things." Naranjo proved instrumental in changing the mathematics department at her new school, as collaborative director Pearl Quigley recalled:

> You get one person like Maria Naranjo from [a school] . . . She learns from her colleagues . . . She's breathing the +PLUS+ philosophy . . . "How can I make this happen at my school?" And just that dynamic, the energy, is what gave her the strength and vision to go to her school, take the worst department in the school and turn it around and now have a $300,000 California partnership grant that's she's involved with.

The details of the story demonstrate how important the collaborative was in this remarkable transition. The mathematics department at the school had rarely met and had not pursued change before Naranjo arrived. Concurrently, she began attending the collaborative's Saturday workshop series, a year-long teacher-led program for teachers to learn about, experiment with, and share teaching ideas on a variety of topics. Through her participation, Naranjo also learned how departments could join the collaborative. She rallied support in her department, which organized itself to write a grant proposal to join the collaborative:

> We wrote the grant. We said, "Look, we've got all these complaints. But what is it that we can do?" And I think that was the beauty of [the collaborative's departmental grants]. Yeah, there's a lot of things . . . No, you can't increase your budget by $50,000. That was an impossible dream. But with what you have, what can you do? And I think that opened the dialog that got us closer, got us talking. And that really made all the difference in the world in that department. Because I think we realized, "Let's not be complainers. There are things that we can change."

Shortly thereafter, Naranjo was elected department chair, and the mathematics program changed considerably in the five years following.

Nate Ortel, a former mathematics teacher and district administrator in Los Angeles, was similarly impressed in the collaborative's early years at how mathematics departments began to change. His words echo the notion of a turning point that Naranjo described:

> [The collaborative] provided a vehicle for . . . math teachers in the district where nothing else existed . . . There was no vehicle for doing that, to mobilize math departments and math teachers . . . It was interesting how just that much money would motivate people who had never talked to each other to start talking to each other, to sit down and write a plan and move forward with it . . . They would get together, they'd write it . . . and then would talk to one another.

When Naranjo's department joined the collaborative, the plan for improvement was, in retrospect, quite simple—a reward system for students' performance. However, it gave the department a sense of purpose and respect it previously did not enjoy: "It kind of brought the math department into focus at that school. We were always the pits . . . If you were in the math department, that was not something to be proud of . . . And we got a lot of compliments, and it was really nice to see." In the ensuing five years, the department's efforts to reform and improve its mathematics program increased substantially. Teaching assignments changed to provide a more equitable distribution to teachers of different levels of mathematics classes. Basic mathematics courses were gradually replaced by courses that offered more algebra and advanced mathematics to all students. More teachers were involved in the collaborative's Saturday workshops, and four attended extensive summer mathematics institutes in Connecticut and Massachusetts with support from the collaborative.

The process of writing a grant to join the collaborative and to improve the department's mathematics program, related Naranjo, was a significant step toward a more active leadership role for the teachers:

> If you get extra money, you get to do some of the extra things that you're always complaining about . . . that you don't have money to do. And as a result of that and all of the other activities . . . in [the collaborative], it kind of opened your eyes . . . I think we made a lot of changes at the high school. Now we have to take it one more step.

Taking the next step led the department to develop, write, and—eventually —receive a California Academic Partnership grant, a major part of which funded reform of the mathematics program with an aim to place more students in colleges and universities. The development, writing, and administration of the grant utilized the leadership of Naranjo, who codirected the administration of the grant: "Seeing other grants that were funded, I guess we were one of the few that actually had a teacher as a codirector . . . I think if you're going to have a good grant, you need someone that's in the classroom that knows what needs to be done." Moreover, Naranjo said, the process of reforming the department's program developed the leadership of the entire department:

> We've talked quite a bit about reform, especially with algebra. So with this grant . . . we got to look at some different curriculum programs. [The collaborative] has kept us in that loop of updated information. You know, without it, I don't think we would have ever learned about Change From Within, IMP . . . the TI-82 calculator . . . Three of us now in the department . . . are going for our master's with VizMath, which is a master's program . . . sponsored by NSF [the National Science Foundation] for math teachers.

As leaders, the teachers in the department took risks with new experiments in the classroom. In the ninth grade mathematics classes, which previously focused on basic skills, Naranjo said, algebra became the emphasis:

> Once we get to the point where they need a calculator, then we bring in the calculators. So I think our main stress is—we're giving them algebra . . . they are doing algebra. And we tell them, "We're not going to waste our time doing basic [mathematics]. We don't want to teach it . . . You've had it for years." You

know, that's our philosophy. "You've had it for three years. Here's a calculator." Help them to do better. Then let's go on; let's learn algebra . . . We're giving them algebra. And we'll keep on giving it to them. So we've embraced that.

Some encouraging results were seen among the students. Naranjo noted:

> Our calculus [enrollment] . . . increased last year, from 10 students to 29, this
> year . . . Two years ago, we really pushed our geometry students to take algebra
> 2. Especially students who were in the ninth grade and in geometry . . . And we
> opened up new calculus. We have [something] like eight students taking the
> second year and we never had that before.

Enrollment in algebra 1 also increased, and a greater proportion of the students
enrolled in higher mathematics courses came from the 10th and 11th grades
rather than from the 12th grade.

Growth in leadership also promoted new roles and responsibilities among
the teachers. For example, Naranjo said, "Part of our job is to keep [the adminis-
tration] informed about what is going on and what's out there, what we need to
be doing." Assuming this leadership role requires that teachers remain informed
with current information and practice, "as teachers, we have to constantly
change," and trust one another as instructional leaders, "when people go to a
[collaborative] workshop or anything like that. They know that it's going to be
worth their while. It's not a waste of time. And it's being done by teachers and
there's that credibility . . . Think of all of the changes that are going on in mathematics right now. We really need it coordinated."

Teacher leadership that combines new responsibilities with classroom duties
is taxing. Leadership requires teachers to find a balance in their professional
lives, which Naranjo found through the collaborative:

> I think with [the collaborative] . . . it kind of keeps you going, keeps you moti-
> vated, keeps you wanting to do more . . . That you don't get bogged down with
> the day-to-day things. Working for [this district] can be very depressing. Things
> don't seem to go right a lot of the time. It's such a big bureaucracy . . . But I
> think with [the collaborative], it keeps you wanting to go on . . . And keeps
> your focus that your job is with students. That it's the students that you're in it
> for, not to make your life easier, to make your job any easier. That you'll really
> get something out of the students that you have in your classroom.

Doreen Eagle, a former principal at another collaborative school in Los
Angeles, attested to a similar change among that school's faculty. Joining the
collaborative gave the mathematics department at the school opportunities and
possibilities it never recognized before. She observed the change most notably
when the teachers began to consider ways to curb the school's high failure rate
in algebra 1:

> What math +PLUS+ allowed the math department to do was give time and the
> incentive to get together to look for solutions. And it did that. The first solution
> that was suggested and tried was a three-semester math course. It didn't work.
> But that was okay, because we were looking for solutions so we went back to

the drawing board. We knew that it didn't necessarily mean that we were fail-
ing, we just hadn't found the key yet. The key was, "Let's keep trying until we
find what we are looking for." I think we found it in the Interactive Math
Project.

Participation in IMP was much more than a matter of discovering the program, especially since the teachers learned about IMP very late in the application process. It took persistence, leadership, and a manageable but convincing
plan—all of which, Eagle observed, emerged from the collaborative teachers:
"They came to me and said, 'Doreen, what can we do? Let's write a letter. Let's
bombard them with phone calls. Let's tell them [to] give us a chance to try this.'
Because they really wanted to do it. They discovered the program. They really felt it
was right for us."

The school was specially admitted into IMP. Eagle and the teachers were
extremely pleased with the engagement and learning of students in the years following.

Teacher leadership developed within the collaboratives. It is further noteworthy that, according to Nate Ortel, leadership translated to other improvement
and change efforts: "Achievement council—which is a state program—people
from +PLUS+ moved into that program and they already knew how to do
things. They knew how to organize. They knew . . . where resources were. They
knew how to tap into resources."

Teachers' Professional Growth

Teachers' professional growth is a process of constructive learning (Simon and
Schifter 1991, and Lambert 1989). Treated as such in the collaboratives and in
this case study, teachers' professional growth required a few fundamental elements. This case study reveals seven elements of professional growth—disequilibrium, exposure, existence proof, modeling, support, experimentation, and
reflection. The contribution of these elements to teachers' growth is further supported by existing research on teachers and reform.

These seven elements of professional growth emerged from analysis of the
case as integral to teacher change. These elements did not occur in a specific
chronological order; they did not exist in a specific hierarchical order; and their
relative importance in each individual's professional growth varied. In relation
to the collaboratives, they were highly interactive and rarely occurred in isolation.

In the brief sections that follow, each element of professional growth is
introduced in the words of one or more collaborative participants or observers.
(All of the elements are illustrated in figure 2, which follows these sections.) A
description of each element and its treatment in existing research is also included. The nature of the seven elements of professional growth, interwoven with
one another and with each individual's experience, complicates the provision of
concise examples from collaborative teachers' experience of how each con-

tributes to professional growth. Rather than attempting to provide such examples, the reader is directed to earlier sections of the case study in which teachers' experiences are treated more comprehensively. Rereading these sections with the seven elements of professional growth in mind will reveal both their individual and collective contributions to teachers' professional growth.

Figure 2.   Elements of Teachers' Professional Growth

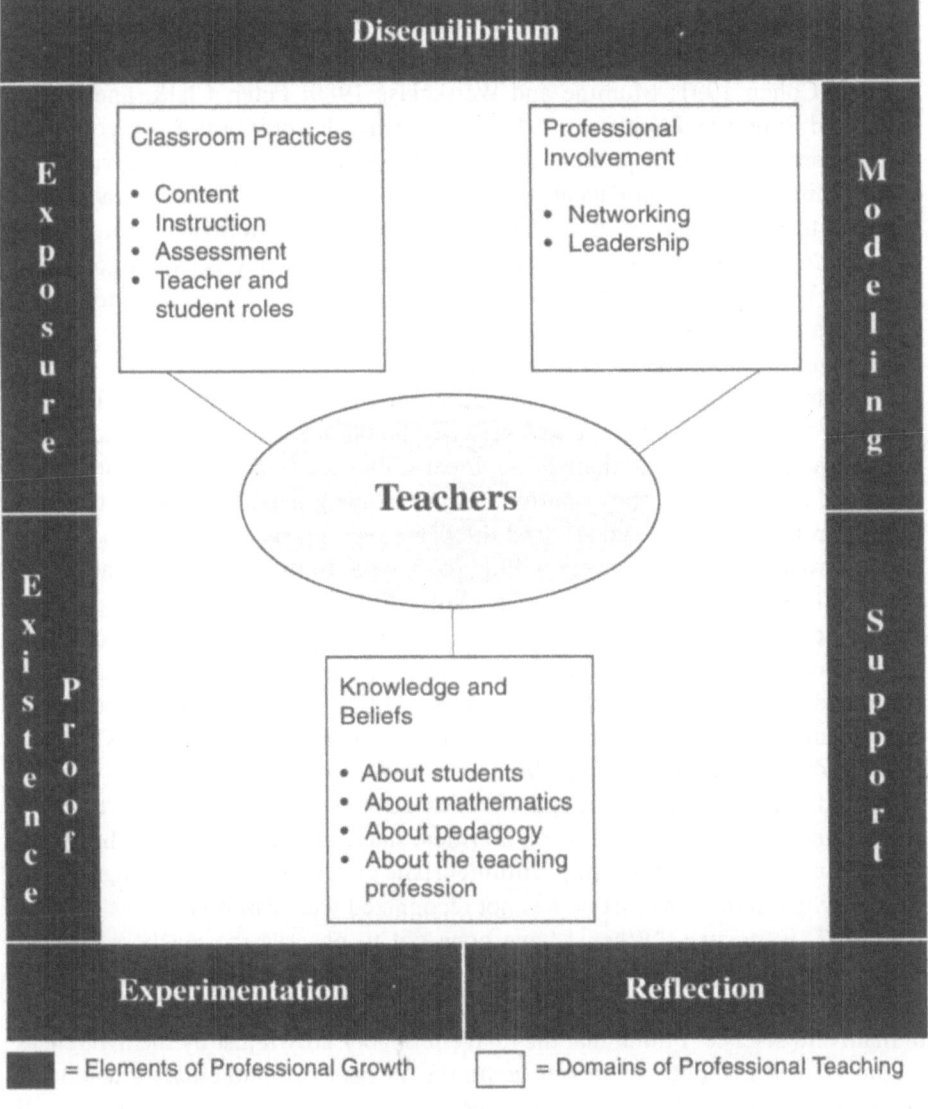

1. Disequilibrium. Chris Erlanger was the professional development coordinator at a high school in Los Angeles. Upon reflection on the growth of mathematics teachers who participated in the collaborative, he said, "You've got to, I guess at one some point, be frustrated to and be open to say, 'Well, I'm aware that maybe I'm not doing as much as I could be, or getting the results that I want, and I'm not quite sure why I'm not getting the results in my classroom.'" Brenda Baston, district administrator in Columbus, concurred: "I think the biggest [change] is the teachers realizing that what they've been doing for many years is not working. The need, just the urge, for something new."

Recent research on professional growth among teachers and reform in education has hypothesized and supported the idea that change begins with disequilibrium (Cohen 1991; Mumme and Weissglass 1989; Peter, Clark, and Carlin 1992; and Simon and Schifter 1991). Specifically, disequilibrium is recognition that current practices or policies do not hold the capacity for an individual to achieve his or her current educational goals. Disequilibrium compels teachers to question the status quo in their classrooms, schools, districts, and professional lives and to search for potential remedies to the discrepancy between goals and realities. When recognized by a teacher, disequilibrium implies at a minimum the desire to improve as an educator, and potentially the willingness to change fundamentally (Simon and Schifter 1991).

Teachers' willingness to change was a cornerstone of the collaborative project. In general, the collaboratives were a valuable asset to those teachers who showed a willingness to change. As local collaboratives evolved, they each expanded their services and opportunities to include greater numbers of teachers from a broader range of schools and districts, drawing their strength and membership from teachers who were willing to change. In individual cases and over time, collaboratives reached teachers whose disequilibrium and willingness to change developed concurrently with, and sometimes in response to, the collaborative. However, the collaboratives were less successful at promoting and harnessing disequilibrium and the consequent willingness to change within the entire mathematics teaching force of schools or districts. Collaboratives faced a peculiar dilemma in this respect: Voluntary participation and expressed willingness to change were powerful and effective motivators for collaborative teachers, but in order to grow and reach a critical mass of teachers, somewhat more explicit methods of promoting volunteerism and fostering a willingness to change where it did not exist or was not recognized were essential.

2. Exposure. "When I have gone out to observe the teachers," stated Kirk Morris, a higher education participant in the Milwaukee collaborative's internship program, "they were including a lot more of the real-life applications in math and science. I think that they have probably grown just by the awareness factor." Chris Erlanger witnessed a similar impact on mathematics teachers' broadened exposure through the collaboratives: "Because of the advent of all those programs, the students had options, but also the teachers had options . . . Being exposed to them made a big difference."

One of the primary conditions the UMC project had hoped to change was the isolation of urban mathematics teachers from ideas, from resources, and from opportunities (Nelson 1994). Reducing teachers' isolation implies that teachers will become aware of ideas, resources, and opportunities; will have genuine access to them; and will have the authority to utilize them. The collaboratives first addressed awareness by exposing teachers to ideas, resources, and opportunities through teachers' local contact with collaborative staff, with participants from business and higher education, and with other teachers. Additionally, the national communication networks sustained through the collaboratives were vehicles by which teachers were connected to the broader mathematics education community. Thereby they obtained exposure to other ideas, resources, and opportunities. Exposure broadens teachers' awareness of possibilities for change and fosters a sense that alternatives to traditional knowledge and beliefs, classroom practices, and professional involvement are available and within their reach. Exposure alone, however, does not ensure teachers' actual or perceived access to or authority over ideas, resources, and opportunities (Joyce and Showers 1980, Lambert 1989, and Wilson 1990).

3. Existence Proof. The Los Angeles collaborative's classroom demonstration program, according to collaborative director Pearl Quigley, was "a strategy to encourage people that weren't coming to the workshops to see that people that were presenting workshops really know what they are talking about."

> And you come to their classrooms and you see the new strategies working with kids. Well, if the new strategies are working with kids, then I as a teacher ought to learn how to use that strategy with my kids, so I'll go to the workshops. So the demos became a recruitment mechanism. It also became a strategy for making real for those teachers who couldn't make the jump from the workshops to seeing the impact it has on kids.

Columbus resource teacher Derrick Duncan echoed Quigley's emphasis:

> Just the fact that there are teachers there who are doing these things and they can hear someone say, "Yeah, I did that," or "I did something like that . . ." I think the interaction of people who are actually doing it . . . that makes a tremendous difference . . . If you took a whole bunch of people who are very traditional teachers, who had not done any of this stuff, sent them off to a workshop somewhere that was done by somebody that they didn't know . . . I don't think they would be convinced and try to change as quick as hearing the person down the hall or the person in the next county . . . I think that makes a difference to teachers . . . to hear people that they can identify with talk about what they're doing in the classroom . . . And they realize that if "so and so" can do it, then I can do it, too!

Teachers' exposure to new ideas, resources, and opportunities often prompts the question: Will they work in practice? Mathematics teachers are sensitive to theoretically sound ideas and materials that do not function well in the classroom or school setting or that do not sustain a workable mathematics program for students. Mathematics education reform movements of the 1960s and 1970s are often cited as examples of this phenomenon (e.g., Cohen 1991, Fey 1979, Fullan and Stiegelbauer 1991). Many teachers first seek substantiation of the applicability of new ideas through examples of their successful utilization—in short, existence proof—in environments similar to their own, usually in another teacher's classroom. Knowledge of the existence of new ideas in practice compelled collaborative teachers to change in two ways. First, an existence proof offers moral support because teachers recognize that the goal of successful implementation of new ideas is realistically obtainable. Second, an existence proof provides a challenge to teachers, because they can test their abilities to make changes in their own classroom, school, or professional life against a known standard. In the context of the collaboratives, the power of existence proofs to foster change was related to the level of similarity between the existence proof's example and the professional situations of other participating teachers. The localization of the collaboratives and the familiarity of participants with one another created convincing existence proofs.

4. Modeling. "When I do a workshop for teachers," stated Derrick Duncan, "I present it to them in pretty much the same way that I would present it with students. To ask the same kinds of questions that I would ask with students . . . I've asked them to justify and explain things. So that they can see how it's presented to the class. I'm modeling for them." Such modeling was commonplace in many collaborative activities. It was most obvious in instances like the one Duncan described—demonstrating classroom activities for teachers in an authentic fashion.

Modeling became a strategy the collaboratives used to promote more systematic change. George Jackson, a higher education representative in the Memphis collaborative, alluded to a use of modeling in a new program in order to influence the school system:

> The project involves providing teachers with resource people as well as materials, and a chance to correspond, to discuss with other teachers who are in the project what is going on. So we are trying to create a supportive collaborative group of people to demonstrate to the system, rather then just tell the system what it is like to do things in new ways.

Most teachers who are exposed to new ideas and aware of the existence of new practices need to experience the idea or practice in an authentic setting to facilitate transfer to their own setting (Joyce and Showers 1982). Modeling is a powerful vehicle to deepen a teacher's understanding of a new idea because it places the idea in a realistic context. It can be a potent extension of an existence proof because the teacher's own experience becomes a part of the proof. The collaboratives provided various types of modeling in their activities and relationships. Workshops, meetings, and institutes sponsored by the collaboratives were frequently designed to mimic a classroom situation in which the presenter modeled the behaviors of a classroom teacher and the participants played the role of students. Some collaborative activities additionally offered teachers opportunities to observe one another teaching mathematics to students in controlled or real classroom settings. Another type of collaborative activity allowed teachers to observe how other professionals used mathematics in their work with the aim of translating applications into classroom activities as realistically as possible. Integral to each of these activities was the modeling of teaching or applying mathematics. Furthermore, such activities provided distinct models of leadership and professionalism. Teachers saw other teachers in leadership roles and observed teachers and other professionals engaged in meaningful discourse, elements of which formed a model of desired professional growth.

5. Support. Victoria Wells was an assistant principal at a high school in Los Angeles where many mathematics teachers were involved in the collaborative. Reflecting on how change occurred among teachers in her school, she related, "I think that the teachers having the time to work on something. That's the most critical factor of all. For teachers to be able to talk with each other, develop programs with each other."

The collaboratives became well-aware of the importance of building collegial support for teachers. Frances Ilps, director of the Memphis collaborative, noted that the teachers became responsible for communicating their need for support: "A lot of teachers express the desire for that opportunity to meet with fellow teachers on a regular basis during the school year . . . When we ask the teachers what they would like to do, a lot of times that is one thing that they say they would like to do."

The collaboratives learned to build the collegial support that teachers desired and required into their programs. Frances Ilps made a general observation about how the collaborative attended to this need: "One of the key principles we've always stuck to is to get pairs of teachers from a school [to participate], so that when they go back they are not isolated in their attempt to implement things. They have a partner." Reliable support among teachers and other participants was critical to many collaborative programs.

Teachers' professional growth and related change in classrooms and schools are evolving processes that require time and support. Exposure, existence proofs, and modeling may plant seeds of growth and change, but substantive and lasting growth and change require sustained support (Cochran-Smith and Lytle 1990; Jaworski 1991; Joyce and Showers 1980; Little 1982; Mumme and Weissglass 1989; Peter, Clark, and Carlin 1992; Wilson 1990; and Wiske and Levinson 1993). The collaboratives variously provided teachers with material, financial, and technical support through resource centers, resource personnel, stipends, and grant programs. Most importantly, the collaboratives provided forums for teachers engaged in change to associate and communicate with one another and with other professionals in a variety of formal and informal settings. In addition to the potential for articulated disequilibrium, increased exposure, solidified existence proofs, and expanded modeling such collegiality bears (Bird and Little 1986, Little 1990, Little and McLaughlin 1991, Mumme and

Weissglass 1989, Simon and Schifter 1991, Wiske and Levinson 1993, and Woodrow 1992), they also offered ongoing and responsive moral, technical, and intellectual support for change.

Anticipated and unanticipated challenges may arise in reaction to any change a teacher attempts to enact. Individually, teachers often do not have—nor can they be expected to have—the resources or experience to remedy all difficulties arising from change (Wilson 1990). Moreover, individual teachers may not be aware of all of the potential positive impacts of properly managed change, thereby missing opportunities to foster student learning (Cohen 1991). Supportive collegiality with fellow teachers and other professionals increases each individual teacher's potential to manage change because it increases the base of problem-solving resources available. Colleagues might have encountered similar challenges and attempted different solutions with varying results that can be shared. Moreover, colleagues who have tried similar changes may have identified a variety of novel positive outcomes in students' learning, in their teaching practice, and in their professional lives which can be further encouraged with appropriate attention. A collaborative's shared, collective knowledge bolsters the capacity of each teacher to manage change; to identify and encourage positive outcomes; to anticipate difficulties and obstacles; and to apply appropriate solutions in his or her own classroom, school, and professional life.

6. Experimentation. In the Columbus collaborative's summer mathematics camps, teachers enjoyed a rare opportunity to teach with very few constraints. Resource teacher Carrie Calloway described this special environment:

> Whatever you want, we provide—whatever you need to teach that lesson that you'd be afraid to teach it in your class. And then [the teachers] succeed . . . We tell them, "Think about it. If this were your class, what would you like to do?" And we try to provide them with everything that they need so that they can try that good situation . . . We try to give them their wish lists.

Suzanne Sanchez was a district administrator who worked with teachers who participated in the camps. She conveyed her observations of the importance of allowing teachers to experiment during the process of change:

> Another thing that I think that they have done that has successfully translated the training back to the classroom is creating a risk-free environment in the summer for teachers to practice the new skills and strategies that they learn. A week of the class instruction, a week of working with some students during the summer, actually practicing what they have learned, and then a week of lesson preparation to prepare the activities to go back into the regular school.

Experimentation, undertaken proactively and deliberately, is a characteristic of schools that demonstrate effective professional development among their staff (Little 1982). It is further argued to be a vital element of an individual teacher's professional growth (Goswami and Stillman 1987; Jaworski 1991; Peter, Clark, and Carlin 1992; Simon and Schifter 1991; Wilson 1990; and Woodrow 1992).

Critics, however, have justifiably cautioned that experimentation by teachers is potentially dangerous when undertaken without sufficient information and deliberation (Hofmeister 1993).

The collaboratives viewed teachers' experimentation as highly variable and personal. As teachers experiment with new ideas and practices, they encounter uncertainties about the reactions of students, fellow teachers, administrators, and parents; and they risk negative reactions (Fullan 1982). The collaboratives provided an environment in which a teacher, as a willing volunteer to change, could experiment at his or her own pace depending on that individual's level of comfort with inherent risks and uncertainties, level of confidence to overcome both anticipated and unanticipated difficulties, and level of competence with desired changes. In support of individual teachers' experimentation, the collaboratives provided information through exposure, existence proofs, and modeling. They offered the support of materials, finances, and resource personnel, which allowed teachers to undertake experimentation that would otherwise be beyond their reach. Most significantly, the collaboratives served for many as a forum for deliberation on the merits of experiments. In some cases, collaborative programs were specifically designed to encourage experimentation and critical deliberation on the outcomes of experiments; in others, a variety of experiences led teachers to seek, through the collaboratives, the resources and support necessary for a particular experiment, as well as colleagues with whom they would consider the results and implications.

7. Reflection. "We have to realize . . . that teachers are at all different stages of development," acknowledged Alice Adams, administrator of the Columbus collaborative, "and if a teacher comes up with this idea, it may be leaps and bounds over yesterday's lesson plan. And you shouldn't discourage a person by immediately making improvements on their plan; you just have to lead them and give them opportunities to reflect . . . and they will grow."

Such opportunities for teachers to reflect freely on their practice and profession are indeed uncommon. Reflection was an important component of change in some collaborative activities. Olympia Pressman, a district administrator and participant in the Los Angeles collaborative, found the collaborative's workshop series an effective professional development mechanism due to the emphasis on shared reflection coupled with experimentation: "After each [workshop], the teachers are developing lessons from what they've learned. They try it in their own classrooms. They bring the lessons back to the next [workshop], and teachers swap and share and they tell about how it worked and what it did and then they develop more."

Reflection is very closely coupled with experimentation. Much of the informative and deliberative substructure vital to experimentation also forms the foundation of reflection. Reflection, however, is delineated as a separate element to underscore its importance in the change process, particularly as it relates to individual teachers. Research on teachers' professional growth has demonstrated in case studies of individuals that change without reflection is shallow and incomplete (Cohen 1991 and Wilson 1990). Moreover, developmental, theoretical, and evaluative reports on professional growth of teachers identify reflection as an essential component of sustained progressive change (Cochran-Smith and Lytle 1990; Jaworski 1991; Little 1982; Peter, Clark, and Carlin 1992; and Simon and Schifter 1991).

Various collaborative activities supported the use of journals or regular meetings of teachers specifically to encourage serious reflection on teachers' engagement in change. More significantly, collaborative teachers described serious reflection on change within the collegial environment they found in collaboratives. Many collaborative teachers uniquely identified with collaboratives the productive examination of ideas, sharing of experiences, and deliberation about solutions, both formally within activities and informally in sustained communication with colleagues.

Collaboratives in the Long View

A Portrait of 16 Collaboratives

An overview of the locations, host agencies, types, governance, and target teachers of the collaboratives is provided in table 2.

Level of Operation. The Ford Foundation, later with the help of EDC, established a total of 16 urban mathematics collaboratives as fully funded original sites or partially funded replication sites between 1985 and 1992. By the summer of 1993, the levels of operation varied considerably among the 16 sites.

Of the 11 original collaboratives, two were well-developed, both organizationally and programmatically—with activities, governance, and participation contributing to a permanent existence. Likewise, of the five replication collaboratives, two seemed to be well-developed in most of the dimensions that were explored for this study. Two of the other original collaboratives at the time supported several strong programs that were exceptional in organizational terms. One collaborative emphasized breadth and continuity across the year in its program offerings, and the other collaborative emphasized highly visible annual events or one-time events.

Although three of the original collaboratives continued to function in terms of activities, governance, and participation, they remained vulnerable to contextual factors that could prove seriously detrimental. Two of the collaboratives had relied on the support of school district budgets to provide personnel and time for critical administrative functions, but both of the collaboratives lost some of this support. The other "functioning" collaborative faced potential financial danger if it could not find financial support for its principal program on assessment.

Two of the original collaboratives were at critical junctures that challenged their existence as formal organizations. One of these collaboratives existed in part to serve as an equalizing force in mathematics education between the city

Table 2. Urban Mathematics Collaborative Sites

Site	Host Agency	Type	Governance	Target Teachers	School Districts
Cleveland	Cleveland Education Fund	Orig.	Corporate	MS and HS mathematics	Urban (1)
Columbus	Columbus College	Rep.	Corporate	K-12 mathematics	Urban, rural
Dayton	Alliance for Education	Rep.	Program specific	K-12 mathematics and science	Urban, suburban
Durham	North Carolina School of Science and Mathematics	Orig.	Corporate	MS and HS mathematics	Urban, suburban
Los Angeles	Los Angeles Educational Partnership	Orig.	Decentralized	MS and HS mathematics	Several urban
Memphis	Memphis State University	Orig.	Corporate	MS and HS mathematics	Urban, suburban
Milwaukee	Marquette University	Rep.	Single body	MS and HS mathematics	Urban, suburban
New Orleans	Metropolitan Area Committee	Orig.	Corporate	MS and HS mathematics	Urban (1)
New York	New York City Fund for Public Education	Rep.	Single body	K-12 mathematics	Urban (1)
Philadelphia	PATHS/PRISM	Orig.	Corporate	K-12 mathematics and science	Urban (1)
Pittsburgh	Pittsburgh School District	Orig.	Grade-level specific	MS and HS mathematics	Urban (1)
St. Louis	Cooperating School Districts	Orig.	Single body	MS and HS mathematics	Urban (1)
San Diego	San Diego State University	Orig.	Corporate	MS and HS mathematics	Urban (2)
San Francisco	San Francisco Education Fund	Orig.	Single body	K-12 mathematics	Urban (1)
Twin Cities	University of Minnesota	Orig.	Corporate	MS and HS mathematics	Urban (2)
Worcester	Alliance for Education	Rep.	Single body	K-12 mathematics	Urban, suburban

and county districts that it served. Financial support from both the city and county districts had already been decreasing when the two systems merged in 1993. The absence of a financial commitment from the newly merged district and the collaborative's long-standing difficulties in obtaining grants were causing the collaborative to reconsider its formal organizational operations. Similarly, the other collaborative that was at a critical juncture had operated within the school district structure, but it too was facing significant organizational changes prompted by district reorganization. The school district at the site had neither a division of mathematics nor a mathematics supervisor. Instead, a general curriculum division emerged to assist school-based decisions regarding curriculum in all subject areas. Lacking the budget and organizational leadership and power it had enjoyed, the collaborative was forced to be redefined in order to persist as an organization. However, if the collaborative were to cease operations as a formal organization, the decision to disband would have to be considered in light of the collaborative's original intentions: The collaborative had always intended to create such fundamental changes in the district that the organization would eventually disappear.

Finally, of the 11 original collaboratives, two in 1993 effectively ceased to operate as formal organizations. One collaborative exhausted its funding in August 1993; approximately one year earlier, its coordinator had left the collaborative's host agency. The other collaborative was unable to fill the position of coordinator outlined in its permanence proposal. The lack of funds to meet the salary requirements of candidates contributed to the collaborative's disbanding as a formal organization. In addition, a lack of understanding among the host agency, advisory board, and teacher members regarding general budgeting of funds and the proper position description, duties, and salary of a coordinator led to the dissolution of the collaborative.

The three remaining replication collaboratives bear separate consideration. Two collaboratives came into existence largely for the sake of single programs. In both cases, the main programs continued to operate, and smaller tangential projects were also emerging. The third collaborative no longer existed as a separate entity, but many of its programs remained in an expanded fashion within the original host agency, with many of the same people participating both in governance and activities.

Host Agencies. In terms of their most recent operation, the 16 collaboratives existed mainly within two categories of host agencies. The most common type of host agency was an educational advocacy organization—in many cases a public education fund.[5] Nine collaboratives had such an agency as host. The other main variety of host agency was an institution of higher education. Five collaboratives had a college or university as host. The other two collaboratives had unique host agency arrangements: One was hosted by the North Carolina

---

[5]Public education funds are nonprofit organizations that are community based, with the primary mission of raising and brokering community resources for the benefit of schools.

School of Science and Mathematics, a public magnet boarding school; the other operated within its city school district, though at one time it was hosted by an educational advocacy organization.

Governing Structures. Two general forms of governance emerged in the 16 collaboratives. One form divided responsibilities between two committees; the other relied almost entirely on a single governing committee. The dual committee structure generally consisted of two bodies: a cross-sectional committee, with business, higher education, school district, and teacher representatives performing some governing functions; and a second committee, entirely or largely composed of teachers, performing other governing functions. Exactly half of the 16 collaboratives used this type of governing structure. In a few of these cases, one governing committee became entirely responsible for decision-making, planning, and steering—despite the formal existence of a second committee.

Collaboratives that were governed by a single committee relied upon the cross-sectional representation and active participation of individuals from different sectors. However, operational details of these organizations varied significantly from site to site. Five collaboratives relied on this administrative structure. At the three remaining collaboratives, governance was essentially a hybrid of the one- and two-committee schemes described above.

As originally conceived, each collaborative was expected to limit its target audience to secondary mathematics teachers only, with permission for expansion. The collaboratives all began with or adopted broader target audiences through vertical expansion to other grade levels or horizontal expansion to other subject areas.

Subject Areas and Grade Levels. Four collaboratives eventually included all teachers of mathematics—secondary, middle, and elementary—in their target audience and membership. Six other collaboratives either established or investigated ways to serve teachers of all grade levels. Although several collaboratives developed or were developing organizations similar to the UMC for local science teachers, or have included science teachers in some activities, only two collaboratives had a joint mathematics and science focus in their target audience and membership.

Of the collaboratives that maintained a focus specifically on mathematics teachers, nine served both secondary teachers and middle or junior high teachers.

Districts Served. Collaboratives were originally intended to serve at a minimum the teachers from schools in one urban district. Some of the collaboratives began with larger or more inclusive constituencies in terms of districts, whereas others grew to include a broader target audience:

• Seven collaboratives mainly served the teachers in a single urban school district.

• Three collaboratives served teachers in more than one urban district.

- Six collaboratives served teachers in both urban and largely suburban or rural school districts.

- Five of the collaboratives also included private and parochial school teachers in their target audiences.

Organizational Collaboration. Most of the collaboratives continued to involve representatives from business and higher education as advisors, through service on governing committees. Many of the collaboratives also relied, in part, on financial contributions from local businesses and institutions of higher education. Participation at these levels varied widely, as did the involvement of partners from business and higher education beyond advisory, governing, or financial responsibilities.

Teacher Participation. Using information provided by collaborative administrators during telephone interviews, it was possible to make very general estimates of total participation by teachers in the work of collaboratives. *Very active* teachers were those who participated in collaborative governance or numerous activities. Estimates of very active teachers range from as few as 11 at one site to as many as 65 at another site. Collectively, the 14 collaboratives that were active as formal organizations reported that they included approximately 350 very active teachers—suggesting that mean teacher participation across the 14 collaboratives involved 25 very active teachers per collaborative. In contrast, *active* teachers were those who participated in collaborative activities on occasion. Estimates of active teachers ranged from a high of 700 at one large city site to approximately 30 at a few sites. When all 16 collaboratives were operating as formal organizations, approximately 2,175 teachers were reported to have been active. As of mid-1993, when 14 collaboratives were operating as formal organizations, approximately 2,000 teachers were reported to be active—yielding an estimated mean of 143 active teachers per collaborative.

It is very difficult to gauge how the numbers of very active and active teacher participants compared to the total potential population of participants. Many of the collaboratives opened their doors to teachers from many grade levels, schools, and districts but did not specifically target those teachers as participants or even consider the number of total possible participants. Furthermore, teachers' level of participation varied considerably over time. The number of very active and active participants at any moment in time represented only a portion of the total number of teachers who participated in the collaboratives during their history.

Higher Education. The five collaboratives that were hosted by a college or university had clear collaborative links with higher education, although the level of actual interaction varied considerably. In each of the five collaboratives the host provided office space, equipment, and clerical support. Two of the collaboratives hosted by a college or university experienced financial frustrations, especially with regard to fundraising, in their relationship with their university host agencies. However, the other three collaboratives with similar host arrangements had notable success in obtaining grants, largely as a result of their close institutional partnerships with higher education.

Several higher education partners hosted professional development workshops, institutes, and other activities for their partner collaboratives; and, in some cases, faculty from institutions of higher education were more directly involved. The following activities illustrate the range of assistance that was supplied by organizational partners from higher education:

- cosponsoring mathematics competitions and offering courses specifically for teachers;

- hosting collaborative summer camps;

- hosting a collaborative's workshop series;

- involving collaborative teachers in university mathematics education programs;

- training teachers in curriculum writing and providing ongoing support for collaborative teachers in a citywide internship program;

- pairing teachers selected for an exemplary practice program with a local university resource person;

- hosting the collaborative's summer institutes, some of which were led by faculty from local universities;

- providing faculty to serve as judges and sponsors of a mathematics fair and contest; and

- hosting an annual fall mathematics conference and other professional development opportunities.

There seemed to be a close correlation between the involvement of colleges and universities as hosts of events and the involvement of faculty members from those institutions in collaborative activities. However, the direction of causality was not clear. One conjecture was that when their institution hosted an event, faculty from colleges and universities became more aware of the collaborative's goals and activities and consequently recognized the possibilities that existed for their own involvement. Conversely, the very effort to foster involvement in collaborative activities by representatives from higher education tended to encourage institutions of higher education to host and support collaborative activities.

Business and Industry. Many of the collaboratives also continued to involve representatives from business and industry on advisory and governing committees. Partners from this sector were also active at several sites in fundraising for the collaboratives. At many sites, however, the involvement of business did not extend beyond this level.

Where there was business involvement in collaboratives, it varied considerably among the sites. Many collaboratives sponsored site visits for teachers to expose them to uses of mathematics in local businesses and industries, though this activity became less frequent over the life of the UMC project. Two collaboratives still sponsored such activities, and a few other collaboratives took advantage of parallel programs that offered this type of activity.

Partners from business also hosted professional development workshops and meetings for a few collaboratives or donated equipment and technical assistance to improve workshops and meetings. In addition, individuals from businesses that sponsored events attended and participated in workshops and meetings with teachers.

A more integral form of collaboration between teachers and business that was explored at several sites was a teacher internship program. Generally, such programs placed teachers in a local business for several weeks during the summer. The programs were designed to expose teachers to actual applications of mathematics. In addition, teachers were expected to utilize their increased understanding of how mathematics was used in various jobs to make school mathematics more relevant and interesting to their students. The success of such programs tended to vary by site. The two most successful business internship programs required teachers to produce, implement, revise, and share curriculum units based on their summer internship experiences.

Another form of collaboration with partners from local businesses recently emerged in three collaboratives. One collaborative attempted to involve representatives from a wide variety of businesses and industries in its Exemplary Practices videotape program. Largely due to the nature of the project, the collaborative was most successful in recruiting representatives from media industries. These partners contributed considerable technical expertise to the program, and they and other interested partners from business also participated in the formation of guidelines defining exemplary classroom practice. A similar collaboration with a media partner occurred at another collaborative. In that instance, a local television station invited collaborative teachers to prepare mathematics questions for students to answer during visits to four local historic attractions. The television station was promoting educational visits to the four sites and used the teachers' questions as the basis for a student competition. Collaborative teachers in a third city participated in a similar venture. The teachers prepared different levels of mathematics questions to guide students through the local science center. Activities such as these represented a cooperative effort to promote mathematics education outside the classroom, to identify local resources that could be used to teach mathematics, to offer customized mathematical guides for class field trips and individual visits to local attractions, and to provide publicity for the collaborative and the attractions.

Two general impressions emerge from an examination of the partnerships that collaboratives formed with businesses. First, the most successful partnerships with business representatives were characterized by the provision of very specific roles for businesses and their representatives to play in collaborative activities and programs. A major difficulty in collaborative partnerships with the business sector seems to have derived from a general lack of understanding about ways that teachers and businesses could work together for common educational goals. The collaboratives that overcame this obstacle to some extent did so by supporting programs and activities that provided a very definite role for all participants. Second, administrators from two collaboratives cited the lack of evidence about student outcomes as a problem in fostering the interest of local businesses, either in supporting the collaboratives financially or in participating in the collaborative's activities. The dearth of either student-centered activities or systematic assessments of outcomes as they affected students was consistently cited by collaborative coordinators as a barrier to effective partnerships with businesses.

School Districts. The 16 collaboratives involved school and district administrators at various levels of their operation. Most of the collaboratives included school district representation on a governing or advisory committee. Most actively engaged current or former district mathematics supervisors in prominent roles.

Another important manifestation of collaborative partnerships with constituent school districts was the provision of released time or restructured time for teachers to participate in collaborative administration or activities.

- At one site, the secondary resource teacher was a classroom teacher on leave for the academic year.

- At another site, the school district provided restructured time for a teacher to serve as collaborative director, as well as leave time for teachers serving on planning committees for the collaborative's mathematics contest and fair.

- One large unified school district provided restructured time for the collaborative coordinator until 1994, when the individual teacher was called back to a classroom teaching assignment.

- Another city system until 1994 permitted teachers to work for the collaborative as part of the state's career ladder program.

- A large urban district assisted the collaborative by matching the host agency's provision of an extra free period each day for several teachers administering collaborative functions.

Other collaboratives enjoyed direct financial benefits from partnerships with their constituent districts. For example, the collaborative in one large city received school district contributions for professional development opportunities. Another form of financial partnership between collaboratives and school districts consisted of shared responsibility for travel expenses and substitute pay when teachers were absent from school for professional development opportunities.

Many school districts utilized collaborative teachers as deliverers and presenters of their in-service opportunities.

- Several districts served by one collaborative allowed teachers to attend collaborative workshops for their required in-service hours.

- Two collaboratives provided the majority of professional development services for their constituent districts' mathematics teachers, and the school districts used collaborative teachers to develop and lead the in-service opportunities they offered.

- One school district invited teachers to assist in delivering in-service workshops and to serve as representatives on district curriculum committees.

- Another collaborative, following district streamlining which reduced the number of mathematics supervisors from three to one, served as the remaining supervisor's informal partner for mathematics education in-service workshops that were offered to middle school and high school teachers.

One additional variety of partnership between collaboratives and local school districts deserves attention. One of the principal programs of one big city collaborative was a Saturday Morning Workshop Series. As a result of reorganization, the local unified school district structured its professional development opportunities in such a way that individual schools and clusters of schools managed their own professional development budgets and selected their own professional development opportunities. The collaborative's Saturday Morning Workshop Series, a number of other collaborative programs, and many programs that developed concurrently with the local collaborative and drew on the leadership of collaborative teachers, were included on the district's list of professional development offerings.

Other Partnerships. Some collaboratives enjoyed partnerships in addition to those with business, higher education, and school districts. Two such partnerships involved relationships with host agencies and their other programs, and relationships with parallel educational reform efforts and organizations.

Examples of the elaboration of previously established relationships with host agencies show the variety of possibilities that existed for this kind of collaboration:

- Two collaboratives used the same outside internship programs as their host agencies.

- Two other collaboratives used their host agencies' newsletters to distribute information rather than producing separate newsletters.

- Two collaboratives shared professional development opportunities with similar organizations that supported teachers in other disciplines.

- At another site, the mathematics collaborative was expected to serve as a model for a science collaborative, and the two collaboratives were to be integral components of a new center for excellence in mathematics and science education.

- Finally, many collaboratives took advantage of fundraising and evaluation offices that existed separately within their host agencies, rather than replicating their functions.

A number of collaboratives maintained formal partnerships with other reform organizations, and many informal partnerships existed through shared membership and goals. One common informal partnership among the collaboratives was one with local, regional, and state affiliates of NCTM. Such organizations offered additional professional development workshops, as well as opportunities for collaborative teachers to preside and present talks at events that attracted audiences that were larger and more diverse than they would be if the collaborative had developed the event alone. Collaborative partnerships with other reform efforts were exemplified by the involvement of collaborative teachers in the Interactive Mathematics Project in Philadelphia, San Francisco, Los Angeles, and the Twin Cities.

One additional class of collaborative partnership deserves mention. Many contacts among teachers and administrators existed nationwide across the 16 collaborative sites. In some cases, this contact led to intercollaborative projects, such as the development of a mathematics-environment curriculum by teachers from the three California collaboratives. In other cases, it took the form of support and assistance to begin new programs, such as the help Philadelphia and San Francisco afforded the St. Louis collaborative in writing a proposal for IMP. Additionally, all five of the replication collaboratives relied in part on the technical assistance and advice of the original collaboratives to develop their administration and activities.

A final significant example of intercollaborative partnership is seen in nationwide UMC links to the CAM project, a program that explored alternative assessment methods, using funds provided by EDC. EDC identified five UMC sites as "hub" cities for the CAM project. The hub cities identified district administrators and a small group of middle schools and teachers from those schools to participate in the development, piloting, and implementation of new classroom assessment strategies in mathematics. The other 11 UMC sites were also invited to send representatives to national CAM meetings annually. The project allowed teachers and other collaborative members from many of the UMC sites to collaborate on a specific project and share experiences, ideas, and outcomes over a sustained period of time. The five hub cities remained involved in the project throughout its existence, and several of the other UMC sites enjoyed significant participation in the CAM project through the endeavors of their members who attended national meetings or who otherwise had substantial contact with participants in the hub cities. EDC began both a new mathematics assessment program and a teacher leadership program built on its experience with CAM and summer leadership institutes as the technical assistance partner of the UMC project. A number of collaborative sites served as founding participants in the new programs.

Up until 1994, the UMCs held annual national and regional meetings. Furthermore, they were all connected by electronic communication. Financial support for intercollaborative meetings and electronic communication, which had been provided by the Ford Foundation through EDC, no longer exist. Though administrators and teachers from many collaboratives maintain informal contact, most recognized the dissolution of formal vehicles for mutual support and sharing of ideas as a significant loss. New vehicles were currently emerging through the newly convened UMC national board, the Leadership in Urban Mathematics Reform project, and the new assessment program built on the foundation of CAM.

Teacher Professionalism. The UMC project had implications for teacher professionalism; we here discuss impacts on professional development, teacher isolation, and teacher leadership.

Teachers' Professional Development. Each of the 16 urban mathematics collaboratives supported professional development activities. In most cases professional development referred to the deepening of teachers' pedagogical knowledge and expertise, although leadership and subject area advancement opportunities were also significant. Many of the collaboratives supported their own professional development workshops, and all 16 encouraged teachers to take advantage of collaborative or other professional development opportunities. Some offered financial assistance for this purpose.

Professional development opportunities ranged from single workshops to thematic workshop series, institutes lasting from one day to several weeks, and from informal discussion groups to dinner meetings with hired speakers. A few collaboratives offered special grant programs to teachers and teams of teachers that supported professional development and implementation of new practices at the school and classroom levels.

Ten of the collaboratives supported ongoing workshops, either as a stand-alone activity or as a component of another program:

- Four collaboratives offered both an independent workshop series and additional workshops that were designed to further specific programs. Thus, an internship program at one site offered ongoing professional development workshops especially designed for project participants.

- Three other collaboratives sponsored workshop series that existed independently of other programs, and an additional two collaboratives offered workshops as part of their respective assessment programs.

- Four collaboratives offered dinner symposia on a regular basis. Such activities generally targeted all collaborative participants: teachers, administrators, and partners from business and higher education.

- Many of the collaboratives had in the past sponsored Woodrow Wilson Summer Institutes, and some continued to do so. A few offered additional long-term summer professional development opportunities.

All of the collaboratives took advantage of parallel professional development opportunities for teachers. Many supported teachers' attendance at regional and national conferences, such as those of NCTM and its affiliates. Several collaboratives offered travel stipends for teachers to attend such events. In addition, most collaboratives actively encouraged their members to present sessions or lead workshops at local, regional, and national meetings and to involve themselves as leaders and participants in other local and state reform programs. Individual teachers and participants from several collaboratives rose to important leadership positions in other local, state, and national reform efforts. Furthermore, of the collaboratives that no longer functioned or were relatively dormant, each identified teachers' extensive participation in other efforts as an important outgrowth of individual professional growth that was attributed to participation in the collaborative.

All of the collaboratives promoted increased contact among their teacher members, and several supported discussion or study groups that provided a forum for open communication among teachers. Sharing of practice through teacher communication in the collaboratives ranged from formal structures, such as teacher presentations at professional development workshops—which many sites employed—and videotaping of teachers using innovative instructional practices, to informal phone calls between collaborative teachers to discuss new practices.

Finally, in some collaboratives professional development extended to the realm of college and university course work. At many sites, individual teachers sought advanced degrees or certification or additional course work as a part of their professional growth. In addition, various workshops and institutes could be taken for college or graduate credit. Two collaboratives offered intensive curriculum writing courses within the context of their internship programs.

Reduced Isolation. One of the clearest and most common outcomes of the UMC project nationwide was a reduction in the perceived isolation of active teachers. As a result of collaborative activity, teachers felt less isolated from each other. The reduction in teacher-to-teacher isolation occurred on many levels. In some cases, participation in a collaborative fostered increased contact among teachers from the same school. Furthermore, teachers chose a variety of means to increase or redefine their contact with teachers from other schools, other grade levels, other districts, and other collaboratives. The newsletters of many collaboratives, a teacher's journal at one collaborative, electronic communication systems, and discussion or study groups were all devices specifically intended to foster professional communication among collaborative members. Moreover, collaboratives viewed year-round follow-up activities as an important means to preserve the professional contacts that teachers made during intensive summer experiences.

Teachers also became less isolated from developments, innovations, and debates in mathematics and science education. Increased local professional development opportunities, support for travel to regional and national conferences, the availability of grants and resources, and the distribution of information regarding mathematics and science education opportunities all represented vehicles by which the collaboratives contributed to the reduction of teachers' isolation from changes in the content and pedagogy of their profession.

Furthermore, some collaboratives reduced teachers' isolation from other professionals who used mathematics. Among the 16 collaboratives, governing committees and professional development and student-oriented activities included—in addition to teachers—representatives from business, higher education, research, and the military. Certain activities, such as mathematics competitions and internship programs, represented specific vehicles for reducing teachers' isolation—not only from other professionals who used mathematics, but also from the applications themselves.

Teacher Leadership. Teacher leadership was explicitly identified as a goal in eight collaboratives, though leadership in collaborative governance or activities and involvement of collaborative teachers as leaders in parallel reform efforts was evident at all 16 collaboratives. In some cases, teachers held the main responsibilities for governance and planning for the collaborative. In other cases, teachers were the chief leaders of certain activities and programs within the collaborative. Furthermore, at some sites, teacher leadership extended beyond the collaboratives' established programs.

All 16 collaboratives had a governing body that included teachers. Teacher participation on governing bodies was initiated differently among the collaboratives: Some held elections; others made appointments; still others accepted volunteers. In four instances, teachers constituted the vast majority of governing representatives and therefore handled the bulk of decisionmaking and administration. In several other collaboratives, administrative and financial decisions resided with a cross-sectional advisory committee or a director, but teachers shouldered the main responsibilities for planning and conducting activities. One collaborative was in the process of attempting to establish a new governing body of teachers, based on leadership training and peer recruitment.

Certain collaboratives also established staff positions for teachers or had involved teachers in administrative positions. Among the positions held by teachers or former teachers in the collaboratives were resource teacher, coordinator, satellite coordinator, and journal editor.

Most, if not all, collaboratives offered professional development opportunities that were chiefly presented by teachers for their peers. Two types of teacher-led professional development activities were most common. First, teachers who had attended a conference or workshop often prepared a presentation to share what they had learned with teachers who did not attend the conference or workshop. Second, teachers who had experimented with a novel curriculum, instructional method, or assessment practice in their own classroom frequently shared their experience, expertise, or support with other teachers who desired to attempt something similar in their classrooms. In many collaboratives, certain teachers became unofficial resource people regarding a particular area of expertise, such as graphing calculators, teaching with manipulatives, and portfolio assessment.

Several collaboratives also provided the means to disseminate curriculum units or other resources that collaborative teachers developed. Some collaboratives used their resource centers or publications for this purpose, while others distributed materials and information at local and regional conferences.

A few collaboratives supported programs that addressed the professional concern of peer evaluation. One collaborative's classroom demonstration series allowed teachers to observe actual classes that were taught by their peers. The internship programs at two other collaboratives allowed participating teachers to share curriculum units they were developing with other participants in order to obtain constructive feedback. Such activities fostered further professional peer evaluation, an often-noted deficiency in teacher professionalism.

Several collaboratives or their host agencies offered grants to teachers or teams of teachers. Generally, teachers received grants to pay for training and supplies necessary to implement an innovative teaching practice. Some collaboratives financed such grants themselves, but most informed their members of the availability of such grants and assisted teachers in applying for them. Grant programs fostered teacher leadership because they placed responsibility for developing and implementing innovative classroom practices directly on teachers.

At a very basic level, the existence of collaboratives gave teachers an opportunity to exercise leadership in very specific ways. Teachers utilized the credibility and support of the collaboratives to seek increased influence in educational decisionmaking. Ten examples illustrate the range of work that was undertaken by collaborative participants:

- Acquisition and use of calculators. Teachers in Pittsburgh were instrumental in the district's decision to provide appropriate calculators to all students, K-12, just as the district provided textbooks to all students. Many other collaboratives provided calculators or the funds to purchase them, and still other collaboratives persuaded districts to provide calculators for mathematics classrooms.

- Curriculum development. Teachers from the three California collaboratives were developing a mathematics-environment unit on water use. Collaborative members in Memphis wrote the curriculum for a new course called Introduction to College Mathematics, which was piloted for three years and approved by the district.

- Textbook adoption. In Los Angeles, a teacher secured district approval for both a textbook and a course on discrete mathematics—which she has subsequently taught. Teachers in Columbus and Durham also worked successfully to promote state or district approval of textbooks that they wanted to use in their classrooms.

- Teaching without a textbook. Based on the experimental work and subsequent writings of one collaborative teacher, a number of teachers in Los Angeles explored the strategy of teaching without a textbook. These teachers relied instead on supplementary materials and software, or on the ideas, conjectures, and conclusions that students derived when exploring mathematical situations with their classmates and teacher.

- Formation of groups to explore classroom issues. In Milwaukee, San Francisco, New York, and St. Louis, collaborative members formed groups that met regularly to discuss such issues as assessment, algebra, or more general issues in mathematics education. Meetings of governing or advisory bodies of teachers often served similar purposes at other collaboratives.

- Classroom observations. In Los Angeles, groups of teachers visited the classroom of a lead teacher who demonstrated the use of methods and materials that were of particular interest to the group. In New York, the classroom methods of "exemplary" teachers were made available to other teachers through a videotaping program.

- Educational policy. Teachers in New Orleans opposed the downgrading of mathematics graduation requirements, supported the inclusion of mathematics as a course required for promotion at certain grade levels, and explored avenues of influence in state-level educational policymaking.

- Departmental leadership. In Los Angeles, two teachers became department heads at their schools and radically improved the atmosphere of departments that had previously been deemed stagnant.

- Site-specific mathematics resources. Teachers in Pittsburgh and St. Louis developed mathematics guides for students who visited local museums and tourist attractions.

- Continued collaboration even when the formal organization of a UMC ceased to operate. In New Orleans and in the Twin Cities area, teachers continued to meet informally or became involved in other reform efforts, even though the collaboratives in those cities ceased their formal operations.

Past and Present Challenges

Reaching Teachers. Numerous examples have been cited throughout the case study of teachers for whom the learning system of the collaboratives had been useful and effective. Important and intriguing questions remain about teachers and schools for which the learning system was not useful or effective. In a small number of instances, a considerable majority of mathematics teachers in a single school were involved in and affected by the collaboratives. The Los Angeles high schools where Maria Naranjo, George Hillman, Harvey Jacobs, and Earl Grant taught and the Georgia high school where Rochelle Robinson, Tonya Trier, Percy Patterson, and Vincent Valentine taught are excellent examples. In these schools, nearly all members of the mathematics faculties were involved in at least one collaborative activity, and each school employed a core group of three to six of the collaboratives' teacher-leaders. Moreover, the school administrators in these schools were extremely supportive of the teachers' initiatives. Their support took the form of allowing teachers time and opportunity to engage in professional development, to meet together, and to experiment in their classrooms. They entrusted teachers with authority and decisionmaking power while avoiding unnecessary assertions of control.

More commonly, however, active collaborative teachers were a minority in their mathematics departments. Although their isolation from other reform-minded teachers in their district or locality was greatly reduced through their involvement in the collaboratives, they still remained somewhat isolated within their own schools and departments. Jo Lamar, a middle grades teacher in a private school in Milwaukee, described the isolation of her situation: "I have no support. I'm a person pursuing the best I can in mathematics education . . . Anybody else who teaches math [in my school] teaches other subjects . . . so I'm alone. And I have needed the support of peers and that's what I found through the collaborative." Yvonne Alberts, a collaborative teacher-leader in a nearby high school, noted similar feelings about the collaborative from other teachers: "It has provided a valuable link to people who do not have a real active school system. Katy Hatfield, who comes from a smaller school system, this is her little lifeline . . . I think it serves some of those people that bother to drive in all the time, as a lifeline." Derrick Duncan, a member of the staff of the Columbus collaborative, observed a similar phenomenon:

> I think the fact that we serve a number of rural systems has had an impact, because a lot of teachers in rural systems are in situations where . . . they are isolated, they are the only math teachers . . . or . . . they . . . are the only ones that want to change or are motivated to implement the NCTM standards. So the collaborative has provided them an opportunity to link with other people and that has supported them in efforts to change and to implement new things in their schools.

Just as teachers in small or rural schools and systems faced serious professional isolation, so also did teachers in large, urban schools. As Los Angeles high school teacher Julie Krauss attested, "It's okay if nobody else is doing this here . . . I teach a lot of cooperative groups and up to this past year nobody else in the department did groups at all and I was just out there on my own, I didn't have anybody to discus it with." Krauss explained how the collaborative alleviated her sense of isolation:

> I attend Saturday workshops that +PLUS+ provides which I think is a really good way to network with other teachers. You have to find out what's going on, to realize that what's happening in your classroom is not in isolation, that it is happening elsewhere . . . Your problems are the same as everybody else's, some are different, but a lot of what you are struggling with is the same across the board.

Krauss enthusiastically reported that a few more teachers in her department began to explore reform-oriented curriculum and instruction, and the department's activity in the collaborative also increased.

Concurrently, in small and large rural and urban schools and districts, many mathematics teachers had little or no contact with the collaboratives; still others had been active in the past but effectively ended their relationships with the collaboratives. In this case study, their stories are, to be sure, as important as those of the many teachers whose professional lives were nearly inseparable from the collaboratives. Most of the teachers who were not active in the collaboratives can be classified in one of three categories.

First, the collaboratives required a voluntary commitment of time and energy. Some teachers simply did not have the time to involve themselves more than occasionally, even if they sincerely believed in the mission of the collaboratives. Time commitments to family, to other school responsibilities such as coaching, and to second jobs—especially during the summer—prevented many teachers from seeking substantial involvement in the collaboratives. Umberto Upshaw, a teacher in Georgia who also coached, typified this scenario. He had attended some collaborative activities at the invitation of "another teacher . . . He's a fellow coach and so on. He kind of convinced me to go with him . . . I haven't been back since but it's not because I don't want to. It's because there's no time, what with coaching and teaching." Although he knew that the collaborative and its resources were openly available to him, Upshaw said, "I haven't used many resources from there. I wouldn't know how to get resources from there because I haven't ever done that. But several teachers here [at this school] have." Upshaw's case was not an uncommon one in the collaborative sites. Most collaborative activities were scheduled after school, on Saturdays, or in the summer. For many teachers, significant participation in the collaboratives was simply not possible because of other meaningful commitments. Although the collaboratives enjoyed many isolated successes in providing paid leave time or worthwhile stipends for learning opportunities, they were rarely institutionalized or widely available. Effective and desirable arrangements sometimes disappeared due to changes in district administration or policy. Understandably, the collaborative's limited resources and opportunities were often largely utilized by those teachers who were already involved.

Second, the collaboratives provided a wide variety of learning and professional growth opportunities, but their activities did not satisfy the needs of all potential constituent teachers. Importantly, this fact applied not only to teachers who never found involvement in the collaboratives satisfying of their needs, but also to teachers who found the collaboratives useful and effective for certain opportunities over a limited period of time, but not fulfilling of their longer term needs. The case of Francisco Hierro, a high school teacher in Los Angeles, is illustrative. Hierro related, "When the +PLUS+ program first came . . . I was one of the charter members involved in the whole process . . . I worked with the +PLUS+ program for two years. We did some really good things, we did some great field trips to industry . . . we were trying to get that connection . . . between us and industry." Two factors of participation in the collaboratives troubled Hierro. The first concerned uses of teachers' limited time:

> What I did not like about +PLUS+ was that we . . . were planning meetings to have a meeting. We had planning meetings to plan that meeting. And I'm an action-oriented person. I don't like to spend 20 hours of work to get one hour of results . . . There's too much more that I can do with those 20 hours in terms of affecting kids in a positive way than spending 19 hours in meetings to have a one-hour fruitful meeting. So that's why I personally . . . dropped out of the +PLUS+ program, after about three years.

A second shortcoming of the collaborative for Hierro concerned the ongoing usefulness of activities. Of the industry site visits the collaborative sponsored in its early years, he said:

> They were very valuable for me at the time . . . We tried to make some lesson plans out of some of the things we saw . . . We weren't real successful . . . It was exciting for us to see what we saw . . . for a while we were able to relate those stories in class, and tell kids that I saw this and I saw that. I still use a couple of the stories . . . in class, but it ended up being just stories that we could tell, not math that we could use because . . . we were never able to sit down with them well enough to see it in action.

In Hierro's collaborative, the deeper connections to local business and industry for which he hoped never materialized to his satisfaction. He, however, found other programs that successfully met his professional needs in this regard.

A third group of teachers had strong reservations about the collaboratives and the reform movement in general, while others had little to no knowledge of either. Considered collectively, this large group was often referred to as the "hard-to-reach teachers," an accurate description from the perspective of advocates of the collaboratives and the reform movement. The attitudes of these teachers toward the collaboratives ranged from indifference to suspicion. Yet such teachers could not be looked upon as adversaries, else the reform would likely never reach them. Many were somewhat informed but remained conservative and skeptical. Wallace Young, a high school teacher for many years in Los Angeles, had participated in a few collaborative activities, but purposefully distanced himself from the collaborative. He explained:

> I've gone to their workshops, but I personally have some reservation because I traditionally am not a joiner . . . I feel like they create an atmosphere there where you're a joiner, and the more enthusiastic you are, the better you are

accepted. It seems very competitive and contrived. I enjoy it if there's a real atmosphere of improvement in instruction and a more mature atmosphere where there are no constraints where you have to be socially accepted or you have to be more than willing to be involved. Then I think it's a very satisfactory idea.

Young's attitudes were grounded in his beliefs about education and the teaching profession. With regard to participation in the collaborative and investment in the reform movement, he said:

> There are some teachers that are in the department that probably feel left out of this general picture. They may feel like I do; they have some trepidation about immersing themselves into a program they're insecure about and they haven't been educated about . . . We're really used to being . . . dependent on leadership. And I feel like there is not the leadership or the vision or the guidance in this department or within the district that is really giving us a concrete direction.

He added some thoughts about mathematics education, "I feel basically that an overhaul is necessary. But I don't feel like the discipline should be overhauled to the extent that we have to eliminate René Descartes because we have a graphic calculator."

Young's reservations and concerns were informed and very realistic. Of the three groups of teachers who were not at the time involved in the collaborative, Young's group represented the greatest challenge. Yet their reservations and concerns served as an important balance in the reform movement. In Young's words,

> Classically, my notion of something experimental is that we have a control group and we have a regular group. The control group is usually smaller than a regular group. We make some observations and rather than immersing ourselves, we try to incorporate or integrate some of the newer ideas into our classroom instruction. Frankly, I'm concerned that we've jumped on a bandwagon and we've gone in a direction that is misdirected.

The control group to which Young referred certainly already existed, though not by design. Eager participants in the collaboratives and the broader reform movement could learn a great deal from those who regarded current reforms with informed conservatism and skepticism.

Finally, a few collaborative participants noted that the collaboratives were not always successful in reaching minority teachers and teachers in largely minority schools. Although this was certainly not a categorical shortcoming, it was a serious concern. The collaboratives were founded on the premise that informing and encouraging urban teachers in mathematics education reform efforts would ensure that students in urban schools would benefit from change. The collaboratives achieved notable success in helping urban teachers not only to keep pace with change in mathematics education, but also to serve as leaders on the forefront of change. Concerns of participants regarding the participation of minority teachers and teachers in largely minority schools represented a microcosmic view of the same concern the Ford Foundation addressed on a national scale.

Evaluation of Collaborative Impacts. Among the 16 collaboratives, systematic self-evaluation of programs, administration, and impact were notably lacking. Several collaboratives made some use of their host agency's evaluative departments or mechanisms, and one used the evaluative services of its local board of education. In general, the collaboratives used such mechanisms to evaluate individual programs. In addition, several collaboratives used surveys, follow-up meetings, informal discussions, and outside evaluators to gauge the success of and explore options for improvement in individual events and programs. However, none of the collaboratives had an established mechanism for comprehensive evaluation.

Several administrators expressed the need for periodic, systematic evaluation of the collaboratives. In addition, the administrators recognized that the existence of an evaluative mechanism and the empirical results of such critical examinations would be useful in soliciting general support from businesses and foundations and in applying for program-specific grants. Furthermore, critical evaluation could assist individual collaboratives in facing challenges posed by transformations in their educational context. For example, one collaborative was seeking its niche in a newly consolidated city-county school district; and several collaboratives were dealing with organizational overhauls in their constituent school districts, generally involving streamlining of district administration and increasing school-based management. Evaluative information that convincingly documented the individual collaborative's impact and importance could be invaluable in securing continued funding and support following such transformations.

Reflections: Teachers at the Center of Reform and Change

Critical to the Urban Mathematics Collaborative Project as a decentralized, voluntary reform effort were teachers and their professional knowledge, beliefs, practices, and involvement. Toward the goal of improving mathematics education in urban areas, the UMC project was initiated to enhance the professional status of teachers; to enable mechanisms by which teachers' awareness of, knowledge about, and access to current developments in mathematics and education might be improved; and to foster interaction among mathematics teachers and other mathematicians. Embedded in the UMC focus on teachers was the conviction that more knowledgeable, better informed, and less isolated teachers would serve as the cornerstones of substantive reform and improvement of mathematics education in urban schools. The UMC project hinged on the experience of individual teachers.

The collaboratives engaged teachers at a variety of levels of professional leadership, involvement, and maturation. Moreover, the collaboratives served some teachers over extended periods of their careers—contributing to considerable professional growth. Each teacher's journey with the collaborative was bound to be unique. However, patterns that consistently emerge suggest that some meaningful generalizations can be drawn regarding how the collaboratives operated with respect to teachers.

Mathematics teachers, classroom teaching and learning, and teachers' professional lives changed in the 10 years of the UMC project. For the most active and involved teachers, change was often dramatic. Largely due to the commitment and perseverance of these teachers, mathematics education in some departments, schools, and districts and relationships between mathematics teachers and other professionals in business and higher education also changed. The closely focused examination of teachers' participation in the collaboratives and related changes, which comprised the majority of the researchers' efforts in this case study, has convinced us that, for many teachers, the UMC project was pivotal in their individual and collective professional development.

Our analysis strongly suggests that voluntary professional development opportunities that are teacher-led and that attend to the realities of teachers' daily work and lives are instrumental in individual teachers' professional growth. Moreover, although such growth is individual, it cannot be considered isolated. In order to achieve the professional growth that teachers attribute to the collaboratives, substantive contact with fellow teachers and with the persons and ideas comprising the current reform movement in mathematics education is essential. The UMC project cannot be solely credited with the many significant changes described in this case study, but the teachers and many others involved in the collaboratives agree that the collaboratives often were the catalyst and the enduring focus of change in mathematics education in their sites. The collaboratives introduced ideas, developed them in a useful and realistic context for teachers, and supported the efforts of teachers to turn ideas into realities in classrooms, departments, schools, and districts.

Could collaboratives reform mathematics education in America's cities to a point at which all students would be learning challenging and appropriate mathematics? Certainly not in their first 10 years. Could collaboratives substantially contribute to such a reform in mathematics education? This case study demonstrates how, in their first 10 years, the collaboratives did so. The collaboratives locally developed new models of professional development for teachers. They opened avenues of information and support through which teachers were able to maintain current and relevant knowledge in mathematics education; experiment with new ideas; and implement those ideas in classrooms, schools, and districts. They developed networks and leadership involving practitioners, which most researchers agree the ongoing reform movement desperately needs.

Research Methods and Personnel

Data Collection

Contact Interviews. The Documentation Project at the University of Wisconsin-Madison documented the development of each of the 11 original urban mathematics collaboratives during the period from 1985 to 1990. Less detailed records exist regarding the five replication sites. During the summer of 1993, a list of coordinators and directors was compiled from the existing records. Although titles and position descriptions varied, the coordinator or director deemed most directly involved at the last previous contact was mailed a letter describing the Organisation for Economic Co-operation and Development case studies research project. The letter also informed the collaborative administrator that a researcher from the Wisconsin Center for Education Research would shortly contact them to schedule a telephone interview designed to last approximately 45 minutes. Between June and September 1993, an individual at each of the 16 collaborative sites participated in a telephone interview with one of four interviewers.

The instrument that guided the telephone interviews consisted of questions and probes designed to elicit responses that would establish the changes that took place in each collaborative during the 1990-93 period, the factors that contributed to such changes, and the current state of administration and activities. Some items were also included to establish what current beliefs and values surrounding constructs such as "teacher empowerment" and "collaboration" existed at each site.

Site Selection. Following the contact interviews, five collaboratives were chosen for further exploration as representative of the group of 16 urban mathematics collaboratives: Columbus, Los Angeles, Memphis, Milwaukee, and Philadelphia. In order to determine appropriate site selection, the data gathered from the contact interviews were used to classify the 16 collaboratives. Classification included categories such as origin, location, history, level of operation, type of host agency, demographics of constituency, and range of activities.

The five sites that were selected for further research offered a cross section of the Urban Mathematics Collaborative project in several ways. First, geographically the five sites were spread across the United States, with most major population regions represented. Second, several of the best models of collaborative activity were found among the five selected sites. Los Angeles and Philadelphia supported two of the largest professional development programs; Milwaukee offered one of the most significant and successful attempts to involve teachers and representatives from business and higher education in a single program; and Columbus maintained one of the most important programs that directly targeted both teachers and students. Collectively, these programs represented an ideal urban mathematics collaborative, which exceeded what any single site had achieved to date. Third, the five selected collaboratives formed a cross section of the categories determined by the nature of the sites' origins and their present level of functioning. Fourth, the Los Angeles collaborative served the nation's second largest school district; the Columbus collaborative served a small urban school district and numerous very small rural school districts; and the other three sites served school districts that fell between these two extremes. Exclusion of one of the two original sites classified as nonfunctioning is the chief shortcoming of the site selection.

Introductory Site Visits. The administrators of the five sites selected for further investigation all agreed to extend their respective collaborative's participation in the case study. A two-day introductory site visit to each collaborative was scheduled with the assistance of the administrator and staff of the collaboratives. The main purposes of the introductory site visits were to identify and initiate contact with key individuals at each site and to determine the most significant themes to be explored at each site.

During the fall and winter of 1993-94, two researchers made an introductory visit to each site, with the exception of Columbus, which was visited by a single researcher. During the site visits, the researchers interviewed collaborative staff and administrators, teachers, and collaborative participants from business, higher education, and school and district administration. Portions of interviews were guided by questions developed from analysis of the contact interviews. The remainder of the interviews was intentionally open-ended in order to explore themes that emerged during the site visit. The participants who were interviewed were pre-identified by the collaborative administrators and researchers as key individuals in the collaboratives. The principal questions were as follows:

- What were the identifiable stages of growth of the collaborative over its existence?

- What strategies did the collaborative employ to enact change?

- Who were the key players in the collaborative in instigating and carrying out change?

- What changes in teachers' roles in or out of the classroom occurred in connection with the collaborative?

Major Site Visits. Reflection on and initial analysis of the data gathered during the introductory site visit identified critical themes to be developed and sources of information to be consulted, both within and across sites, during a major site visit to each collaborative. Interviews and observations were scheduled with individuals either directly or through the assistance of those contacted during the introductory site visit. The main purposes of the major site visits were to collect data via interviews and observations that would inform the themes emerging in the ongoing analysis and to immerse the researchers in the context of the collaborative sites.

During the spring of 1994, one researcher made a major site visit of three to five days to each of the five collaboratives. The visits variously included data collection by interviews with teachers, school and district administrators, collaborative staff and administrators, and collaborative participants from business and higher education; as well as observations of mathematics classrooms and collaborative activities. Interviews were guided by universal questions and probes related to themes emerging within and across sites, and questions and probes specific to certain individuals. Most interviews also intentionally included discussion of issues that arose in the course of observations or during the interview. Most of the research participants were pre-identified as sources of relevant information by the researchers. Additional information was gathered through informal conversations and interactions with the pre-identified individuals and other individuals in the schools or at collaborative functions. The principal questions were as follows:

- How did mathematics education in the classroom (content, instruction, assessment, teacher and student roles) change in relation to the collaborative?

- What educational goals did teachers and others hold for students with regard to mathematics? How were those goals related to the collaborative? How were those goals manifested?

- What professional goals did teachers hold for themselves or did others hold for teachers? How were those goals related to the collaborative? How were those goals manifested?

The principal points of observation were:

- objectives, goals, and purposes of mathematics lessons;

- discourse in the classroom between teacher and students and among students;
- engagement of students with challenging mathematical content; and

- activities and roles of teacher and students in mathematics learning.

Exit Interviews. The researchers identified critical points of information that were missing from existing data as well as themes that were further articulated or that emerged through analysis and writing. Points of information and clarification of themes or response to themes formed the substance of brief exit interviews with one or two administrators or teacher-leaders at each of the five selected collaborative sites. The main purpose of the exit interviews was to obtain, update, and verify specific information regarding the constituencies of the collaboratives; the specific goals of each collaborative; and the strategies utilized to achieve those goals.

Exit interviews were conducted via telephone during the spring of 1995 by the researcher who had made both the introductory and major site visits to each collaborative. Interviews were specifically guided by a set of developed questions and probes designed to solicit precisely and succinctly the desired information. The research participants were identified by the researchers as those most likely to have knowledge or documentation of the required information.

The principal questions were as follows:

- What percentage of teachers in the target population participated in the collaborative in any way over the course of its existence?

- What emphasis was given to developing teacher leadership? How successful was the collaborative in developing teacher leadership?

- What was the collaborative's vision or definition of collaboration?

Procedures. Interviews were tape recorded with the permission of the respondents and later transcribed into electronic and printed text for the researchers' examination. Observation notes were handwritten. Additional printed materials (reports, articles, meeting agendas, classroom handouts) were occasionally provided by research participants to supplement the data collected during interviews and observations. Source tapes, transcripts, observation notes, and additional materials were consulted extensively throughout the processes of analysis and writing of the case study.

Analysis

Coding of Data. The text of all interviews was electronically processed by a single researcher using HyperRESEARCH (Hesse-Bieber et al. 1993), a computer application designed to facilitate qualitative research. Among its uses for qualitative research, HyperRESEARCH allows the user to read a text file and associate passages of text with one or more code words or phrases for later retrieval. Each interview was coded twice.

The principal purpose of the initial round of coding was to help the researchers identify important areas of interest and emerging themes in the case study. The coding scheme utilized for the initial round of coding consisted of 4 primary codes and 17 secondary codes based on an early outline of the case study. No changes were made to the coding scheme during the first round of coding.

The principal purpose of the main round of coding was to extract evidence from the data relevant to each area of interest and theme. The coding scheme utilized for the main round of coding consisted of 4 primary codes, 14 secondary codes, and between 63 and 66 tertiary codes. The codes were derived from a list of areas of interest and themes generated by the researchers during examinations and discussions of the data during the initial round of coding. The primary and secondary codes did not change during the coding process. Several tertiary codes were added, deleted, divided, or collapsed during the main round of coding depending on the researchers' determination of their necessity and usefulness to aggregate information relevant to the case study's areas of interest and themes.

Identification and Coding of Themes. During the periods of data collection, analysis, and writing the UMC case study research team met two to four times each month to articulate themes and areas of interest in the case, as well as to handle technical details of the case study.

During the data collection periods, the researchers who conducted each site visit offered reports, impressions, and insights regarding the visit at subsequent meetings. Identification of themes and areas of interest was generally focused on individual sites during data collection.

During periods focused on analysis, the researchers discussed themes and areas of interest emerging within and across sites. A series of matrices (site × theme) was generated to aggregate information within and across sites. As the matrices evolved, they were utilized to generate and modify the case study's coding scheme and to generate outlines for the writing phase of the project. Validation of themes and areas of interest focused on both the individual sites and the entire case during periods of intensive analysis.

During early periods focused on writing, the researchers divided the case study according to themes and areas of interest. Each researcher assumed primary responsibility for developing a written analysis of one major area of interest and all underlying themes and reporting to the team his questions, insights, and observations. Investigation of each area of interest and substantiation of each theme focused on the entire case during phases of writing.

Evidence of Change. A major concern for the researchers of the case study of the UMC project was the need to establish connections between the goals and activities of collaboratives and changes that were described by participants in the collaboratives or observed by the researchers during data collection. The researchers attended to this need during data collection by seeking corroborating evidence of change from several sources, including interview data from individuals with different roles in collaboratives' operations, observation data, and printed material, as appropriate.

The processes of coding and analysis also demanded attention to evidence that particular changes had actually taken place and that a relationship existed between identified changes and the activities of collaboratives. The links between change and the collaboratives that the researchers conjectured generally arose from impressions and insights following data collection. The coding of interview data and examination of observation notes and printed material provided potential evidence to substantiate the researchers' conjectures. Conjectures were initially tested by generating matrices of evidence within sites—and subsequently across sites—for validation of the proposed connection between identified change and activities of collaboratives. Importantly, the discussions and deliberations of the research team also served as a forum for testing the validity of conjectures about change in relation to collaboratives. The combined perspective of the research team was viewed as a critical filter for eliminating the biases and preconceptions of each individual researcher.

Ultimately, evidence of change and of the connection between change and collaboratives depended on the reports of collaborative participants and observers regarding the changes that had taken place, why particular changes had taken place, and how the activities of the collaboratives were related to particular changes.

The Writing Process. Writing was a vital component of the analysis and construction of the case study of the UMC project. Several hundred pages were written for publication and for internal use, in the forms of reports, memos, and draft documents, during the three years of the case study. The writing process served as a vehicle to record questions, insights, impressions, and results of analysis. Moreover, the writing process was a vital component of the analysis of the case.

Each report, memo, or draft written for publication or internal use provided new information and often a new perspective on the case study. Writing and the researchers' discussions and deliberations about written text were utilized extensively to examine the relative importance and applicability of ideas and themes generated during the course of the project. Furthermore, the nature and strength of evidence of change for particular propositions, as well as the lack of sufficient evidence for certain claims, only became fully apparent during the writing process.

The final report of the case study of the UMC project presented here is a consolidated presentation of the results of three years of visiting, interacting with, thinking about, writing about, and attempting to make meaning of the collaboratives and the people they comprised. It represents what the researchers consider the most vital aspects and most important contributions of the UMC project as a living example of innovation in mathematics education.

Case Study Team

Norman Webb brings to this study of the Urban Mathematics Collaborative project an extended history with the project. From 1986 to 1990, he served on the UMC Documentation Project that recorded, analyzed, and described the development of the collaboratives from their initial funding by the Ford Foundation. He, along with Thomas Romberg, culminated that work by editing the book *Reforming Mathematics Education in American Cities* (1994). This long history with the project gives Webb a strong historical, but possibly somewhat biased, perspective on UMC. He became professionally acquainted with collaborative staff and continues to interact with them in other professional contexts. His background is mathematics and mathematics education, with an major emphasis on assessment and evaluation. His participation on the writing team of the NCTM *Curriculum and Evaluation Standards for School Mathematics* (1989) signifies his support of the current reform efforts in mathematics education with a tempered view—as is the case with the other team members—between traditional and constructivist conceptions of learning.

Daniel Heck, a social constructivist and moderate traditionalist, comes to case study work with a background in history, mathematics, and education. Along with teaching both high school and pre-elementary education, he has a strong interest in curriculum and technology.

William Tate brings to this study a strong interest in the economics of mathematics education. He has worked as an elementary teacher, a high school teacher and coach, and a district supervisor of mathematics. His work in policy analysis has trained him to identify organizational structures as they impede or expand the opportunity of students to learn mathematics.

This study represents the collective thinking and analysis of all three researchers, achieved through weekly meetings and even more frequent discussions of data and observations.

References

Ayers, W. 1990. Rethinking the profession of teaching: A progressive option. *Action in Teacher Education* 12(1): 1-5.

Battista, M. T. 1994. Teacher beliefs and the reform movement in mathematics education. *Phi Delta Kappan* 75(6): 462-70.

Bell, B., and J. Gilbert. 1994. Teacher development as professional, personal, and social development. *Teaching and Teacher Education* 10(5): 483-98.

Berlak, H. 1992. Toward the development of a new science of educational testing and assessment. In *Toward a new science of educational testing and assessment*, ed. H. Berlak, F. M. Newmann, E. Adams, D. A. Archbald, T. Burgess, J. Raven, and T. A. Romberg, 181-206. Albany: State University of New York Press.

Bird, T., and J. W. Little. 1986. How schools organize the teaching occupation. *The Elementary School Journal* 86(4): 493-511.

Bruckerhoff, C. E. 1991. The Cleveland collaborative and the pursuit of mathematics curriculum reform. *Educational Policy* 5(2): 158-77.

California Department of Education. 1991. *Mathematics framework for California public schools: Kindergarten through grade 12*. Sacramento, CA.

Clark, D. L., and T. A. Astuto. 1994. Redirecting reform: Challenges to popular assumptions about teachers and students. *Phi Delta Kappan* 75(7): 513-20.

Cochran-Smith, M., and S. L. Lytle. 1990. Research on teaching and teacher research: Issues that divide. *Educational Researcher* 19(2): 2-11.

Cohen, D. K. 1991. Revolution in one classroom (or, then again, was it?). *American Educator: The Professional Journal of the American Federation of Teachers* 15(2): 16-23, 44-48.

Education Development Center, Inc. (EDC). 1991. *Policy statement on equity in mathematics education*. Newton, MA.

Ernest, P. 1991. *The philosophy of mathematics education*. Bristol, PA: Falmer Press.

Fey, J. E. 1979. Mathematics teaching today: Perspectives from three national surveys. *Mathematics Teacher* 72(7): 490-504.

Fullan, M. G. 1982. *The meaning of educational change*. New York: Teachers College Press.

Fullan, M. G., and S. Stiegelbauer. 1991. *The new meaning of educational change*. New York: Teachers College Press.

Goswami, D., and P. Stillman. 1987. *Reclaiming the classroom: Teacher research as an agency for change*. Upper Montclair, NJ: Boynton-Cook.

Heaton, R. M. 1992. Who is minding the mathematics content? A case study of a fifth grade teacher. *Elementary School Journal* 93(2): 145-52.

Heck, D. J., N. L. Webb, and W. Martin. 1994. *Case study of Urban Mathematics Collaboratives: Status report.* Madison, WI: Wisconsin Center for Education Research.

Hesse-Bieber, S., T. S. Kinder, P. R. Dupuis, A. Dupuis, and E. Tornabene. 1993. HyperRESEARCH [computer software]. Randolph, MA: ResearchWare.

Hofmeister, A. M. 1993. Invited rejoinder: Innovativeness is not a synonym for effectiveness. *Remedial and Special Education (RASE)* 14(6): 33-34.

Jaworski, B. 1991. Develop your teaching. *Mathematics in School* 20(1): 18-21.

Joyce, B., and B. Showers. 1980. Improving in-service training: The messages of research. *Educational Leadership* 37(5): 379-85.

———. 1982. The coaching of teaching. *Educational Leadership* 40(1): 4-8, 10.

Keedy, J. L., and K. Rodgers. 1991. Teacher collegial groups: A structure for promoting professional dialogue conducive to organizational change. *Journal of School Leadership* 1(1): 65-73.

Lambert, L. 1989. The end of an era of staff development. *Educational Leadership* 47(1): 78-81.

Leinwand, S. J. 1992. Sharing, supporting, risk-taking: First steps to instructional reform. *Mathematics Teacher* 85(6): 466-70.

Lichtenstein, G., M. McLaughlin, and J. Knudsen. 1991. *Teacher empowerment and professional knowledge.* CPRE Research Report Series RR-020. Stanford, CA: Consortium for Policy Research in Education.

Little, J. W. 1982. Norms of collegiality and experimentation: Workplace conditions of school success. *American Educational Research Journal* 19(3): 325-40.

———. 1990. The persistence of privacy: Autonomy and initiative in teachers' professional relations. *Teachers College Record* 91(4): 509-36.

———. 1993. Teachers' professional development in a climate of educational reform. *Educational Evaluation and Policy Analysis* 15(2): 129-51.

Little, J. W., and M. W. McLaughlin. 1991. *Urban Math Collaboratives: As the teachers tell it.* Stanford, CA: Center for Research on the Context of Secondary School Teaching.

Lytle, S. L., and R. Fecho. 1991. Meeting strangers in familiar places: Teacher collaboration by cross-visitation. *English Education* 23(1): 5-28.

Mathematical Sciences Education Board (MSEB). 1989. *Everybody counts: A report to the nation on the future of mathematics education.* Washington, DC: National Academy Press.

McCarthy, C. 1994. Being there—A mathematics collaborative and the challenge of teaching mathematics in the urban classroom. In *Reforming mathematics education in America's cities: The Urban Mathematics Collaborative Project*, ed. N. L. Webb and T. A. Romberg, 173-95. New York: Teachers College Press.

Middleton, J. A., N. L. Webb, T. A. Romberg, S. D. Pittleman, G. M. Richgels, A. J. Pitman, and E. M. Fadell. 1989. *Characteristics and attitudes of frequent participants in the Urban Mathematics Collaboratives: Results of the Secondary Mathematics Teacher Questionnaire.* Madison, WI: Wisconsin Center for Education Research.

Middleton, J. A., N. L. Webb, T. A. Romberg, and S. D. Pittleman. 1990. *Teachers' conceptions of mathematics and mathematics education.* Madison, WI: Wisconsin Center for Education Research.

Mumme, J., and J. Weissglass. 1989. The role of the teacher in implementing the *Standards. Mathematics Teacher* 82(7): 522-26.

National Council of Teachers of Mathematics (NCTM). 1980. *Agenda for action.* Reston, VA.

———. 1989. *Curriculum and evaluation standards for school mathematics.* Reston, VA.

———. 1991. *Professional standards for teaching mathematics.* Reston, VA.

———. 1995. *Assessment standards for school mathematics.* Reston, VA.

Nelson, B. S. 1994. Mathematics and community. In *Reforming mathematics education in America's cities: The Urban Mathematics Collaborative Project,* ed. N. L. Webb and T. A. Romberg, 8-23. New York: Teachers College Press.

Peter, A., D. Clark, and P. Carlin. 1992. Facilitating change for secondary mathematics teachers. *Journal of Science and Mathematics Education in Southeast Asia* 15(2): 67-79.

Pitman, A. 1994. Professionalism. In *Reforming mathematics education in America's cities: The Urban Mathematics Collaborative Project,* ed. N. L. Webb and T. A. Romberg, 67-82. New York: Teachers College Press.

Popkewitz, T. S. 1988. Institutional issues in monitoring of school mathematics. *Educational Studies in Mathematics* 19(2): 221-51.

Popkewitz, T. S., and S. Myrdal. 1994. The "urban" in the mathematics collaboratives: Case studies of eleven projects. In *Reforming mathematics in urban schools: The Urban Mathematics Collaborative Project,* ed. N. L. Webb and T. A. Romberg, 129-50. New York: Teachers College Press.

Porter, A. 1989. External standards and good teaching: The pros and cons of telling teachers what to do. *Educational Evaluation and Policy Analysis* 11(4): 343-56.

Prawat, R. S. 1992. Are changes in views about mathematics teaching sufficient? The case of a fifth grade teacher. *Elementary School Journal* 93(2): 195-211.

Putnam, R. T. 1992. Teaching the "hows" of mathematics for everyday life: A case study of a fifth grade teacher. *Elementary School Journal* 93(2): 163-77.

Remillard, J. 1992. Teaching mathematics for understanding: A fifth grade teacher's interpretation of policy. *Elementary School Journal* 93(2): 179-93.

Richert, A. 1994. Knowledge growth and professional commitment—The effect of the Urban Mathematics Collaborative on two San Francisco teachers. In *Reforming mathematics education in America's cities: The Urban Mathematics Collaborative Project*, ed. N. L. Webb and T. A. Romberg, 151-72. New York: Teachers College Press.

Roberts, H. 1992. The importance of networking in the restructuring process. *NASSP Bulletin* 76(541): 25-29.

Romberg, T. A., and J. A. Middleton. 1994. Conceptions of mathematics and mathematics education held by teachers. In *Reforming mathematics education in America's cities: The Urban Mathematics Collaborative Project*, ed. N. L. Webb and T. A. Romberg. New York: Teachers College Press.

Romberg, T. A., and N. L. Webb. 1992. The Urban Mathematics Collaborative Project: C^{2}ME as a case study on teacher professionalism. In *Science and Mathematics Education in the United States: Eight Innovations*, Organisation for Economic Co-operation and Development, 183-207. Paris.

Sato, N., and M. W. McLaughlin. 1992. Context matters: Teaching in Japan and the United States. *Phi Delta Kappan* 73(5): 359-66.

Serra, M. 1989. *Discovering geometry: An inductive approach*. Berkeley, CA: Key Curriculum Press.

Simon, M. A., and D. Schifter. 1991. Towards a constructivist perspective: An intervention study of mathematics teacher development. *Educational Studies in Mathematics* 22(4): 309-31.

Stake, R., and J. Easley, eds. 1978. The case reports. *Case studies in science education*. Vol. 1. Urbana, IL: University of Illinois Press.

Stanic, G. 1991. Social inequality, cultural discontinuity, and equity in school mathematics. *Peabody Journal of Education* 66: 57-71.

Stodolsky, S. 1988. *The subject matters: Classroom activity in mathematics and social studies*. Chicago: University of Chicago.

Sykes, G. 1991. In defense of teacher professionalism as a policy choice. *Educational Policy* 5(2): 137-49.

Talbert, J. E., and M. W. McLaughlin. 1994. Teacher professionalism in local school contexts. *American Journal of Education* 102(2): 123-53.

Thompson, A. G. 1984. The relationship of teachers conceptions of mathematics teaching to instructional practice. *Education Studies in Mathematics* 15(2): 105-27.

Webb, N. L., and T. A. Romberg, eds. 1994. *Reforming mathematics education in America's cities: The Urban Mathematics Collaborative Project*. New York: Teachers College Press.

Webb, N. L., S. D. Pittleman, T. A. Romberg, A. J. Pitman, and R. J. Williams. 1988. *The Urban Mathematics Collaborative Project: Report to the Ford Foundation on the 1986-87 school year*. Madison, WI: Wisconsin Center for Education Research.

Webb, N. L., S. D. Pittleman, T. A. Romberg, A. J. Pitman, E. M. Fadell, and J. A. Middleton. 1989. *The Urban Mathematics Collaborative Project: Report to the Ford Foundation on the 1987-88 school year.* Madison, WI: Wisconsin Center for Education Research.

Webb, N. L., S. D. Pittleman, T. A. Romberg, A. J. Pitman, J. A. Middleton, E. M. Fadell, and M. Sapienza. 1990. *The Urban Mathematics Collaborative Project: Report to the Ford Foundation on the 1988-89 school year.* Madison, WI: Wisconsin Center for Education Research.

Webb, N. L., S. D. Pittleman, M. Sapienza, T. A. Romberg, A. J. Pitman, and J. A. Middleton. 1991. *The Urban Mathematics Collaborative Project: Report to the Ford Foundation on the 1989-90 school year.* Madison, WI: Wisconsin Center for Education Research.

Weiss, I. R. 1995. *Mathematics teachers' response to the reform agenda: Results of the 1993 National Survey of Science and Mathematics Education.* Paper presented in April at the American Education Research Association Annual Meeting, San Francisco. CA.

Wilson, S. M. 1990. A conflict of interests: The case of Mark Black. *Educational Evaluation and Policy Analysis* 12(3): 309-26.

Wiske, M. S., and C. Y. Levinson. 1993. How teachers are implementing the NCTM standards. *Educational Leadership* 50(8): 8-12.

Woodrow, D. 1992. Learning from experience: Some principles of inset practice. *Mathematics in School* 20(4): 11-13.

Yerushalmy, M., and J. Schwartz. 1985. Geometric Supposer [computer software]. Pleasantville, NY: Sunburst Communications and Education Development Center, Inc.

Acronyms and Abbreviations

+PLUS+	Professional Links With Urban Schools
AMATYC	American Mathematical Association of Two-Year Colleges
AP	Advanced Placement
ARISE	Applications Reform in Secondary Education
ASSM	Association of State Supervisors of Mathematics
C²ME	Cleveland Collaborative for Mathematics Education
CAM	Classroom Assessment in Mathematics
CBMS	Conference Board of the Mathematical Sciences
CEEB	College Entrance Examination Board
CGI	Cognitively Guided Instruction
COMAP	Consortium on Mathematics and Its Applications
CPAM	Council of Presidential Awardees in Mathematics
DMC	Durham Mathematics Council
DMP	Data, Models, and Predictions
EDC	Education Development Center, Inc.
ICM	Introduction to College Mathematics
IIAC	Instructional Issues Advisory Committee
IMP	Interactive Mathematics Program
MAA	Mathematical Association of America
MIT	Massachusetts Institute of Technology
MSEB	Mathematical Sciences Education Board
NACOME	National Advisory Committee on Mathematical Education
NAEP	National Assessment of Educational Progress
NCEE	National Commission on Excellence in Education
NCSM	National Council of Supervisors of Mathematics
NCSSM	North Carolina School of Science and Mathematics
NCTM	National Council of Teachers of Mathematics
NRC	National Research Council
NSB	National Science Board
NSF	National Science Foundation
OECD	Organisation for Economic Co-operation and Development
OERI	Officeof Educational Research and Improvement
PR	public relations
PRIME	Positive Reinforcement in Mathematics Education
PTA	Parent Teacher Association
RAC	Research Advisory Committee
SIMS	Second International Mathematics Study
SMP	School Mathematics Project
SMSG	School Mathematics Study Group

TIMSS	Third International Mathematics and Science Study
TORCH	Teacher Outreach
UCSMP	University of Chicago School Mathematics Project
UK	United Kingdom
UMC	Urban Mathematics Collaborative

366

368